MECHANISMS IN INSECT OLFACTION

MECHANISMS IN INSECT OLFACTION

Edited by

T. L. PAYNE
Department of Entomology, Texas A & M University

M. C. BIRCH
Department of Zoology, University of Oxford

and

C. E. J. KENNEDY
Department of Zoology, University of Oxford

Contributions from the NSF Western Europe Regional Seminar and NATO Advanced Research Workshop entitled Mechanisms in Perception and Orientation to Insect Olfactory Signals

CLARENDON PRESS · OXFORD
1986

Oxford University Press, Walton Street, Oxford OX2 6DP

Oxford New York Toronto
Delhi Bombay Calcutta Madras Karachi
Kuala Lumpur Singapore Hong Kong Tokyo
Nairobi Dar es Salaam Cape Town
Melbourne Auckland

and associated companies in
Beirut Berlin Ibadan Nicosia

Oxford is a trade mark of Oxford University Press

Published in the United States
by Oxford University Press, New York

British Library Cataloguing in Publication Data
Mechanisms in insect olfaction.
1. Insects—Behavior 2. Smell
I. Payne, T. L. II. Birch, M. C. III. Kennedy, C. E. J.
595.7'01826 QL496
ISBN 0-19-857616-1

Library of Congress Cataloging in Publication Data
Main entry under title:
Mechanisms in insect olfaction.
Bibliography: p.
Includes indexes.
1. Insects—Physiology. 2. Smell. I. Payne, Thomas L. II. Birch, Martin C. III. Kennedy, C. E. J.
QL495.M43 1986 595.7'01826 85-18839
ISBN 0-19-857616-1

Set by Spire Print Services Ltd., Salisbury, Wilts
Printed by St Edmundsbury Press,
Bury St Edmunds, Suffolk

Foreword

Sir Richard Southwood, FRS

The role of olfactory signals in an insect's perception of its environment has long been appreciated by naturalists: Henri Fabre's observations on assembling moths are but one example. However, for a long time, and within the memory of some of us participating in this volume, the picture was relatively simple: there were those odours that attracted insects to their food (e.g. flies to carrion), those that repelled enemies (e.g. the stinkgland secretions of bugs), and those that attracted the opposite sex. The first two could be detected by the human nose; the existence of the third was generally a matter of inference from the behaviour of the insects. Where the scent could be detected, classifications were sometimes developed: 'pineapple scent', 'banana-like scent', 'bug-type scent', but these were of little value.

In the last few decades the development of new methods of chemical analysis that have enabled the detection and identification of minute quantities of odours has revolutionized our approach to this subject. It has been shown that the functions of these odours are very diverse; indeed it is clearly a profound mistake to seek a single evolutionary driving force for the development of many of these materials. Some workers have responded by devising elaborate classifications of various -mones depending on function; but I suspect that in the long run these will prove of little more value in themselves than those based on the human perception of the scent! The organizers of the workshop from which this volume developed have, I believe, shown wisdom in eschewing this jargon and referring straightforwardly to 'insect olfactory signals'. I have no doubt that the following papers will show the timeliness of this review by those active in research in this rapidly advancing field of entomology and point the way to further significant research.

Preface

During the last two decades a wealth of information has been developed from research on the chemistry, sensory physiology, and behaviour of insects with respect to olfactory stimulation, particularly by pheromones. This research has for the most part, however, been circumscribed by the boundaries of the individual disciplines (behaviour, physiology, etc.). Also the body of information in each of these areas has not, until now, been developed to a state where it could contribute to a co-ordinated overall concept of the mechanisms underlying insect responses to olfactory signals. The state of knowledge, although far from complete, is now conducive to integration of the available information into a coherent analysis of the mechanisms underlying the behaviour and physiological events taking place in olfactory perception.

The state of the art in the chemistry of olfactory signals is now such that chemical identifications are often routine. For example, the pheromone output of individual insects can be quantified and biosynthetic pathways are being investigated. A range of compounds are used by insects as olfactory signals; some of these are simple one-compound systems, others are complex mixtures. Until recently, analytical equipment and techniques were not developed to a stage where these compounds could be defined fully with respect to geometric and other stereoisomerism, stability concentrations, and output ratios, all of which have been found to be critical in initiating normal behavioural responses. In contrast to the early work stemming from the identification of the pheromone of the silkworm (*Bombyx mori*) by Butenandt *et al*. (1959) as a single compound, it is now also generally understood that simple pheromone systems, where they exist, are a rarity, and that most pheromones are complex mixtures. Many laboratories are now equipped for pheromone identification, largely following the procedures developed by W. L. Roelofs, R. M. Silverstein, and others in the USA, and more recently, for example, K. Mori in the area of stereo-specificity. The result is that pheromone components, whose precise configurations are determined, are now available for stimulation of receptors and for challenging behavioural hypotheses.

In the area of sensory physiology, a depth of understanding of the process of perception has been developed largely under the initiation of D. Schneider at the Max Planck Institut in Seewiesen, West Germany, using *Bombyx mori* as the model. This field of research now continues from the level of the structure and function of olfactory receptors to include elucidation of the mechanisms of central integration of olfactory information and its translation into the motor responses of behaviour.

Peripheral receptor mechanisms are now relatively well characterized, at least in one or two model species, but the study of central processing mechanisms is still in its infancy.

From the behaviour perspective, there is little comparison between the early studies of attraction (yes or no) to isolated chemicals, and what is now known about responses to olfactory signals. For example, behavioural responses are now known to be influenced, often critically, by concentration effects, by pulsed versus uniform airstreams in which odours are conveyed, by the influence of abiotic factors such as temperature, by the probability of one behavioural response leading to the next, and by the pheromonal stimuli. Much of this work has built on the analytical approach established by H. H. Shorey during the 1960s, and now greatly extended by laboratories in the USA, United Kingdom, and West Germany. In particular, ideas on the ways in which insects orient to odour sources have generated hypotheses about the integrated role of chemical, tactile, and visual-pattern stimuli in orientation, in which divergent approaches (e.g. analysis of tethered flight versus free flight in wind tunnels) appear to be converging on common orientation mechanisms.

Superimposed on these approaches, the potential use of insect olfactory signals in replacing or augmenting insecticides in pest control practices world-wide has stimulated a great deal of research on olfactory communication in economically important species. Pheromones are being considered for direct manipulation of pests, while kairomones are being studied for use in manipulating natural enemies against pests. The practical use of these advances in pest management has, however, been only slowly implemented. In the case of pheromones, for example, this must be due, at least in part, to the fact that, whether pheromone applications use trap-out or disruption (confusion, inhibition, etc.) strategies, their effectiveness depends on manipulating or blocking the insects' mechanisms of perception and orientation to pheromones. This manipulation can only be as effective as is our understanding of the underlying mechanisms.

This volume is the result of an NSF–NATO-sponsored seminar, held at Lady Margaret Hall, Oxford, England, in August 1984. The seminar was organized to bring together leading research scientists in these overall fields, with the aim of presenting recent research results, debating and discussing areas of questions or controversy, and of producing a synthesis of how insects respond to olfactory signals from the molecular level to the overall behavioural level. It is our hope that the volume will act as a catalyst for future directions in this active research area.

College Station, Texas
and
Oxford
August 1985

T. L. P.
M. C. B
C. E. J. K.

Acknowledgements

We gratefully acknowledge the financial assistance of the US National Science Foundation Division of International Programs, Western Europe (Dr Marilyn Rurak, Director) and the NATO Scientific Affairs Division, Advanced Research Workshop Program (Dr M. diLullo, Director), without which the seminar and this volume would not have been possible. We thank E. E. Davis, SRI International, and W. E. Seabrook, University of New Brunswick, for their assistance in development of the seminar. We thank the participants in the seminar for their contributions to the meeting and this volume. Also we thank the discussants for their participation in the seminar. For their invaluable assistance in local arrangements, accommodations, dining and much more, we are most grateful to Mrs J. Bell and the staff of Lady Margaret Hall where the seminar was held. We thank Wendy Laffer who helped with correspondence and Mike Amphlett who arranged the photography. Further help with the running of the seminar was provided by Huw Edwards, Steve Simpson, Pete White, and Dorothea Lucas. We also thank Nancy Andryszak and Janet Delorme, Texas A&M University, for their assistance and most specially we thank Carol Russell, Texas A&M University, for her tireless and outstanding assistance in the many clerical needs to develop the seminar and this volume.

Contents

Contributors

T. C. Baker, Department of Entomology, University of California, Riverside, California 92502, USA.

R. J. Bartell, P.O. Box 1700, Canberra City, ACT 2601, Australia. (Correspondence to E. R. Rumbo).

L. Barton-Browne, CSIRO, Division of, Entomology, P.O. Box 1700, Canberra City, ACT 2601, Australia.

W. J. Bell, Department of Entomology and Department of Physiology and Cell Biology, University of Kansas, Lawrence, Kansas 66045, U.S.A.

M. C. Birch, Department of Zoology, University of Oxford, Oxford, OX1 3PS, U.K.

J. Boeckh, Institute of Zoology, University of Regensburg, Universitätstr. 31, D-8400 Regensburg, West Germany.

J. -L Boevé, Département de Biologie Animale, Faculté des Sciences, Université Libre de Bruxelles, 50 av. F. -D Roosevelt, 1050 Bruxelles, Belgium.

J. H. Borden, Department of Biological Sciences, Simon Fraser University, Burnaby, B.C. V5A IS6, Canada.

J. W. S. Bradshaw, Departments of Biology and Chemistry, The University, Southampton, SO9 5NH, UK.

R. T. Cardé, Department of Entomology, Fernald Hall, University of Massachusetts, Amherst, Massachusetts 01003, USA.

C. Cearley, Department of Biological Sciences, University of Maryland Baltimore County (UMBC), Catonsville, Maryland 21228, USA.

C. T. David, Silwood Park, Ashurst Lodge, Ascot, Berks. SL5 7DE, UK.

E. E. Davis, Neuroscience Program, SRI International, 333 Ravenswood Ave., Menlo Park, California 94025, USA.

J. -L Denenubourg, Service de Chimie Physique II, Faculté des Sciences, Université Libre de Bruxelles, 50 av. F.-D. Roosevelt, 1050 Bruxelles, Belgium.

V. G. Dethier, Department of Zoology, University of Masschusetts, Amherst, Massachusetts 01003, USA

J. C. Dickens, USDA Boll Weevil Research Laboratory, P.O. Box 5367, Mississippi State, Mississippi 39762–5367, USA.

F. E. Hanson, Department of Bilogical Sciences, University of Maryland Baltimore County (UMBC), Catonsville, Maryland 21228, USA.

J. G. Hildebrand, Arizona Research Laboratories Division of Neurobiology, 603 Gould-Simpson Science Building, University of Arizona, Tucson, Arizona 85721, USA.

P. E. Howse, Departments of Biology and Chemistry, Unit Building 3, The University, Southampton, SO9 5NH, UK.

D. W. A. Hunt, Department of Biological Sciences, Simon Fraser University, Burnaby, B.C. V5A IS6, Canada

R. L. Jones, Department of Entomology, Wildlife and Fisheries, University of Minnesota, Minneapolis, Minnesota 55455, USA.

K. E. Kaissling, Max Planck Institut für Verhaltensphysiologie, D-8131 Seewiesen, West Germany.

C. E. J. Kennedy, Department of Zoology, University of Oxford, Oxford, OX1 3PS, UK.

J. S. Kennedy, Department of Zoology, University of Oxford, Oxford, OX1 3PS, UK.

S. Kogge, Department of Biological Sciences, University of Maryland Baltimore County (UMBC), Catonsville, Maryland 21228, USA.

E. Kramer, Max Planck Institut für Verhaltensphysiologie, D-8131 Seewiesen, West Germany.

D. M. Light, Western Regional Research Center, Agricultural Research Service, U.S. Department of Agriculture, Albany, California 94710, USA.

C. E. Linn, Department of Entomology, New York State Agricultural Experiment Station, Geneva, New York 14456, USA.

J. C. Lisk, Departments of Biology and Chemistry, The University, Southampton, SO9 5NH, UK.

J. Löfqvist, Department of Animal Ecology, University of Lund, S-223 62 Lund, Sweden.

C. Löfstedt, Department of Animal Ecology, University of Lund, S-223 62 Lund, Sweden.

J. R. McLaughlin, Insect Attractants, Behaviour and Basic Biology Research Laboratory, Agricultural Research Service, U.S. Department of Agriculture, 1700 S.W. 23rd Drive, P.O. Box 14566, Gainesville, Florida 32604, USA.

D. R. Miller, Department of Biological Sciences, Simon Fraser University, Burnaby, B.C., V5A IS6, Canada.

R. A. Montague, Department of Biological Sciences, Fairchild 913, Columbia University, New York, New York 10027, USA.

J. Murlis, Tropical Development Research Institute, College House, Wrights Lane, London, W8 5SJ, UK.

H. Mustaparta, Department of Zoology, University of Trondheim, 7055 Dragvoll, Norway.

R. J. O'Connell, The Worcester Foundation for Experimental Biology, Shrewsbury, Massachusetts 01545, USA.

J. M. Pasteels, Département de Biologie Animale, Faculté des Sciences, Université Libre de Bruxelles, 50 av. F.-D. Roosevelt, 1050 Bruxelles, Belgium.

T. L. Payne, Department of Entomology, Texas Agricultural Experiment Station, Texas A&M University, College Station, Texas 77843, USA.

J. N. Perry, Rothamsted Experimental Station, Harpenden, Herts. AL5 2JQ, UK.

R. Preiss, Max Planck Institut für Velhaltensphysiologie, D-8131 Seewiesen, West Germany.

E. Priesner, Max Planck Institut für Verhaltensphysiologie, D-8131 Seewiesen, West Germany.

R. J. Prokopy, Department of Entomology, University of Massachusetts, Amherst, Massachussetts 01003, USA.

Y. Quinet, Département de Biologie Animale, Faculté des Sciences, Université Libre de Bruxelles, 50 av. F.-D. Roosevelt, 1050 Bruxelles, Belgium.

L. M. Riddiford, Department of Zoology, University of Washington, Seattle, Washington 98195, USA.

W. L. Roelofs, Department of Entomology, New York Agricultural Experiment Station, Geneva, New York 14456, USA.

E. R. Rumbo, CSIRO, Division of Entomology, P.O. Box 1700, Canberra City, ACT 2601, Australia.

C. J. Sanders, Great Lakes Forest Research Centre, Environment Canada, P.O. Box 490, Sault Sainte Marie, Ontario P6A 5M7, Canada.

D. Schneider, Max Planck Institut für Verhaltensphysiologie, D-8131 Seewiesen, West Germany.

K. N. Slessor, Department of Chemistry, Simon Fraser University, Burnaby, B.C., V5A IS6, Canada.

J. N. C. Van Der Pers, VDP Laboratories, P.O. Box 1547, NL-1200 BM Hilversum, The Netherlands.

J.-C. Verhaeghe, Département de Biologie Animale, Faculté des Sciences, Université Libre de Bruxelles, 50 av. F.-D. Roosevelt, 1050 Bruxelles, Belgium.

R. G. Vogt, Institute of Neuroscience, 219 Science III, Eugene, Oregon 97403, USA.

C. Wall, Rothamsted Experimental Station, Harpenden, Herts, AL5 2JQ, UK.

Discussants

Dr G. Bergström, Department of Ecological Chemistry, Göteborg University S-400 33 Göteborg, Sweden.

Mr L. E. Browne, 3555 Timberline Drive, Easton, Pennsylvania 18042, USA.

Dr P. Guerin, Swiss Federal Research Station, Eidgenossiche Forschungsanstalt, CH-8820 Wädenswil, Switzerland.

Dr A. K. Minks, Instituut voor Plantenziektenkundig Onderzoek, P.O.B. 9060, 6700 GW Wageningen, The Netherlands.

Dr D. B. Pinniger, Ministry of Agriculture, Fisheries and Food, Slough Lab., London Road, Slough, Berks, SL3 7HJ, UK.

Dr N. Rachman, Chevron Chem Co., Ortho R & D, Registration and Reg. Affairs, 940 Hensley Street, Richmond, California 94804, USA.

The Honourable Miriam Rothschild, Ashton Wold, Peterborough, PE8 5LZ, UK.

Dr E. Städler, Swiss Federal Research Station, Eidgenossische Forschungsanstalt, CH-8820 Wädenswil, Switzerland.

Dr R. A. Steinbrecht, Max Planck Institut für Verhaltensphysiologie, D-8131 Seewiesen, West Germany.

Dr M. Stubbs, Ministry of Agriculture, Fisheries & Food, Slough Lab., London Road, Slough, Berks. SL3 7HJ, UK.

Dr J. P. Vité, Forstzoologisches Institut, Universität Freiburg, Bertoldstrasse 17, 7800 Freiburg i.Br., West Germany.

Dr L. J. Wadhams, Department of Entomology, Rothamsted Experimental Station, Harpenden, Herts. AL5 2JQ, UK.

1 Chemoreception and behaviour from an evolutionary and comparative perspective

V. G. Dethier

There is a classic article in the literature of psychology entitled 'The snark was a boojum'. The article, referring to Lewis Carroll's poem *The hunting of the snark*, was written by Frank Beach as an allegory depicting the status of comparative studies in his discipline. Beach compared the comparative psychologist to a member of a hunting party seeking a Snark known as Animal Behaviour. In his hunt the comparative psychologist found the Snark, but the Snark turned out to be the white rat and proved to be a Boojum, whereupon the comparative psychologist suddenly and softly faded away.

I direct this article to your attention because the message that it transmits is relevant to the field of chemoreception where genuine comparative approaches do not flourish. A perusal of the literature reveals that scientists who study the chemical senses of vertebrates seldom compare their results with those from invertebrate studies; students of insect chemoreception are equally neglectful of the literature on vertebrates; few followers in either camp pay much attention to synaptic and central nervous system receptors; and still fewer, unless they are microbiologists, are familiar with chemoreception in bacteria.

As a consequence of this parochialism, seminal ideas and hypotheses generated in one corner of the field are seldom perceived in other corners. Also, few of the remarkable technological advances made in other areas of biology are put to useful work in the general field of chemoreception. Electrophysiology, for example, found its way more slowly into this field than into others; application of modern genetic analyses has lagged. Only very recently have learning theory and experimental techniques been applied with sophistication to the study of insects. Too long have we been seduced by the dogma that insects are 'instinctive' animals. Radioimmunological, immunofluorescent, and mono- and polyclonal antibody techniques are adopted here more slowly than elsewhere. Furthermore, when these techniques are finally employed, they are more often applied to peripheral problems than to fundamental ones.

Admittedly there are sound reasons for this state of affairs, especially with respect to insect chemoreception. Investigators in general direct their efforts to those problems that capture their enthusiasm; people conducting their research with insects are generally more interested in understanding the insect itself than in using it as a white rat to assist them in the pursuit of answers to elusive basic phenomena. Furthermore, insects are not the most tractable of 'white rats'; in fact, some problems of chemoreception and related neurophysiology and behaviour are fiendishly difficult to solve with insects as experimental animals. Nevertheless, even if the primary goal is to understand insects, we should raise our eyes to wider horizons.

If we expect to understand those patterns of behaviour that are initiated, terminated, or otherwise modulated by olfactory and gustatory stimulation, we gain by having some knowledge of: the precise nature of the controlling stimulus, its magnitude, and temporal and spatial dynamics; the manner in which it is transduced by receptors; the nature of the neural signal generated; and how signals are integrated in the central nervous system. In short, we should have at least a working knowledge of the concatenation of events from stimulus input to behavioural output in order to be able to evaluate limits, errors, noise and meaning. Finally, we should strive to characterize behaviour as accurately as we attempt to characterize the stimulus.

We can certainly profit by technical advances in other fields, but to what extent are phenomena discovered in other animals applicable to insects? To what degree can comparative awareness actually serve the ends of the entomologically oriented investigator? I believe that some of the answers to these questions are to be found in evolutionary considerations. The following brief evolutionary survey is presented, therefore, to establish probable common denominators of chemoreception. It provides some rationale for espousing a more vigorous comparative approach and for viewing chemoreception in the broadest possible perspective.

Chemoreception, the capacity of a cell to react with foreign molecules without transporting them into itself for metabolic purposes, is one of the earliest hallmarks of life. It may have antedated the abilities to trap light by way of photosynthesis and photoreception, themselves ancient abilities. From this primitive chemosensitivity taste and olfaction evolved, the latter being the richest and most acute of chemical senses and, in human beings, the most evocative.

Whether one espouses Oparin's (1957) theory of coacervates as protolife or Fox's (Fox and Nakashima 1980) theory of proteinoid microspheres, the precursors of organisms were undoubtedly complex aggregates with the property of selective permeability. These membranes served water molecules and surface-active substances that endowed the aggregates with the property of selective permeability. These membranes served as barriers maintaining the integrity of the aggregates and as collections of gates that selectively permitted the entry of molecules destined to meet energetic and structural needs. In addition, however, to sites for transmembrane transport, there evolved sites that reacted specifically with molecules destined neither to serve as building blocks nor to provide energy. They were the first chemoreceptors.

These speculative steps in the evolution of chemoreception find expression today in the prokaryotes as exemplified by the bacterium *Escherichia coli*. Probably in no other organism is the mechanism of chemosensory transduction so well understood and comprehensive knowledge of its relation to behaviour so close to realization. In the first step the stimulus molecules are bound to a protein. At this point, chemoreception and transport are related, but after the initial binding the subsequent steps leading eventually to behavioural response are distinct (Adler 1976).

A partial list of chemoreceptors includes galactose, glucose, ribose, aspartate, and serine receptors. Each is a different 30,000 mol. wt. protein of characteristic specificity and breadth of action spectrum. The serine receptor, for example, binds L-serine, L-cysteine, L-alanine, and glycine; the aspartate receptor binds L-aspartate and L-glutamate, and so on (Adler 1969). The molecules first pass through the outer bacterial membrane. Some of the proteins with which they bind are in the periplasmic space, and others are bound in the inner membrane. Details of subsequent steps in the process of transduction are given by Adler (1969, 1976) and Koshland (1983). In essence, chemosensors influence the degree of methylation of membrane proteins, and the degree of methylation determines the direction of rotation of the flagella. The direction of rotation alters the frequency of tumbling, which in turn determines whether a bacterium is attracted or repelled by the stimulating chemical.

A provocative suggestion is that the chemoreceptive proteins of bacteria may not differ greatly from the gustatory proteins of insects and mammals. There is support for this idea in evolutionary studies. As Schwartz and Dayhoff (1978) have pointed out, many proteins and nucleic acids are 'living fossils'. Their structures have been conserved successively in living organisms for billions of years. For example, crucial proteins involved in photosynthesis were already established about 1 billion (thousand million) years ago. Their probable evolution has been worked out by protein and nucleic acid sequence data derived from living bacteria and blue-green algae. It is, therefore, quite possible that the study of bacterial chemoreceptor proteins might yield clues to the identity of insect and vertebrate chemoreceptor proteins, the more so because efficient mechanisms that have evolved early tend to be preserved in time. In terms of threshold, discrimination (Weber fraction), specificity, and breadth of action spectra, bacterial carbohydrate and amino acid receptors appear to be as effective as comparable gustatory receptors in metazoans.

With the advent of chemoreception the behavioural capabilities of organisms, cells, and metazoans alike were greatly expanded. An organism could 'recognize' others of its kind, thus setting the stage for sexual reproduction; it could differentiate among foreign organisms and compounds, thus permitting the establishment of prey/predator relationships, the selection of substrates for attachment or locomotion, and the exercise of selective feeding. Despite these advantages, however, chemosensitivity alone could not realize its full potential.

As long as cells remained small, their passivity could be compensated for by Brownian movement and streaming, convection, turbulence, and diffusion in the fluids in which they lived. They could be moved randomly from one part of their environment to another. Once they exceeded a critical size beyond which Brownian movement was ineffective and only gross fluid movement translocated them, immobility imposed severe handicaps. This has been demonstrated by comparative studies of the selective advantage of motile bacteria over non-motile ones. Without motility there could be no emigration from a place of nutiritional aridity to one of nutritional bounty, from a physiologically inimicable area to a felicitous one, from a loveless environment to one abounding with other cells. The coupling of chemoreception and motility and chemokinesis and chemotaxis possible.

Chemotaxis undoubtedly evolved in ancient prokaryotes in a form observed today in contemporary bacteria where the mechanisms connecting sensing to specific behaviors are fairly well known. It is not surprising that chemotaxis at the cellular level is a ubiquitous phenomenon of striking biological importance (Lackie and Wilkinson 1982). The basic machinery for chemoreception was perfected in the prokaryotes, coupled with motility, and developed to a remarkable state of sophistication in the Protista, especially the slime moulds, and in metazoan leucocytes and cells concerned with immunological events.

The nature of the linkage between chemosensitivity and contractile mechanisms is not known. Doetsch (1972) proposed that responses in bacteria were triggered by bioelectric changes in the membrane. In Protista, especially in *Paramecium*, significant advances have been made in relating molecular, electrical, and behavioural events (Eckert *et al*. 1976). Receptors generate an inward current, presumed to be carried by Ca^{2+} This current, propagating electrotonically, depolarizes the entire cell membrane. An increased frequency and a counter-clockwise shift of the power stroke of the cilia follow. This causes the ciliate to retreat. When the posterior end of the cell is stimulated, the membrane is hyperpolarized, and there is an influx of Ca^{2+}. An increase in frequency and a clockwise shift of the power stroke of the cilia follow. Locomotion is then in a forward direction.

These bioelectric properties resemble those of metazoan nerve, muscle, and epithelium. Thus, two major components of the machinery for behavioural responses to chemical stimulation emerged very early in evolutionary history. The next problem was how to adapt these capabilities to a multicellular mode of life.

When multicellularity occurred many cells lost their ancient talents. Sensing of the chemical richness of the external world, and the internal milieu as well, with its hormonal tides and its complex patterns of neurotransmitters and neuromodulators, became specialized skills relegated to a few strategically situated cells. Cells in general did not lose the capacity for internal motility (this has been preserved in all cells, where it plays major roles in mitosis, embryogenesis, neuronal physiology, immunological responses, and a host of other essential functions). What was lost to a large extent was the coupling of motile mechanisms directly to individual cell movement.

The motility of selected cells was drafted to the common cause of moving the multicellular organism as a whole. Chemoreception was now cast in the grander role by association with specialized transmitting systems. Transitional stages between the two functional levels, self-contained unicellular systems and neurally linked multicellular systems, are preserved in contemporary coelenterates. Here are found the earliest metazoan chemoreceptors: specialized epithelial cells in an all-encompassing sheet of tissue. Simple functional linkage is still present in some dual-purpose cells, independent effectors such as nematocysts; but adumbrations of higher function are already in evidence.

As organisms became larger and intercellular relationships more complex the nascent nervous system added to its role of transmittal that of integration. Sense organs became concentrated anteriorly; masses of interneurones were added to the complex; ganglia became brains (but compare the view of Koopowitz and Keenan (1982), who suggested that bilateral symmetry rather than cephalization was the prime morphological feature necessitating the development of a brain). From the point of view of chemotaxis, translating information about chemical gradients, direction of sources, and temporal and spatial patterns of concentration discontinuities into directional progression required more complex processing of sensory input than the rather direct and simple linkages that sufficed in unicellular organisms. Intricate ganglionic circuitry evolved even before animals emerged onto land.

Terrestrial living brought additional problems. Chemoreceptors were originally concerned with molecules in an aqueous medium, in which the organism was totally immersed. Now receptors were exposed to an entirely different set of circumstances, circumstances that challenged organisms on many other fronts as well. Terrestrial exposure to a chemically and physically thin, non-supportive, and desiccating atmosphere forced novel evolutionary responses. Among these were: rigid external and internal skeletons designed to bear the full weight of the body, copulation and internal fertilization to provide an aqueous medium for union of egg and sperm, embryogenesis in hard-shelled eggs or in a uterus, and elaborate mechanisms to conserve water and thwart desiccation.

Discussions of these changes often overlook the challenge to the chemical sense. In the new environment receptors had to detect molecules existing in a gaseous phase, and often at extraordinary dilutions. Different thermodynamic constraints existed because molecules had to be partitioned between an aqueous phase and a gaseous phase.

In the service of olfaction, previously existing structures that had served chemoreception in the sea were modified. In vertebrates the first cranial nerve, relaying information from chemoreceptors in the epithelium of the nasal cavities after it had been integrated by a collection of interneurons (the future olfactory bulb), became the principal olfactory system. The system was already internalized, and air-breathing mammals incorporated the bellows of the new respiratory system to bring odours to the receptor area. A further innovation was behavioural: the sniff. Other innovations are hinted at by comparative studies of the olfactory systems of salamanders before and after metamorphosis from aquatic larvae to terrestrial adults. It is noteworthy that sensitivity to amino acids is greatly enhanced in the adults.

Marine and aquatic arthropods evolved large populations of receptors on the antennules. These receptor fields were well designed to react to low concentrations of compounds. Amino acids, for example, can be detected in the nano and pico molar range. Behavioural adaptations for enhancing the sensitivity of the system include antennule flicking, the analogue of the sniff, and, in numerous organisms, directional currents produced by ciliary activity.

This brief foray into evolutionary history indicates that fundamental principles of chemoreception and its attendant behaviour were established early, that basic mechanisms are essentially the same throughout the animal kingdom, that modifications are probably variations on a theme, and that comparative studies are not only valid but potentially profitable. The questions to be answered are everywhere the same. They relate to transduction (de Kramer 1983) encoding and transmission of information, central decoding and integration, and command and execution of appropriate behaviour.

The most spectacular advances toward understanding transduction have been made with bacteria, various vertebrate central nervous system receptors such as those for acetylcholine and opiates, rate-sensitive immune systems (Rosen 1976), and neurotransmitter systems at synapses. It would be surprising, but gratifying, if any of these were identical to olfactory transducing systems in insects; however, considering the evolutionary conservatism of chemosensitivity and our ability to untangle protein genealogies it is not unreasonable to suppose that valuable clues can be found by comparative studies.

More especially, powerful analytical techniques have been invented in the fields mentioned, and they have not yet been exploited to their fullest. One of the problems with insects is the near impossibility of obtaining enough material for direct molecular analysis. Here is where some of the probes employed by molecular biologists might prove valuable. Even more exciting is the possibility of cloning primary receptory cells in order to obtain sufficient material for analysis (Lancet 1984) and also to facilitate intracellular electrophysiological recording.

Another tool that has not been exploited aggressively is genetics. In the insect gustatory system some progress has been made toward elucidating molecular sites on the dendrite. A more extensive employment of mutants in the realm of olfaction might be profitable.

With respect to coding of sensory information, insects may be the 'white rats' par excellence because the receptors are identifiable primary neurones. Technical problems still remain, especially in the gustatory system where intracellular and single axon recording has for the moment been unachievable. Conditions are better in the olfactory system. Yet, dependence on spike amplitude and wave form as sole criteria for identifying receptors is not without its risks. Matters are complicated by peripheral interactions, inhibition, and synergism, which can alter the characteristics of action potentials. Here, greater use of computer analysis is becoming indispensable. It might be added that information theory may have valuable contributions to make to this phase of chemoreception (Smith *et al.* 1984).

In the final analysis it is how the central nervous system reads and interprets spike trains at any moment that determines behavioural output. A fair number of investigations of central projections of gustatory and olfactory pathways have been carried out in vertebrates, but few sallies have been made into the insect brain. At a conference on the insect brain held in 1979 (Altman 1980) it was clear that, aside from sensory processing in optic and antennal lobes, surprisingly little is known about insect brains.

At the behavioural level, studies of insect chemoreception are still predominantly descriptive and ecological. Since most behaviours involving chemoreception (e.g. oviposition, feeding, etc.) are orchestrated by multiple modalities (orientation of flying insects of pheromones is a fine example), the role of chemoreception can best be elucidated by dissecting the behaviour. It should be emphasized that the nature of behaviour above the simple reflex level is linked and hierarchical. If understanding of a particular behaviour is to proceed beyond the descriptive level, or beyond the identification and assessment of goals, analysis must be in terms of proximate causes (Dethier 1984). To this end, genetic analyses, coupled with neurophysiological analyses, are invaluable. The emphasis that is now being given to genetic analysis of learning in insects (e.g. Quinn *et al.* 1978; Tully and Hirsch 1982) might be extended with profit to olfactory systems. It has been done with some success with gustation (Falk and Atidia 1975; Ishikawa *et al.* 1963; Isono and Kikuchi 1974; Tanimura *et al.* 1983; Rodrigues and Siddiqu 1978; Tompkins *et al.* 1979).

The present volume addresses most of the basic questions that I have discussed: transduction, coding, central nervous system pathways and integration, perception, and orientation. We shall in the following chapters of this book be presented with the latest advances, from which we shall be able to assess the state of the field and promising vistas into the future.

Part I
Behavioural mechanisms

2 Some current issues in orientation to odour sources

J. S. Kennedy

Since the aim of this book is to set the stage for future research it seemed that the most useful way to introduce this behavioural section would be to raise for discussion some of the issues that are the subject of current concern and debate.

2.1. The incidence of any kind of distant olfactory attraction

There are now so many established cases of both walking and flying insects homing in on a distant source of animal or plant odour, that it is time to look again at the cases where that component of resource-finding behaviour appears to be lacking.

For instance, recent work has greatly extended Wood's original field observations of host finding by the bark beetle *Dendroctonus brevicomis* (Birch 1984). Wood interpreted these observations as evidence that there was no olfactory attraction to susceptible but as yet uninfested trees, despite their known odour emission. A tree began to attract the beetles only after some had already invaded it and started emitting their 'aggregation pheromone'. The recent work shows clearly that some species in four genera do, after all, respond to host odour in flight because they land on sticky screens enclosing beetle-free tree trunks more frequently than they do on similarly screened non-host trees. Nevertheless, Wood's original findings were confirmed, because some other species in two of the same genera landed in equal numbers on the two kinds of screened tree. These differences within a genus seem very odd. And if it does really mean there is no olfactory attraction to the host itself in some species, are we to expect that many more insects rely on the simple stay-or-leave mechanism of host selection after contact, as Thorsteinson (1960) and Beck (1965) once suggested most insects do?

Now the bark-beetle evidence of indiscriminate landings on hosts and non-hosts recalls our old records of three aphid species doing the same (see Kennedy 1977). This was likewise taken to mean there was no oriented flight towards their hosts in response to the host odour, an unexpected conclusion at that time which has often been quoted since. However, the interpretation of those field counts of aphid landers was later questioned by Taylor (1965; Taylor and Taylor 1983); moreover it has recently been shown that host plant odour does attract another aphid over at least 1 metre, either anemotactically or by inducing visual orientation (Chapman *et al*. 1981).

Migrating insects may become ready to respond to the *visual* stimuli from plants by indiscriminate approaches and landings, before they have become ready to respond discriminately to host odours by upwind-oriented flight. Taylor's point was that the conditions during our field counts were exceptional in that great numbers of aphids were flying within a few metres of the ground, instead of being dispersed upwards by turbulent convection as usual. In these conditions most landers have probably come from nearby sources. Any that have flown long enough to respond discriminately to host odour (if they do) may then be too few for statistical significance, so that the landings may give a false impression of being quite indiscriminate.

The evidence Taylor offered in support of this hypothesis can be questioned, but it has not yet been tested in aphids, or in bark beetles which are in fact known to need to get through a period of migratory flying before they become responsive to attractive odours. But it is evidently necessary to allow for the migratory phase that frequently intervenes before an insect proceeds to find a host or mate or some other odorous resource (Kennedy 1985), before we can say that an insect does *not* use odour cues in finding that resource.

2.2. The incidence of odour-conditioned anemotaxis in flight

This optomotor anemotactic mechanism (David (Chapter 5)) is also now known to be widespread, but there is growing evidence that it need not be used even in distant olfactory attraction by flight (Miller and Strickler in Bell and Cardé 1984). Some Diptera evidently use *mechanoreceptive* anemotaxis instead. As first described in the adult cabbage root fly, the approach to a source of host odour is made in a series of very short, low flights ('hops'), without the zigzagging typical of the sustained flight of male moths nearing a sex pheromone source. Each time the fly lands it turns to face into wind on the ground where it can detect the wind direction mechanically. When it takes to the air again it can no longer 'feel' the wind and would then need visual cues from fixed objects to correct for deviations due to the wind changing direction. But the wind is most unlikely to do that significantly while the fly is airborne for one brief hop, so that airborne anemotaxis hardly seems necessary.

This would be the 'open-loop' technique of oriented movement not dependent on feedback while in progress: 'first aim, *then* shoot', like the female treefrog's hops towards a vocalizing male (Feng *et al*. 1976; Rheinlaender *et al*. 1979), or the famous raptorial strike of the praying mantis. Bursell (1984) has obtained field evidence that tsetse flies are very likely using this same technique as they approach a source of animal odour in a series of brief and very fast flying 'hops'.

Gillett (1979) speculated about another variant of mechanoreceptive anemotaxis which would not even require landings, in flying mosquitoes. They have been shown to use optomotor anemotaxis in flying upwind towards a human odour source in the laboratory, but Gillett says that in the field they can fly upwind to the source perfectly well when there is not enough light for visual cues to be used. They fly very low where the vertical gradient of wind speed at its steepest, and they frequently dip even lower as they fly. Gillett's idea is that they manage to detect the wind direction by feeling the wind *shear* mechanically at each dip. The dips might also involve actual contacts with the ground or plant stems, with or without actual landings, which would better enable the flier to detect the wind direction with its mechano- and proprioreceptors. These ideas deserve following up.

2.3. Self-steering in olfactory attraction

The *self*-steered component in olfactory attraction is another issue of rather recent appearance on the agenda for most workers west of the Rhine. A manoeuvre is described as self-steered when it is initiated and modulated by the olfactory stimulus, but not steered by it (Bell and Tobin 1982; Havukkala and Kennedy 1984; Kennedy 1978). Instead, the manoeuvre is controlled by an internal programme utilizing current and stored information about the insect's own steering activity ('idiothetic' cues).

That is the strict sense of self-steering—course control by 'dead reckoning'. Bell (in Bell and Cardé 1984) calls it 'indirect chemo-orientation based on internally stored information'. Often, however, completely internal control has not been established with certainty and then the term may be employed in a looser sense to cover cases where the programme utilizes non-olfactory *external* directional cues that are independent of the insect ('allothetic' cues); Bell calls this 'indirect orientation based on a second modality'. For example, the wind provides mechanical directional cues for an odour-stimulated insect on the ground, and visual ones for it in the air.

German workers have been way ahead of the rest of us on this. They have a much stronger tradition of research on the mechanisms of *Orientierung*, and for them it has gone without saying for at least 30 years that every manoeuvre appearing as a response to some external stimulus is more than a *re*action to the stimulus: it is an '*action-and-reaction-in-one*'. The fact that the rest of us, save Bell and Tobin (e.g. 1982), have been so slow to assimilate that simple but fundamental point is explained in great part by the extraordinary longevity outside Germany of Fraenkel and Gunn's readable and rigorous book *The orientation of animals*, published in 1940. For its pervading weakness is precisely that it fails to bring out the self-steered component in the adaptive manoeuvres of animals. It presents every class of manoeuvre as a straight *re*action, a stimulus-and-response, as if the stimulus always steered the responding animal's manoeuvre as well as initiating it.

Even in the class *klinokinesis*, where it was implicit in the description and plain to see that only the animal itself could be steering the observed path, and the external stimulus could not be doing more than modulate this self-steering—even then, the self-steered component was not brought out as such. The class *klinotaxis* has been transformed by recognition of the self-steered component. Fraenkel and Gunn recognized only cases where the animal samples the external stimulus intensity successively to the left and right of its path by small alternating swings of its antennae, head or whole body, a form of scanning described as 'wig-wagging'. The animal turns when it detects a left–right difference until the difference disappears, and this servo-mechanism steers it along a gradient of odour concentration. But a number of authors subsequently accepted the first diagnostic feature—turning in response to a stimulus difference detected by *successive* sampling of the environment over time—and ignored the second one that restricted this successive sampling to side-to-side movements.

'Klinotaxis' thus came to include cases where the insect turns in response to a stimulus difference detected during sampling *along* its path, without obvious scanning movements. The two sampling modes were in due course distinguished and qualified explicity as *transverse* and *longitudinal* (Kennedy 1978). For brevity's sake one could then speak of 'transverse klinotaxis' and 'longitudinal klinotaxis', although this is confusing because the two qualifications refer only to the mode of sensory sampling, the input side, and not at all to the motor output, the direction taken by the insect in response to that sensory input (see Table 2.1 and Section 2.7 for a new name).

The distinction between the 'transverse' and 'longitudinal' cases is nevertheless crucial. In transverse klinotaxis the odour stimulus both initiates and resultantly steers the manoeuvre, restricting it to movement along the odour gradient. This places it unambiguously in the *taxis* group of 'directed', feedback-controlled reactions along with *tropotaxis*, where again the insect samples the odour on its two sides, but now simultaneously with bilateral receptors instead of successively, and again proceeds along the gradient (Table 2.1). Longitudinal klinotaxis, on the other hand, is not in fact a directed reaction or taxis because here the initiating stimulus does not also provide feedback control of the subsequent path. That path is self-steered, so there is no theoretical limit to the variety or sophistication of the manoeuvres that can be triggered by a simple change in the sensory input.

This brings longitudinal klinotaxis close to klinokinesis, for the manoeuvre is entirely self-steered in both cases (Table 2.1). The difference is that in klinokinesis it is a simple increase or decrease in the mean turning rate (a change only in the directional variance: Havukkala 1980), with no change in the mean direction, while in longitudinal klinotaxis the mean direction does change one or more times in a definite pattern ('a change in the location and/or number of peak frequencies in the angular distribution of turns': Havukkala and Kennedy 1984).

TABLE 2.1

Categories of odour-induced manoeuvre resulting in displacement towards an odour source

Name	Sensory input: Odour concentration differences detected by sampling:	Sensory input: Minimum receptors needed	Locomotory output = manoeuvre: Form	Locomotory output = manoeuvre: Steering
Orthokinesis			Change in linear speed or frequency of locomotion	
Klinokinesis	Successively along path		Change in rate or frequency of turning	Self-steering throughout
Schemakinesis ('longitudal klinotaxis')*		One	A particular pattern	
Klinotaxis ('transverse klinotaxis')*	Successively, by small swings left & right ('wig-wagging')		Turning to the more or the less stimulated side	By the odour gradient, but 'wig-wags' probably self-steered
Tropotaxis	Simultaneously left & right	Bilateral pair		By the odour gradient
	Directional cues generated by wind			
Menotaxis (now including 0 ° & 180 ° angles)	Wind direction may be sampled successively left & right by serial counterturning	Multidirectional	Maintaining a set angle to the wind direction = *anemotaxis*†	By wind-generated cues: mechanical for walkers, visual for fliers

* See text.

† 'Positive' anemotaxis = upwind; 'negative' = downwind; across-wind was called anemo-*meno*taxis in the original Kühn sense.

The two major groupings in Fraenkel and Gunn's system were the directed reactions or *taxes* on the one hand, and the undirected reactions or *kineses* on the other. 'Directed' meant directed relative to the stimulus direction only, and not relative to the animal's previous path, since the idea of self-steering was missing. So, instead of their clear division of reactions into taxes and kineses, we now have an almost continuous spectrum of manoeuvres differing in the role of the self-steered component.

2.4. Mechanism of counterturning

Counterturning is the most elaborate manoeuvre yet shown to be self-steered in insect olfactory behaviour. It consists of a succession of turns and excursions alternately to left and right which are regular enough for the resultant displacement (if any) to be roughly straight. The interval between successive counterturns varies from about one to many seconds, and the distance between them from several to many times the length of the insect. 'Zigzagging' is counterturning with significant resultant progression (or regression), as in the well-known zigzagging track of male moths and various other insects flying or walking towards a source of wind-borne pheromone. 'Casting' is counterturning with no significant net progression or regression, and the excursions are typically wide sweeps, notably longer than zigzagging, as when a flying moth loses contact with a pheromone plume or enters a uniform cloud of pheromone (Baker (Chapter 4); David (Chapter 5); also Kennedy 1983, but note that his distinctions based on the *angles* turned were inappropriate).

What is new here is the demonstration (Baker (Chapter 4)) that flying male moths respond to pheromone by zigzagging *even in still air*. Counterturns are not necessarily wind-oriented (i.e. not 'anemomenotactic', contrary to what we used to think), being basically self-steered. They are driven by a central nervous turn generator, which alternates between left and right at a frequency governed by the concentration and pulsing of the pheromone input (David, Kramer, and Baker (Chapters 5, 6 and 4).

When there is a wind, the concurrent anemotactic turning responses 'polarize' that self-steered counterturning into a resultantly upwind direction in the presence of suitable pheromone input (Baker, Willis and Phelan 1984; Baker (Chapter 4); David (Chapter 5)). Thus the female's chemical signal switches on *two* programmes of manoeuvre by the flying male: anemotaxis *and* self-steered counterturning.

Baker (Chapter 4) presents the extensive evidence that—apart from the anemotaxis—'locking on' to a pheromone plume is due simply to the high-frequency counterturning induced by appropriate pheromone stimulation, and not to turning back up the odour concentration gradient on reaching the plume's edge. If the flier strays over the edge into clean air, the counterturning frequency drops. The lateral excursions between turns therefore lengthen until one of them reaches the lost plume, whereupon they shorten again and the flier is detained there once more by its high-frequency, narrow counterturning, and so on.

On the other hand, Cardé and Charlton (1985) have come to the conclusion that flying gypsy moths straying out of a plume are indeed brought back to it by a counterturn provoked by cessation of the pheromone stimulation: that is to say, by transverse chemo-klinotaxis and not self-steered counterturning. Tobin and Bell (1982, and unpublished; Tobin 1981) have espoused the same idea for walking and flying insects. Thus the zigzags of three flying moths have been recorded as narrower along a discrete plume than in a uniform cloud of pheromone, and Cardé and Charlton (1985) interpret this to mean that the flier turns back when it detects a plume border. An alternative explanation yet to be ruled out depends on the rapid habituation recorded in two of those moths under the continuous pheromone stimulation that they received in a uniform cloud but not when flying along a discrete, fenestrated plume (Kennedy *et al*. 1981; Willis and Baker 1984). For the moth, habituation of the initial narrowing-down response to entering pheromone would be like passing into clean air again, with the well-known result that the zigzags widen. See Baker (Chapter 4) for a third possible explanation.

The second piece of evidence adduced by Cardé and Charlton (1985) for the transverse chemo-klinotactic theory of plume following is that a flying gypsy moth that was presented with the abnornal situation of a narrow but slightly *meandering* pheromone plume in a straight wind, zigzagged along the meandering path of the plume, not straight upwind. Again there is an alternative explanation consistent with the above idea of anemotactically 'polarized' self-steered zigzagging. The illustrated tracks show that the zigzags often took the flier a short distance out of the plume and these excursions were biased to the left of the plume when it deviated to the right, and vice versa, that is to say *towards the wind-line*. This suggests that anemotactic correcting responses were occurring, but could not draw the flier away from the plume into line with the wind because the frequent counterturns took it back to the plume, which was always within range because the meander was slight. The frequent contacts with the plume maintained the high frequency of counterturns. But the issue remains to be resolved.

2.5. The function of counterturning

The function of self-steered counterturning is again an unresolved issue for future research. One generally accepted function in the flying male moth has been mentioned already: the maintaining and regaining of contact with the airborne pheromone, a highly elusive requisite under natural conditions (Kennedy 1983).

Counterturning may also improve an insect's aim. In the first place, without any involvement of anemotaxis or other orientation by an external cue, regular alternation of excursions to left and right can produce a straighter *resultant* track than an irregular series even of quite small and infrequent deviations. For example, a foraging thrush gets out of an area where food items have become scarce by straightening out its path, and it does this partly by alternating more regularly between left and right turns (Smith 1974).

More to the point when there is anemotaxis, as with the flying insects, there must also be a threshold size of deviation from the due upwind course below which no optomotor correcting turn is provoked. The threshold has not been measured in flying anemotaxis, but in the walking phonotaxis of female crickets, the 'sector of ambiguity' extends as much as 30° to each side of the midline (Bailey and Thomson 1977; Oldfield 1980; Rheinlaender and Blätgen 1982; Thorson *et al*. 1982). The sector of ambiguity is the insect's frontal sector of directions of arrival of the sound signal, within which the insect is no more likely to turn towards than away from the sound source.

The ambiguous sector is presumably smaller for visual course-correcting responses. Nevertheless, when a flier is still at some distance from an odour source but in an unobstructed wind blowing more or less straight from the source, then even quite a small uncorrected deviation from due upwind could take the flier well off to one side of that line into more obstructed and disorderly airflow, and so delay its approach to the source. But *self*-steered turns across wind which are large enough to take the directional input (the image flow over the insect's eyes) out of the ambiguous sector, would provoke frequent *extra* course corrections (Cardé and Charlton 1985; Kuenen and Baker 1983). In this way the distance travelled along any one deviant track would be cut short and the *resultant* track would stay closer to the unobstructed wind line, so long as pheromone was there. By counterturning the insect is thus conceived as constantly correcting its aim into the wind bearing the scent. In effect, it is *tracking* the wind direction.

A human observer has the immediate impression that a moth flying in zigzags along a pheromone plume is 'feeling for the edges' of it. This is a vivid teleological way of saying—or rather implying, once again—that the moth is not only detecting the odour gradient across the plume edge but is also steering by that gradient in turning back into the plume: in other words following the plume chemotactically (transverse chemo-klinotaxis). But there is laboratory and field evidence that clearly contradicts this idea and no conclusive evidence for it (Section 2.4 above).

Putting the matter into similarly teleological language, it is more likely that the flier is 'feeling for the error sign' (left or right) of its image flow, in order to bring the flow back into line with its body axis. By zigzagging it quickly detects to which side its track deviates from upwind; the resulting optomotor correcting responses keep its resultant track aligned on the odorous wind, its 'pointer' to the female. On this view the female cricket or treefrog zigzagging toward a calling male is 'feeling for' an interaural sound pressure difference, the walking insect zigzagging upwind towards an odour source is 'feeling for' off-axis wind pressure, and the flying one is 'feeling for' off-axis image flow.

Admittedly, no one seems to have suggested that there is any self-steered component in cricket zigzagging, which is described as '*corrective* meandering' (Thorson *et al*. 1982) as if the counterturns were due solely to course-correcting responses to off-axis sound input, although the published tracks hint that there is also a self-steered component in the meanders (e.g. Bailey and Thomson 1977, Figs. 6 and 7). But in treefrogs, self-steered counterturning has been described on two distinct scales (Rheinlaender *et al*. 1979): side-to-side head swinging ('wig-wagging') while stationary before a leap, and zigzag leaps, thought by the authors to have the function of provoking phonotactic turns, as in the moth anemotactic model above.

To sum up, in the case of the moth it now appears likely that odour-modulated, self-steered counterturning has at least two functions. It enables the insect to track both of the essential sensory cues: (i) the female's non-directional chemical signal, needed to identify her and turn on both the anemotaxis and the self-steered counterturning, and (ii) the wind direction, needed to home in on her. But here again the question is still open.

2.6. Odour dispersion

All the above work on zigzagging notwithstanding, Perry and Wall (1984b, this volume) have observed no clear horizontal zigzagging in many years of field work with pheromone traps for pea moth. Significantly, they suspend their (strong) source of synthetic pheromone in the wheat fields where the moths emerge, *at plant level* where the real calling females perch. Whereas the source was suspended unnaturally higher up in the free, smoother-flowing air above the plant (grass) canopy in our field experiments with gypsy moths and bubbles (David *et al*. 1983), and likewise in Murlis and Jones's (1981) field study of wind-borne plume structure using ionized air (Murlis (Chapter 3)).

The pea moth pheromone was absorbed by and re-released from the plants in male-attracting amounts, and appeared to lose its plume formation among them in a matter of metres, becoming a diffuse, blanket-like cloud; and in this cloud the males were seen flying upwind everywhere. It remains to be established whether a pheromone source emitting at the natural female rate would also produce secondary emission from plants in male-attracting amounts. But we certainly should have been expecting this, remembering the dramatic male-attracting power of J.-H. Fabre's *empty* boxes that had contained a female moth some hours before.

In any case, leaving such secondary emission and its diffusive effect aside, the airflow within and just above the plant canopy (be that grass sward, wheat field or forest) will commonly be more devious, disrupted and diffusive than higher up (see also Elkinton *et al*. 1985; van der Kraan and van Deventer 1982). Yet the pheromone sources used in wind-tunnel experiments have been set up routinely in unobstructed, smooth-flowing air. Perhaps we have been brainwashed by the physicists into feeling that any respectable wind-tunnel must have laminar flow, and have consequently gone on working with discrete and crisply fenestrated pheromone plumes and correspondingly unnatural sensory inputs for moths not very close to the source. It looks as if the field role of the behavioural mechanisms that have been established in the laboratory may need reappraisal, along with the mechanism of mating suppression by multiple pheromone sources.

Our information to date on the incidence of counterturning is indeed quite confusing. In line with the pea moth observations, gypsy moths in the field did less zigzagging at 10 or more metres from the pheromone source than they did closer to it (David *et al*. 1983; cf. larch case-bearer moths: Witzgall and Priesner 1984). Yet, in wind-tunnels, zigzagging or casting persists in uniform pheromone clouds; and with pheromone plumes, zigzagging becomes *more* pronounced when the concentration is reduced (Baker (Chapter 4)).

Clearly, both the dispersion of odour plumes and the behaviour of attracted insects need more field investigation with the pheromone source in the natural female position in her habitat, along with more appropriate experimental simulations in wind-tunnels. A number of chapters in both the behavioural and the sensory parts of this volume converge on this central area of odour structure affecting sensory inputs and behavioural outputs.

2.7. Categorization of behavioural mechanisms

This more general issue of how—and indeed, whether—to categorize olfactory manoeuvres is also a current one, judging from Bell and Cardé's (1984) remarks about terminology in the preface to their book. There is an urge for a purge of the many special terms that plague this field. Bell and Cardé find a measure of consensus 'towards more practical, functionally related terms, rather than classic Greek stems or teleological terms'. They condemn 'terms that are teleological, poorly defined, non-mechanistic, non-probabilistic and difficult to spell'. The references to Greek stems and spelling articulate a reaction against the array of terms that proliferated for some years, particularly in the German literature, and are best known as the names of the categories in Fraenkel and Gunn's (1940) system. Bell and Cardé offer no alternative frame of reference for workers in this field, as if none were needed. Bell (in Bell and Cardé 1984) proposes that researchers should instead (or in addition) give a detailed characterization of each case including (i) the information available to the animal, (ii) the type of information processing, (iii) the motor output, and (iv) the nature of the guidance system.

It will be a real step forward if authors do provide as much of this information as they can. If they stop at that, however, it will become difficult to see the wood for the trees. Actual cases will differ not only in each of those four properties, but also in the particular way they are combined. Every case will stand more or less by itself in a category of one. The resulting multiplicity of cases will be a catalogue, not a frame of reference capable of replacing Fraenkel and Gunn's. This has been the difficulty with Jander's (1975) logical and erudite survey of the subject. He presented a manageable list of six categories of ecological *function* for spatial manoeuvres (object orientation, geographic orientation, etc.), but 70 generalizations about the manoeuvres and mechanisms themselves 'and only space limited the establishment of many more'. Schöne's (1984) new treatise presents much the same problem.

For the purposes of thought, cases have to be grouped into a limited number of categories, and the categories into a system. For the purposes of communication the categories have to be coded into brief, indicative terms. For example, 'self-steering' codes for Bell's (in Bell and Cardé 1984) 'indirect orientation based on internally stored information or referenced to another modality'. Lacking anything better, this chapter (following Kennedy 1977, 1978, 1983) uses some of Fraenkel and Gunn's terms with amendments requiring revision of the system. They did, after all, define their terms as rigorously as anyone could at the time, adapting, pruning, and improving on Kühn's system and even at that early date avoiding most of the terminological sins listed by Bell and Cardé (above). Out-dated though their system is, its terms continue to appear in research papers and even new textbooks.

This is not mere Anglo-Saxon conservatism. It is due, first, to the continuing convenience of the system terms as a communication code *at a particular level of input–output analysis.* The level is quite superficial. being the organization of two systems: (i) the stimulating environment with respect to one or more sensory modes, and (ii) the associated manoeuvres of a mobile 'black box'. This is a legitimate level of study in its own right. It forms an important part of causal and functional behavioural ecology (Havukkala and Kennedy 1984). That point is overlooked when the Fraenkel–Gunn terms are criticized on the grounds that finding one that fits a given reaction becomes an end in itself, an easy substitute for lower-level, cybernetic and physiological analysis. The deeper analysis is obviously desirable, but it is a different pursuit. It is not the bounden duty of those working at a higher analytical level or a substitute for their necessary work.

In the second place, the survival of the kinesis–taxis terms is due to the fact that the system was robust enough to stand repeated revision in the light of experience. Thirdly, and perhaps most important, although the system is still faulty, there is no alternative system in sight useful enough to achieve a consensus. When one is forthcoming it can be relied on to displace the old warhorse by natural selection.

There have been a series of recent, detailed critiques of the Fraenkel and Gunn system, culminating in Schöne's (1984) monumental treatise and, as far as insect olfactory behaviour is concerned, in the reviews of Bell and Tobin (1982) and Bell (in Bell and Cardé 1984). They include the basic criticism elaborated above in Section 2.3, that 'the role of internally stored information is not considered' (Bell and Cardé 1984). But these criticisms are concentrated on the original system without considering how far past and future revisions might meet the criticisms, at any rate for a while.

So we find ourselves in a most unsatisfactory situation. Everyone is dissatisfied with Fraenkel and Gunn's terms, but that is as far as consensus presently goes. Some authors see only the weaknesses of their system and favour simply abandoning it to die of neglect. But this alone cannot achieve a better consensus because others feel the need for some such fairly simple system of categories and so far can see no better alternative.

However, both parties may well come to agree on the limitations of the Fraenkel–Gunn system, which no amount of revision can escape. It is limited in the range of behaviours it can embrace even at its own superficial, near-ecological level. It is also limited *to* that level, that is to say in its explanatory power. It cannot incorporate, for example, the underlying complexities of the optomotor control system in odour-modulated anemotaxis (David (Chapter 5); Preiss and Kramer (Chapter 7)). Once these limitations are generally recognized, then it may be more generally felt to be worthwhile continuing to revise it for its own proper purposes. It is in the very process of doing this that a better alternative seems most likely to emerge and replace it entirely.

Thus my stop-gap term 'longitudinal klinotaxis', a product of the history sketched in Section 2.3, seems already to have outlived its usefulness. It does not in fact refer to a taxis and is clumsy and confusing. Risking much obloquy, a new name, *schemakinesis*, is proposed in its stead. The prefix is familiar in the English (and German) word *schema* and comes from the Greek word meaning a pattern of various kinds (including, felicitously, 'a dance figure'). Given that replacement, the term 'transverse klinotaxis' can revert to plain 'klinotaxis' in the original sense, tidying the system considerably. Table 2.1 then summarizes the position currently reached in the revision process as far as olfactory attraction is concerned, omitting odour-induced visual attraction (Prokopy (Chapter 8); Lanier, and May and Ahmad, in Ahmad 1983; Miller and Strickler in Bell and Cardé 1984).

The word 'manoeuvre' has been used instead of 'orientation' throughout this chapter, advisedly. Orientation has a strictly directional connotation in English (being derived from the Latin *orior*, to rise, and referring to the direction of the rising sun: hence *oriental*, and the *orientation* of a church).

It is therefore confusing in English to include the kineses under 'orientation' since these manoeuvres have no directional relation to the stimulus direction. The confusion arises from mistranslation of the German word *Orientierung* which unlike 'orientation' does embrace all kinds of adaptive manoeuvre—like the English word for a modern sport, *orienteering*. It is indeed unfortunate that there is no English word as comprehensive as *Orientierung*, which Jander (1975) defined as active 'maintenance or change of body position relative to the environmental space'.

2.8. Terminology and teleology

Finally, the current concern over terminology raises an ancient philosophical issue with a direct bearing on future research. The recommendation that both teleological and Greek-stemmed technical terms should be avoided in favour of 'more practical, functionally related terms' is fine except for one thing: teleology can hardly be avoided without the discipline of special technical terms.

Naturally enough, technical terms with strange Greek roots cannot expect a ready welcome, especially in a new and strongly pragmatic field like chemical ecology. The fact remains that in the development of every other scientific discipline it has been found necessary to adopt some specially defined and standardized technical terms, often deliberately derived from dead languages, for the express purpose of avoiding, as far as possible, the ambiguities, individually variable overtones, and conscious or unconscious assumptions of the familiar terms used in everyday speech. It is therefore both interesting, and important, to ask *why* it is felt that this particular branch of science, insect behaviour, should be the single exception to the general rule. What is there so special about behaviour?

It cannot possibly be the strangeness or the spelling of the terms—'klinokinesis' is not harder to memorize or to spell than 'enantioisomer'. What is special about behaviour is simply that we 'know all about it' already, and that is a dangerous reason for making it the exception. Each of us is intensely aware of behaviour and has a ready vocabulary for it in ourselves and our conspecifics. In this respect no other domain of scientific enquiry comes anywhere near it. We have no comparably direct knowledge of molecules or cells or X-rays or quarks, no ready-made vocabulary for them in everyday discourse. Any anthropomorphic terms for them such as 'Maxwell's Demon' are unmistakably metaphors only.

Not so with behaviour, because we know it so well subjectively and nothing comes more easily than applying this knowledge and its terminology to the behaviour of animals too. Subjectively, behaviour is purposeful; it requires a deliberate mental effort *not* to assume that insects are likewise driven by purposes. That of course is teleology: the explanation of behaviour by its end-result—one of those 'hand-waving' explanations that is immediately satisfying and waives the need for closer scrutiny of the animal's actual behaviour. The literature of olfactory behaviour is still littered with teleological terms such as 'searching' and 'appetitive'. Being teleological, what they really describe is the animal's presumed state of mind, not what it is actually doing. What it is actually doing goes without saying—or even looking.

If that were not still a real problem there would be no objection to teleological terms now that nearly everyone recognizes, at any rate in principle, that a teleological description is only a handy metaphor for the *function* of an item of behaviour, which presumably gave it survival value in the past (its 'ultimate' cause), and tells us nothing about its present ('proximate') causes.

One sure result, then, of a general shift towards 'more practical, functionally related' terminology would be more use of the familiar language of our everyday speech and hence more teleological, anti-analytical language and thought. Such a trend would be a sad retrogression, for there has been enormous progress in the opposite direction since the unbridled anthropomorphism of Fabre and other early students of insect behaviour.

Many teleological terms will doubtless remain in daily use because they are handy and it is taken for granted that they have now been stripped of their anthropomorphism. At the same time, trying to find an objective substitute for a teleological term will always pay off in research because it forces a mental break-out and a closer look at the components of the behaviour as actually observed, components which the handy teleological term leaves unnoticed or at best unformulated in the back of the mind.

For example, the activity of a male moth that is ready to mate but has not yet picked up the scent of a female, is described teleologically as 'appetitive'. The same behaviour has been described objectively as 'random flight' by Perry and Wall (1984b), a term that pinpoints the locomotory component of the behaviour. But Cardé (Bell and Cardé 1984) wished to bring out, in addition, a second observed component of this behaviour: the male's especially low response threshold for the female's chemical signal, on receipt of which he promptly switches into different behaviour. To convey that, Cardé resorted to the teleological term 'appetitive' for lack of an objective alternative. In fact Jander's term 'ranging' as he defined it (1975, p. 180) does include this component, as well as the other two, *viz*. locomotion, and long stretches of straight track between turns. Bell (in Bell and Cardé 1984) therefore applies it to the moth behaviour in question.

The risk of falling into teleology at the expense of close scrutiny of the actual manoeuvre makes special care over terminology not less, but *more* necessary in animal behaviour than in any other branch of science. And that means some standard technical terms, not excluding Greek stems. Animism has always been with us in the ambient culture and is not about to disappear even among scientists (e.g. Baker 1982; Thorpe 1954).

Acknowledgements

Thanks are due to Drs W. J. Bell, C. T. David, R. Dawkins, J. N. Perry and C. Wall for helpful comments, and to several authors cited in the text for permission to cite their unpublished work.

3 The structure of odour plumes

J. Murlis

3.1. Introduction

This chapter is about the form an odour signal takes, what information it contains and how this information could be extracted. The discussion is limited to small odour sources positioned just above the ground, for example a calling female moth sitting on a leaf at the top of a plant. It is therefore strictly relevant only to aerial trails and flying insects, although some of the conclusions apply equally to sources at ground level and walkers.

3.2. A description of odour plumes

An odour signal is the product of the structure of the odour trail that is made as odour molecules are taken from their source by the wind. This aerial trail is known as an odour plume.

Odour molecules are released from their source by evaporation. They immediately begin to move about randomly, which is known as Brownian motion, and as a result they move apart from each other, a process which is known as diffusion. In doing so the molecules become less concentrated, but they also migrate through the thin layer of still air surrounding the source until they enter the freely moving wind. Once in the free stream, the molecules are carried away from their source; this process is called dispersal. In the very early stages of dispersal there is a coherent trail of odour molecules, gradually expanding, still by molecular diffusion, as it moves from the source. There is, however, another mechanism which produces diffusion, and that is the disorderly movements of the air known as turbulence. In the atmosphere, turbulence is generated by mechanical forces and by buoyancy; it starts as motions on a scale of up to the atmospheric boundary layer thickness, about a kilometre. These, the largest eddies, transfer their energy to successively smaller eddies until the whole layer of air next to the ground contains eddies of all sizes moving randomly in all directions. The smallest eddies have the least energy, and viscosity is continually removing what little they have; this is called viscous dissipation. Unlike the Brownian motions, which cause molecular diffusion, the turbulence therefore has a significant lower size limit, known as the Kolmogoroff scale, which is roughly (and empirically) a function of wind speed and height above ground (Miksad and Kittredge, 1979). In the atmosphere the Kolmogoroff scale is very approximately a centimetre. (In a typical wind-tunnel airflow it is a fraction of a millimetre.) There are two consequences of this. Firstly, the odour trail has to grow to roughly the Kolmogoroff scale in cross-section before turbulent diffusion can take place. Secondly, the turbulent plume will contain parcels of clean air roughly this size. Once turbulent diffusion does start, it works far more rapidly in spreading the odour molecules than molecular diffusion does. Cigarette smoke shows this: the plume of smoke rises at first in a gradually expanding uniform stream, but within a few centimetres it suddenly breaks down into a cloud of unevenly distributed smoke filaments and expands far more rapidly.

A smoke source in the atmosphere, a chimney for example, illustrates the next stage: the turbulent plume waves about vertically (undulation) and horizontally (meandering) as it moves further from the source. This is due to the action of the larger turbulent eddies. The undulation scale depends on the source height, but the scale of meandering can be as large as the largest eddies present. The length of the initial, molecular diffusion stage of plume growth depends on the source size and the Kolmogoroff scale. For a source of a millimetre in diameter in light winds, Miksad and Kitteredge (1979) estimated it would be some metres in extent. Aylor, Parlange and Grannett (1976), however, concluded that for practical problems the effective source size would be considerably larger, a centimetre or so, and that the initial growth stage was unlikely to exceed a few centimetres. This discussion concentrates on the turbulent stage of plume development.

3.3. The general form of an odour signal

Fig. 3.1 shows how a typical plume would look in plan view if the odour molecules within it could be visualized and the pattern frozen at an instant in time. A small odour receiver moved from downwind up to the source would, assuming perfect response, experience and register a fluctuating, intermittent odour signal, a series of bursts of odour separated by odourless patches, like that shown in Fig. 3.1, trace 2. In the atmosphere the duration or length of the bursts would range from a few tenths of a second up to a second. The gaps between the bursts, which cause the signal inter-

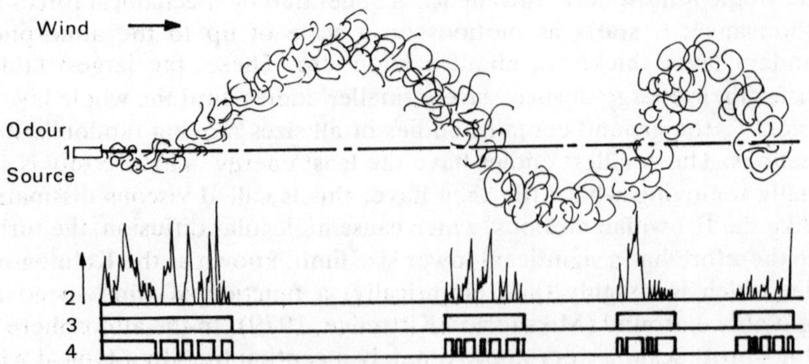

Fig. 3.1. A schematic drawing of an odour plume and an odour signal.

mittency, would cover a wider range; three factors contribute to them, which in order of increasing scale are:

(1) imperfect mixing within the plume, the 'filamentous nature' first recognized in this context by Wright (1958);
(2) the undulating and meandering of the whole plume;
(3) synoptic scale movements of the mean wind direction.

In practice because such a wide spectrum of sizes is involved it is not possible to separate precisely the contributions of each of these to overall intermittency. However, in the atmosphere, small-scale intermittency comes from gaps between bursts of a few tenths of a second, undulating and meandering give rise to gaps of up to a few seconds, and synoptic scale motions can cause gaps in the signal of many minutes.

In the special case of wind-tunnel plumes, undulating and meandering are limited (by the wind-tunnel dimensions) and synoptic scale motions are clearly absent. Burst lengths and gaps have similar values and can be very short, down to a millisecond.

A perfect receiver would detect intermittency from all sources without being able to determine with certainty which source was involved. Even if by adroit manoeuvring the receiver could stay inside the plume, the signal would still appear to be intermittent.

The other important characteristic of the odour signal is its fluctuation in strength. The individual bursts contain varying amounts of odour and the concentration within them varies, as shown schematically in Fig. 3.1, trace 2. The next section deals with methods of measuring this odour signal.

3.4. Estimating odour plume structure

Detection of odour molecules themselves is only possible over very long sampling times. So in order to measure the odour signal in all its detail the odour molecules must be simulated by some other more readily detected material (such materials are known as tracers). Various chemical tracers have been used in conjunction with fast flame photometric detectors to analyse plumes. Resolutions of up to 30 ms were obtained by Barynin and Wilson (1972) and Hadjitoffi and Wilson (1979) using sulphur dioxide and phosphorus pentafluoride. These materials could be detected more than 5 km from their sources. Fackrell (1978) used a dilute propane tracer and obtained a resolution of 3 ms in wind-tunnel experiments.

Another approach is to use ionized air as a tracer and to detect it electrically. Jones (1977) described a method using unipolar ions and giving a resolution of a few milliseconds. The range of ions in the atmosphere depends on their lifetime, so that it is the time of flight which determines range. In moderate winds, ions can be detected some hundreds of metres from source. Corrections can be made for the slight repulsion which occurs in a unipolar ion cloud and which exagerates the expansion of the plume (Jones 1983), but a bipolar ion source avoids this problem. Boreham and Harvey (1984) describe such a system for use in wind-tunnels. It has a resolution of 1 ms, and the detectors can be as small as 5 mm in diameter. This system is now under development for field use (C. D. Jones, personal communication). In the case of a bipolar source, corrections must be made for the reduction of concentration caused by progressive recombination as time of flight increases.

Ionized air methods are straightforward in use: ion generators and detectors (called ion collectors) can be simply made and need only battery power. The collector output is stable and is in the form of an electrical signal which can be directly amplified and recorded.

A set of typical ion collector signals is shown in Fig. 3.2. These were recorded at a height of 1 m, at fixed positions downwind of a 1 m high source. The amplifier gain has to be increased with distance from source, so these traces are not to the same scale. There is a difference of two orders of magnitude between the strength of the closest and furthest signals and, although they are closely similar to the schematic traces of Fig. 3.1 the further downwind signals are relatively more noisy.

These signals are the closest simulation available to us of how an odour plume is presented to an insect. Their analysis is described in the next section.

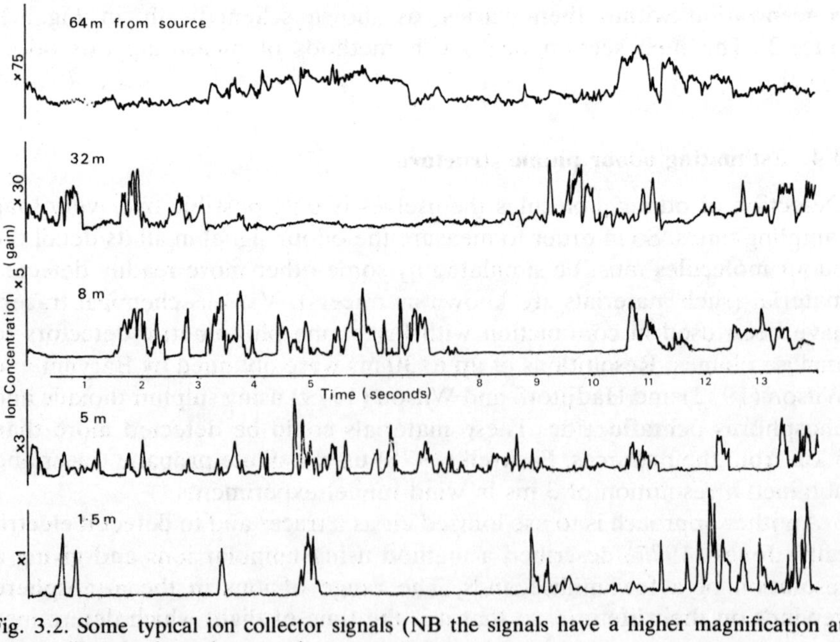

Fig. 3.2. Some typical ion collector signals (NB the signals have a higher magnification at greater distances from the source).

3.5. Analysis of plume structure

It is usual in the analysis of physical processes to deal first with averages and then with a detailed breakdown of the signals into their various parts. In this case, however, I believe that it will be more useful to deal first of all with the pure instantaneous signal and then with the various ways or processing it to extract information, including averaging.

The relevant properties of the instantaneous signal are the lengths of odour bursts, the times between them, their strengths, and how these are related to the position of the receiver. In Fig. 3.1, trace 4 shows a signal which takes the value 1 when there is odour present and 0 otherwise. This binary trace is the intermittency function for trace 2. The same method could be applied to the ion collector signals in Fig. 3.2 and it would then give the intermittency function at a fixed point relative to the source. The analysis of the intermittency function yields information about burst lengths and the rate at which bursts arrive at the receiver.

Murlis and Jones (1981) made an analysis of intermittency functions of ion signals by measuring the lengths in time of a sequence of bursts and the times between the arrival of successive bursts, and then showing these as a frequency distribution. Fig. 3.3 shows a similar analysis of ion signals recorded at night during trials with moth sex attractants. Burst lengths and return periods are approximately log-normally distributed, lengths over a range of a few milliseconds up to a second with a modal value of about 100 ms, and return periods over a range of 30 ms to 30 s with a mode at about 0.5 s. Fig. 3.3 also shows the effect of higher thresholds for detection on the probability distributions. The range of burst lengths is successively reduced and the modal value decreases. For return periods the range remains similar but higher values become relatively more frequent.

There is recent evidence that these time-scales are of biological significance. Insects can detect odour signals as short as 20 ms in duration, see Kaissling (Chapter 22). Short burst return periods are important in maintaining orientation; at periods in excess of about 1 s orientation is considerably weakened (see Priess and Kramer (Chapter 7)). However, Jones (1983) showed that these distributions do not depend greatly on the distance between the source and the receiver, a surprising result which implies that the time-dependent pattern of an odour signal would not convey information to a receiver about its position relative to the source.

More likely sources of positional information are the strengths of bursts, which diminish on average as the distance between source and receiver increases. Murlis and Jones (1981) took the maximum tracer concentrations within bursts as a measure of the bursts' strengths. They found that mean strength did in fact diminish, but that there was a wide distribution of values for individual bursts (Fig. 3.4). They concluded that a flying insect would be unlikely to obtain consistent positional information by comparing the strength of a burst it encountered with the previous burst's strength as

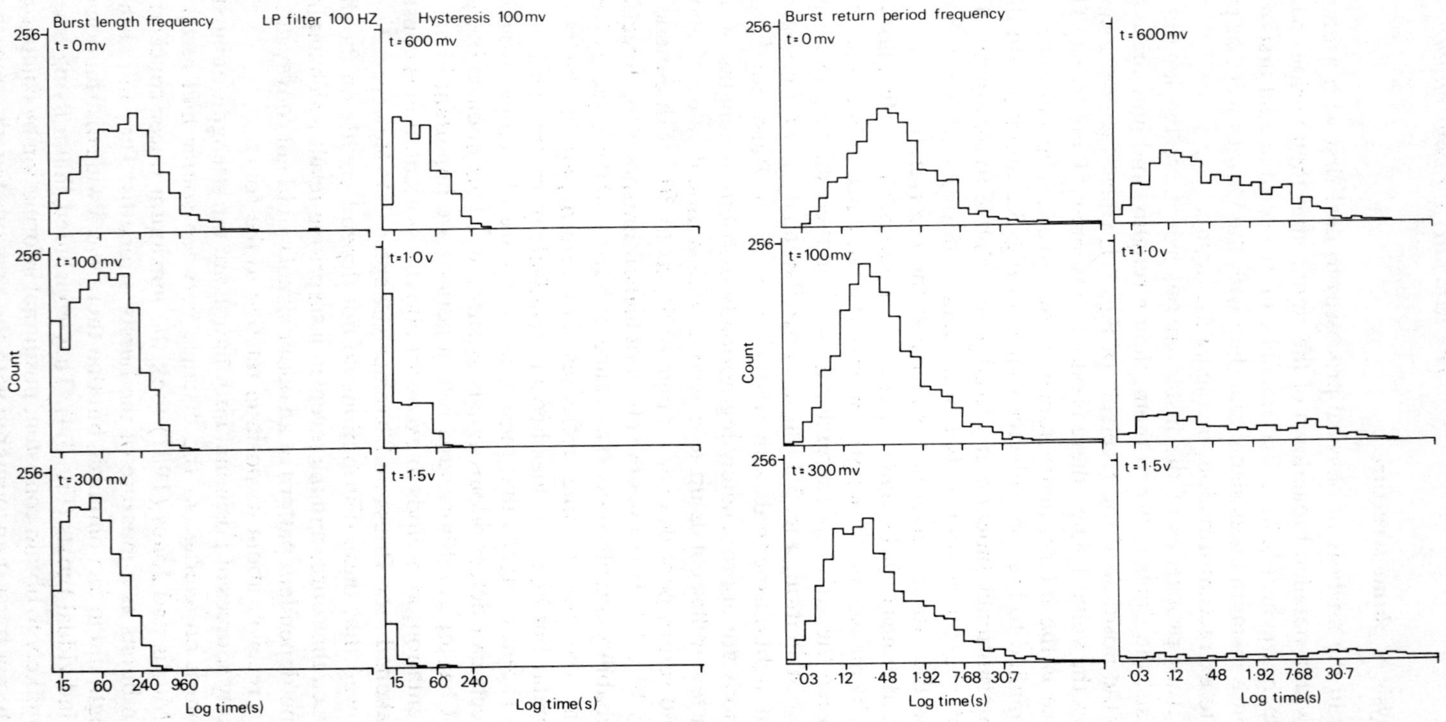

Fig. 3.3. Frequency distributions of burst lengths and return periods from an ion signal recorded by a collector 1 m high at 5 m downwind from a 1 m high source during night-time pheromone trials (t = threshold for detection of bursts).

the stronger of the two would not reliably be located nearer to the source. The modal burst strengh at 10 m, for example, occurred with a finite probability at 15 m from source (Fig. 3.4).

The odour signal can be processed by short-period averaging to smooth the signal input. This is the equivalent of signal filtering. Fig. 3.5 shows the effect of averaging on the ion collector signal analysed in Fig. 3.4. It can be seen that the signal becomes noticeably smoother as the averaging time increases, but that even at an averaging time of 3.3 s it is still intermittent (which may be important in delaying the onset of adaptation or habituation in a receptor system with a slow response time) and the major bursts of the original can be clearly distinguished. At a moderate averaging time of 1 s, most of the original information is preserved intact. The influence of averaging on the variation in burst strength, however, is not great. Jones (1983) shows that an increase of averaging time from 0.03 to 3 s causes the

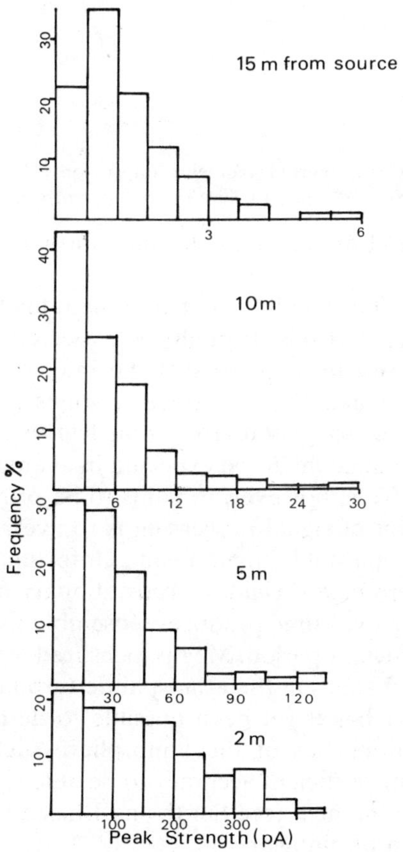

Fig. 3.4. Frequency distributions of peak strength of ion bursts at set distances downwind of an ion source (collector and generator are at 1 m height above ground).

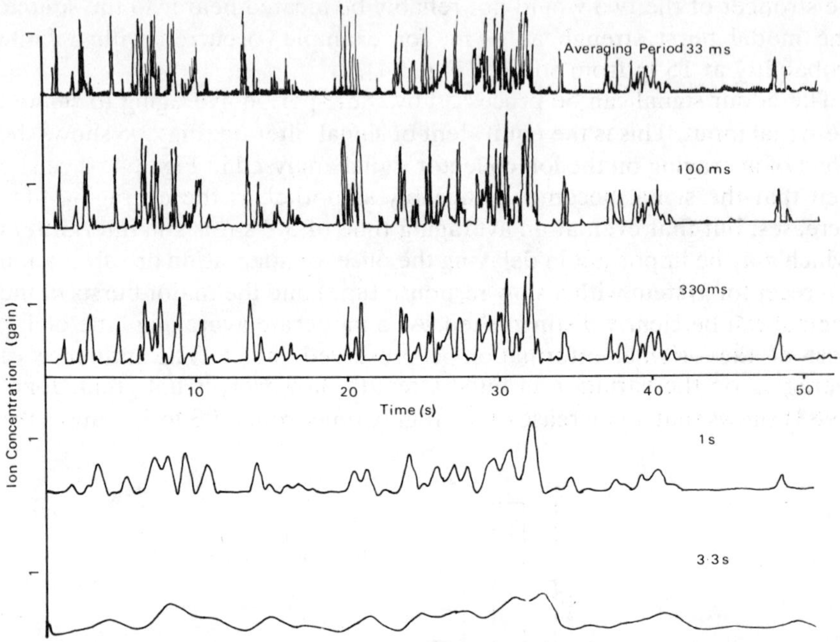

Fig. 3.5. The effect of short period averaging on an ion collector signal. The averaging periods are 33 ms, 100 ms, 330 ms, 1 s and 3.3 s. Collector signals are analysed in Fig. 3.3.

standard deviation of an ion signal to fall by a half (the mean, of course, remains constant).

One further possibility exists: that burst lengths would be affected by averaging over short periods. It might be expected that averaging would tend to fill in the smaller gaps so that the intermittency function would begin to look more like Fig. 3.1. trace 3. Jones (1983) shows that the plume radius expands roughly tenfold from 1 m to 15 m downwind of the source. A similar change might be expected in averaged burst lengths over a similar distance. This, however, remains to be proved.

The simplest form of signal processing is to average over a sufficiently long time to produce a stable 'time-mean'. Unfortunately atmospheric dispersion does not produce variables with stationary means, and this is not possible. Averaging over time produces 'ensemble-means', describing conditions for that particular period. Means measured in different periods vary considerably with changes in the atmospheric turbulence and the synoptic wind direction. Nor has it yet been possible to describe the influence of easily measured properties of the atmosphere such as temperature on plume structure with sufficient accuracy to be able to extrapolate structure from one period to the next (see Smith and Blackall 1979 for a discussion of parameterization of plumes).

Despite these problems, mean values, usually of concentration, have traditionally played an important part in the study of plume structure.

It has been recognized that mean concentrations would be of little value in orientation (Bossert and Wilson 1963; Schwink 1958). However, the use of mean concentration in defining the region of action of an odour source is common (see for example Bossert and Wilson 1963; Fares *et al.* 1980), and such measurements therefore have considerable historical importance in the study of odour orientation.

Measurements of mean concentration in a wide range of meteorological conditions have been made and are summarized in Pasquill (1974). The downwind distribution and cross-wind profiles of concentration for any particular set of conditions can be approximated to simple power laws, but as these cannot be interrelated other than by empirical and unreliable functions of the meterorological conditions, they have no general validity.

Measurements of mean values of the characteristics of the intermittent plume structure are less common: Murlis and Jones (1981) give a series of ensemble-means of ion collector signals (see Table 3.1). They show that the mean burst length is about 0.3 s, the mean return period is about 1 s, and that they do not significantly change as the source to receiver distance increases. The mean signal intermittency, the average value of the intermittency function over a given period, varies unsystematically but is typically about 20 per cent. In a highly intermittent signal there will be a considerable contrast between mean values and the peak signal value. Murlis and Jones (1981) calculated the ratio of the mean signal peak to the overall mean concentration to be more than 20. The peak value within an ensemble can be much greater; Storebø *et al.* (1983) found parcels of undiluted ions more than 30 m from the source. Thus even where mean concentration is below a response threshold there may be intermittent stimuli sufficently strong to trigger a response. Elkinton, Cardé and Mason (1984) found moths responding to pheromone where concentrations were predicted to be well below the response threshold.

Because of the importance of peak values it may be better to give mean concentrations in the form of probability distributions. Mean signal strength is given in Murlis and Jones (1981) as the mean concentration at a

TABLE 3.1

Average characteristics of an ion detector signal (generator and detector at 1 m above ground level)

	Downwind distance from source			
	[2 m]	[5 m]	[10 m]	[15 m]
Mean burst length	0.25 s	0.27 s	0.27 s	0.34 s
Mean burst return period	0.91 s	1.36 s	1.19 s	1.28 s
Mean signal intermittency	23%	15%	20%	16%
Peak : mean ratio	22	26	20	21

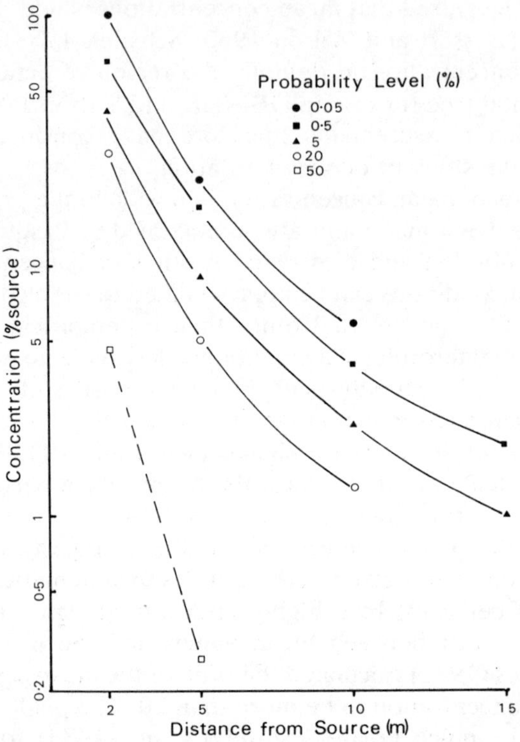

Fig. 3.6. Concentration probability of ions downwind of an ion generator (collector and generator are 1 m above ground).

series of set probability levels. Fig. 3.6 shows these data for up to 15 m from the ion source. It can be seen that undiluted material is expected 2 m downwind about once every 3 min, assuming a median burst length of 100 ms. At 10 m downwind the 3 min peak is nearly 10 times the peak which would be measured on average in a 1 s sample. This method could be used to produce a 'map' of odour distribution on a probability basis, but more understanding of the processes causing the non-stationary nature of atmospheric diffusion would be needed before it could have general value.

A more comprehensive set of tracer trials using groups of ion collectors has been carried out both in the day and at night in order to improve information about concentration probability. Four ion collectors were evenly spaced along a 90 degree circular arc centered on an ion generator placed roughly upwind. Three-hour-long recordings of the collector signals were made at distances from the generator of 2 m to 64 m. At any one distance, the time and amplitude structures of the four collector signals, averaged over 3 min, were statistically indistinguishable. Thus active spaces may be far greater in extent than generally considered.

Further analysis is being carried out on these data.

3.6. Plume calculation methods

Most existing methods of calculating odour plumes are designed to predict mean concentration distributions. Such methods are summarized by Elkinton and Cardé (1984), and it is concluded that they have limited value for our present purposes.

There are, however, new methods which show greater potential. Hall (1975) described a method of simulating the movement of tracer molecules by a numerical random walk. Legg (1983) used the method for problems of biological interest and achieved a better agreement with experimental results than could be obtained using more classical methods. The critical innovation in random-walk methods is that the processes of atmospheric turbulence are modelled in the simulation, rather than being described by simple functions.

The majority of random-walk models, such as those above, describe the movement of a single particle for which many paths are averaged. They therefore give mean concentrations. However, there are now newer methods being developed for treating particle pairs (see for example Sawford 1983). These can model the separation between the particles and can therefore give higher moments of concentration and other characteristics of the instantaneous structure of a turbulent plume. Although two-particle models presently use considerable computer resources, they do represent the necessary way forward for odour plume calculation.

3.7. Conclusions

Odour plumes can be modelled using suitable tracers, and simulated odour signals can be recorded at fixed points. Analysis of such signals strongly suggests that in an unprocessed form an odour plume would not carry information which an insect could use for orientation to the odour source.

Smoothed signals, averaged over up to a second or so, may carry positional clues in their time structure, but more work is needed to confirm this. At present, however, it must be concluded that the main role of the odour signal is to act as a switch to activate orientation behaviours. The fine scale structure is therefore of crucial importance in maintaining an oriented response, and in defining the region of influence of the source. Mean concentrations do not provide helpful definitions of the region of influence because of the large peak to mean ratios within an odour signal. A map of concentration probabilities would be a more useful basis.

Calculation of concentration probabilities could be possible in the near future using random-walk methods of plume modelling. Tracer experiments are needed to provide experimental data for probability maps of concentrations of odour downwind of sources in various weather conditions and environments. A study by tracer of odour distributions in forests and above crops would be particularly valuable (see Cardé (Chapter 20) and Perry and Wall (Chapter 9)).

Acknowledgements

I am happy to acknowledge the collaboration of Dr C. D. Jones of CDE Porton Down in the TDRI Pheromone Plume Structure project. In particular I am grateful for his advice and encouragement during our field trials.

I wish to thank my colleagues at TDRI for their advice during the preparation of this chapter, particularly Mr D. E. Pedgley, Dr. J. R. Riley, Mr M. R. Tucker, Dr D. G. Campion and Dr W. A. Sands. I am indebted to Professor J. Kennedy and Dr C. David for further suggestions. I also wish to thank my wife Helen for typing the first drafts and for editing the final version.

4 Pheromone-modulated movements of flying moths

T. C. Baker

In the earlier days of sex pheromone research, when it was thought that moth species each used one compound to communicate, it was known that different concentrations of pheromone must be affecting the movements of moths because too much or too little pheromone could result in reduced captures of males in traps. When in the early 1970s, it became clear that the majority of species used blends of compounds, the same trap capture-based inferences on movement were made, only now the issue was blend quality. Then we actually began to watch the moths; we advanced from previous inferences about 'synergism' and 'inhibition' (many authors), and 'primary' and 'secondary' pheromone components (Roelofs and Cardé 1977) to statements about 'close-range' and 'long-range' effects of chemical blends. From here we advanced to 'attraction' and 'arrestment', terms actually describing the gross movements of moths (Baker and Roelofs 1981). During this evolution of thought, J. S. Kennedy (1972) cautioned that we did not in fact know that pheromone chemicals directly affected the movements of moths, and that all the evidence had thus far indicated that the chemicals merely switched on a steering response to wind, not to the chemicals themselves.

Wright (1958) earlier had described a way that attractant compounds could be used directly for steering. The frequency of pheromone filaments in a plume could code for the towards-source direction. Later he abandoned this idea when evidence from his experiments showed the overwhelming influence of the optomotor anemotactic response in odour-mediated insect flight behaviour (Wright 1962, 1964). Optomotor anemotaxis was first demonstrated by Kennedy (1940) in the flight of yellow-fever mosquitoes. Farkas and Shorey (1972) brought chemical steering to the fore again when they claimed that their pink bollworm males steered only with respect to the chemicals, not to the wind at all, in successfully flying up a pheromone plume to the source in zero wind. In their view, the wind merely served to draw out the pheromone into a linear trail which was 'followed' chemotactically by flying moths in a manner similar to the way ants follow terrestrial trails. Kennedy and Marsh (1974) apparently neatly negated this idea when they demonstrated with another species how Farkas and Shorey had overlooked the powerful optomotor anemotactic response that had influenced displacement of the moths before the wind had been stopped. Kennedy and Marsh (1974) and Marsh *et al*. (1978) maintained that two purely anemotactic mechanisms, positive anemotaxis in the presence of odour and reversing anemomenotaxis upon odour loss, still best explained the ability of moths to locate a pheromone source in wind. Onset and offset of odour merely served to switch on these steering responses to wind.

If it were true that pheromone merely switched on anemotaxis, I would have very little to talk about from here on, because this talk is supposed to be about how pheromone chemicals modulate the movements of flying moths. Happily, I can continue on much longer because experiments by J. S. Kennedy and colleagues using clouds of pheromone (Kennedy 1982; Kennedy *et al*. 1980, 1981) and by our lab using zero wind (Baker and Kuenen 1982; Kuenen and Baker 1983) showed that, apart from triggering anemotaxis, sex pheromone does directly influence the movements of flying moths by altering the frequency and angular magnitude of zigzags. Unlike Farkas and Shorey (1972) who proposed that all turns were steered with respect to pheromone gradients (both lateral and longitudinal) Baker and Kuenen (1982; Kuenen and Baker 1983) invoked the concept of self-steering (Kennedy 1978, 1983). Pheromone stimulation would initiate and maintain a programme of movements, which for the oriental fruit moth, *Grapholita molesta*, consists of relatively narrow and regular 'zigzags' (Kennedy (Chapter 2), or counterturns (Bell and Tobin 1982; Kennedy 1983). As in *Plodia interpunctella* (Marsh *et al*. 1978), the turns are mainly performed in the horizontal plane with only about half as much movement occurring vertically (Willis *et al*. unpublished). Because the movements are part of a 'programme' originating from within the moth, they are therefore self-steered and rely on idiothetic cues for their execution (Kennedy 1978, 1983, Chapter 2). They would not, therefore, be the type of direct reactions to an odour gradient envisioned by Farkas and Shorey, but would fall into the class of indirect reactions to odour (Kennedy 1978), as more recently outlined by Bell and Tobin (1982).

From here on I will present further evidence, mainly using the oriental fruit moth, that supports the idea that zigzagging flight in orientation to pheromone is primarily self-steered, not gradient-steered. I will also present evidence that this self-steered zigzagging needs to be integrated with optomotor anemotaxis in order for the optimum efficiency of source location to occur. Finally I will argue that the self-steering programme in some moths can be maintained only through intermittent, not continuous, chemical stimulation.

We have observed the counterturning programme, uncomplicated by wind, by releasing male oriental fruit moths in zero wind into a stationary pheromone plume (Baker *et al*. 1984). Such moths initiated a series of left–right turns of about 180°–200° magnitude, each performed with about 60° accuracy, on average. Because the turns alternated left and right and had a high degree of angular regularity, they fall into the category of 'counterturns' (Kennedy (Chapter 2)). Another form of 'regularity', the *temporal* regularity of consecutive reversals, is not considered here, although they also had this type of regularity (Willis *et al*. unpublished). With no pheromone present (again beginning in zero wind) turning magnitudes were more varied, with many more shallow turns; the difference between consecutive turns, one measure of counterturning regularity, was also more varied than with pheromone present. Importantly, without any wind or in-flight experience in wind the counterturns meandered around in the plume and the tunnel. In pheromone, they were performed with high angular accuracy with or without wind, but without it there was no significantly preferred counterturning direction with respect to the source or the plume, as measured by interreversal track angles. Instead of being oriented with respect to the 'edges' of the stationary plume, the zigzags were as likely to run parallel to them, which is inconsistent with the idea that each counterturn is steered with respect to the plume's lateral gradients (Farkas and Shorey 1972) or otherwise 'related to' the plume's boundaries (Tobin and Bell 1982).

As visualized by smoke, the stationary plume diffused slowly outward from the central axis, resulting in the time-averaged boundaries expanding and reaching the walls of the tunnel after a minute or so. Therefore even many of the zigzags performed perpendicular to the edges were well within the plume's 'edges'. Thus the counterturns appear to have been self-steered according to a motor programme, the angular regularity of the consecutively performed turns being a result of an unknown internal mechanism or perhaps also due to external, visual feedback (Ludlow 1984; David (Chapter 5)).

The counterturning tracks had a consistent alignment to anything other than to the immediately preceding turns only when wind was present, or in zero wind when males had already taken flight before the wind stopped (Baker and Kuenen 1982; Baker *et al*. 1984). The latter males already had a chance to experience wind optically and to compensate for wind-induced drift through optomotor anemotaxis. Their success at locating the source was significantly greater than that of males that had experienced wind only mechanically while on the ground (Baker *et al*. 1984). Thus the counterturns were polarized in a direction generally across the wind-line by the optomotor response to wind-induced drift (Ludlow 1984; David (Chapter 5)). It is apparently only a coincidence that they are *also* oriented consistently with respect to the edges of the plume and that the moths thus appear to be steering by them.

Other previous experiments in wind with this species (Kuenen and Baker 1982b) and with others (Cardé and Hagaman 1979; Farkas *et al*. 1974) indicate that the concentration of pheromone inversely affects linear velocity of flight and directly affects the frequency (and therefore amplitude) of the counterturns. At the highest concentrations, the moths' upwind zigzags are the narrowest and are executed very often entirely within the boundaries of the time-averaged plume, not what one would expect for chemotaxis in such plumes that have theoretically wider behaviourally 'active' boundaries than weaker plumes.

Oriental fruit moth males flying to the most concentrated sources eventually bring themselves to a point where in-flight arrestment of progress toward the source occurs. This cessation of progress is not brought about by a further concentration-mediated change in linear velocity or counterturning frequency, but rather by an apparent sudden increase in the course angle between each reversal, resulting in 90° cross-wind tracks where such tracks used to be about 70° (Kuenen and Baker 1982b). Thus it appears that the concentrations causing arrestment do not significantly affect the self-steered programme, but somehow they modify the optomotor response.

Blend quality alterations also can result in in-flight arrestment of oriental fruit moths, among others (Baker and Cardé 1979b; Baker *et al*. 1981; Linn and Roelofs 1981, 1983; Linn *et al*. 1984), and it may be that similar anemotactic changes also account for arrestment to off-ratios. The movements resulting in upwind flight and arrestment to blend ratios need to be examined in the same detail as those for blend emission rates. One thing is clear from existing data for several species of moths: upwind flight effects can usually and most sensibly be ascribed to the *blend* of components, not to individual components in the blend. For the oriental fruit moth, Baker and Cardé (1979b) and Baker *et al*. (1981) argued, with clear experimental evidence, that the blend of three components acted as a unit to influence all behaviours associated with upwind flight, despite erroneous citations of this same work declaring that individual components each evoked different behaviours (Boeckh and Ernst 1983; Boeckh *et al*. 1984). As soon as the blend is dissected, and only partial combinations or individual components are emitted, attraction to the source is reduced, especially at close range (Baker and Cardé 1979b; Baker *et al*. 1976, 1981; Linn and Gaston 1981; Linn and Roelofs 1983). Authors who try to calculate 'nested' upwind flight active spaces for successive individual compounds in a blend (e.g. Buda 1980) inevitably ignore the data which show significantly lower thresholds for the blend compared with any component alone or with incomplete blends, which would translate in the field into larger active spaces for upwind flight to the complete blend than to any partial mixture. A final mistake is that the behaviour involved in nested upwind flight active spaces is often imprecisely defined. What is forgotten is that *upwind flight* must be evoked by the chemical(s). I would agree with the possibility sometimes raised by neurophysiologists (E. E. Davis, personal communication) that at some point far downwind an emitted blend may be perceived as only one component, for example, the one predominating in the emissions or the one for which there are the most receptor sites. All the others may be below receptor thresholds at that distance. But the critical question here would then be whether the major component alone could elicit upwind flight. If it only elicits activation or take-off, as *cis*-8-12:Ac does for

G. molesta (Linn and Roelofs 1983), then it too is below threshold for upwind flight, and it is still the blend that defines the *upwind flight* active space (Baker *et al*. 1981; Linn and Roelofs 1983). It does not matter whether the moth perceives or is excited by this compound, unless one wishes to redefine the active space of interest as that for *perception* or *wing fanning* or *taking flight*. Partial blends or individual components at close range may be able to evoke responses other than upwind flight, such as copulatory attempts (Bradshaw *et al*. 1983). More experiments with more species certainly are needed to further our understanding of how blends affect behaviour.

The time-period immediately following arrestment is characterized by widening counterturns at right angles to the wind with a cessation of progress towards the source ('casting', see Kennedy 1983), concurrent with an increase in altitude that takes the moth up and out of the plume, and then with increased displacement in the downwind direction (Kuenen and Baker 1983). Casting flight also occurs when the pheromone source is lost, as when the source is removed, but there is a slight but significant delay following loss of the plume before this begins (Kuenen and Baker 1983; Marsh *et al*. 1978, 1981). During this interval, narrow upwind zigzagging continues in clean air, often for two or more turns. Because there are no lateral gradients present in clean air, this behaviour is consistent with a self-steered programme of counterturns, not a single gradient-steered counterturn.

The switchover to casting demonstrates the concentration-dependent continuity of the counterturning programme (Kennedy 1983; Kennedy *et al*. 1980, 1981). In addition, the self-steered nature of the casting counterturns was revealed in experiments in zero wind where males flew out the end of a truncated, stationary plume (Baker and Kuenen 1982; Kuenen and Baker 1983). The longer intervals between reversals upon flight into clean still air, plus the change in magnitude and direction of the turns, were similar to what occurs during casting in clean wind. If wind is present to polarize the casts and it then shifts direction, as it often does in the field, the cast shift with it and the moth continued to maintain 90° reversing tracks across the wind-line (David *et al*. 1983). Then the result is often that the plume is re-intersected at a point closer to the source than where it was lost by the moth (David *et al*. 1983). The wide casts thus may not only aid in relocating pheromone, but may also contribute to progress toward the source in a shifting wind-field.

The narrow counterturning programme during more intense pheromone stimulation and wide counterturning during weak stimulation appears, therefore, to function in keeping the moth in contact with pheromone (Kennedy 1983). But an additional function may be to allow the moth to detect the wind direction quickly, either following a lull, or as it shifts (Baker 1984; Cardé 1984; Kuenen and Baker 1983). Kennedy (Chapter 2) has called this 'tracking the wind direction', and the moth may be constantly 'feeling for the error sign' of image flow across its eyes by zigzagging quickly. This function of counterturning remains to be demonstrated experimentally, however, especially since not all flying insects exhibit zigzagging movements in response to odour (such as *Drosophila*, although upon odour loss they do counterturn across the wind; C. T. David, personal communication).

Perhaps the most interesting evidence for the self-steered counterturning programme in wind comes from experiments in which flying *Adoxophyes orana* males were engulfed in a cloud of pheromone of near-uniform concentration (Kennedy 1982; Kennedy *et al.* 1980, 1981). Upon entering such clouds, males previously casting in clean air briefly narrowed their reversals and made a 'surge' of upwind progress. No excursion across a lateral gradient could have triggered each narrow counterturn because no lateral gradient existed. The males made only brief upwind progress, although the narrow cross-wind counterturning continued for quite a while. A mean of 27 such counterturns was recorded from individual males after they were engulfed by the cloud, each occurring without any change in concentration or the presence of lateral gradients (Kennedy *et al.* 1981).

The lack of sustained upwind progress was not due to the somewhat low concentration of pheromone in the cloud relative to a plume (Tobin and Bell 1982) because a cloud generated by a different technique and having a side corridor of uniform mean concentration twice as high as that of a plume that had evoked upwind zigzagging (and over 100 times higher than the first tunnel-filling cloud) also failed to elicit sustained upwind flight. Likewise, the lack of upwind progress was not due to an excessively high concentration of pheromone, because a point-source plume placed within both low- and high-concentration clouds caused significant levels of upwind narrow zigzagging in the plume (Kennedy *et al.* 1980, 1981).

This brought up the very important point that perhaps males require intermittent, not continuous, stimulation with pheromone, such as occurs in a structured plume, in order to maintain the narrow zigzagging programme and upwind progress. Willis and Baker (1984) confirmed and extended these findings using the oriental fruit moth. Males would not take off and fly upwind in a uniform cloud of pheromone, regardless of its concentration, but would if a point-source plume was within the cloud. Higher cloud concentrations reduced the likelihood of zigzagging up the plume. As in the study of Kennedy *et al.* (1980, 1981), males entering a side corridor uniformly permeated with pheromone would usually narrow their zigzagging at the pheromone–clean air boundary, but here oriental fruit moth males would also progress upwind along the edge and not just zigzag in its vicinity. Higher frequency of zigzags and lower linear velocity of flight occurred along the most concentrated side corridor compared with a weaker one. Males would also zigzag upwind if a corridor and its edge were positioned horizontally along the floor, not vertically. Willis and Baker (1984) concluded that the self-administered, intermittent stimulation resulting from flight in and out of pheromone aided in maintaining the narrow counterturning programme along the edge, and each turn was not steered by the gradient. In addition, because the edge was likely to be somewhat heterogeneous in a zone several centimetres wide owing to turbulent mixing, males could have been receiving slightly fluctuating stimulation from it and this too could have aided in promoting counterturning.

We recently tested the effects of continuous versus intermittent pheromone stimulation in a more direct way in the wind-tunnel, by giving males both a uniform continuous cloud, and one that was pulsed either 1 or 0.5 times per second (Baker *et al.* unpublished). We checked our ability to produce the pulsed clouds by means of smoke plus an infra-red photometer, and also by using an electroantennogram (EAG) sensor placed in the cloud of pheromone at the concentration used for evoking upwind flight. It became clear rather quickly that we could not make our cloud's amplitude and speed of oscillation come anywhere close to what a moth experiences in a plume. A male's receptors register deep, quick stimulation to a plume, with sharp onsets and offsets, whereas we were able to give them only slow, shallow fluctuations. Nevertheless, significantly more males took off and zigzagged farther upwind when the cloud was pulsed at either rate than when it was presented continuously. The males zigzagged upwind without the aid of lateral gradients to steer the counterturns. Kramer (Chapter 6) found that fewer *Bombyx mori* males would walk directly upwind when presented with continuous pheromone compared with pulses at a narrow range of frequencies.

It is not surprising that the zigzags of males flying upwind to the pulsed clouds averaged about twice the width of those flying upwind to the plume, considering the relatively slow and shallow amplitude of stimulation we were able to give them. In previous experiments, weak point-source plumes elicited zigzags approximately double the width of those performed in plumes from sources loaded with 10 to 30 times the amount of pheromone (Kuenen and Baker 1982b). Murlis and Jones (1981) showed that at 15 m downwind from a source of ionized air, the mean concentration of ions was about 50 times lower than at 2 m. The mean of burst peaks (peak-to-trough amplitude) was about 75 times lower at 15 m than at 2 m, although the peak-to-mean ratios stayed relatively constant at a value of about 20, regardless of distance. Therefore, it is conceivable that our pulsed cloud not only mimicked the shallower amplitude of stimulation a male would receive tens of metres downwind of a moderately emitting source, but it may also have mimicked the longer durations and decay times of the pheromone bursts that would occur at such distances (Murlis and Jones 1981).

Considering what has been observed about pheromone-mediated attraction and arrestment in flying moths, it is apparent, as pointed out by Kennedy (1983), that peak-to-trough amplitude fluctuations in pheromone concentration appear to be a key, for some species, to initiating and maintaining narrow counterturning and males' progress toward the source. I suggest additionally that an optimal blend ratio and emission rate may be one that on the one hand maximizes fluctuations in neuronal firing (flickering) at locations far from the source where the plume's filaments are less concentrated, but also minimizes *fusion* of the neuronal fluctuations when the peak amplitudes of the plume's filaments become high very close to the source. The moth's own zigzagging movements may help prevent fusion near the source. However, to an off-ratio or excessive emission rate, the male may eventually reach a zone where even these excursions into clean air can no longer enhance flickering, and arrestment will occur. Kaissling (Chapter 22) has found that receptor cells can respond to rapid, experimentally administered pulses of pheromone such as would occur in a plume, and we were able to record plume fluctuations with the EAG.

For moths requiring fluctuating pheromone stimulation, the traditional way of defining the active space in terms of a time-averaged concentration (Bossert and Wilson 1963) would need to be revised, but in a way that takes into account more than peak pheromone flux (Elkinton and Cardé 1984). Peak-to-trough amplitude (Kennedy 1983) would be sufficient for describing a threshold stimulation for many moths, but would not be adequate for those species with an upper threshold such as the oriental fruit moth (Baker and Roelofs 1981). Here, fusion of the fluctuating signal may also need to be invoked, and the thresholds of flickering and fusion would thus play a part in defining the upwind flight active space for a particular blend and emission rate. The plume's structure would now play a much larger role in the active space and would depend on the wind speed and the amount of large- and small-scale turbulence, among other factors. The small-scale turbulence would be generated from foliage or from other surfaces such as pheromone traps (Lewis and Macaulay 1976).

Recently, interneurones that respond phasically to pheromone bursts have been discovered in *B. mori*, and interestingly some also seem to respond to transverse visual pattern movement (Olberg 1983a). The interneurones exhibit two distinct firing frequencies, 'high' and 'low', and alternate between there states upon repeated presentations of pheromone. Usually, increases (but sometimes also decreases) trigger changes in state. If such interneurones were present in *G. molesta* they would require fluctuating pheromone stimulation to change stages, and the frequency of state changes may be what modulates the counterturning program.

EAG responses of males of many species have often revealed that one pheromone component presented as a brief 30 ms puff will evoke a high-amplitude, fast-decaying response, whereas a second component (e.g. *trans*-11-14:Ac in the red-banded leafroller moth) will elicit a lower-amplitude and more slowly-decaying EAG response (Baker and Roelofs 1976; Roelofs and Comeau 1971). The *trans* isomer of the oriental fruit moth pheromone blend has been implicated as a key cause of arrestment of upwind flight, not only when present in a high ratio at natural concentrations (Baker and Cardé 1979b; Baker *et al*. 1981; Linn and Roelofs 1981), but also at excessive concentrations of the optimal ratio (Linn and Roelofs 1981).

The neuronal events that would result in arrestment due to fusion of the neuronal fluctuations may be the following. First, close to the source the highly concentrated filaments may saturate the receptors. They not only may be unable to continue to reflect the actual peak concentration as it increases past the receptors' capacities (Baker 1985), but the surplus molecules also may then take longer to locate unoccupied receptor sites and be cleared by enzymatic degradation (Kaissling 1971). The receptors, therefore, would be more likely to fire continuously at significant rates even during intervals when the moth is actually in a pocket of clean air or 'outside' the plume during a reversal. For an off-ratio of normal concentration (e.g. too much *trans* in the red-banded leafroller moth blend), the slower time-course apparently required for clearing the *trans* component would be more likely to 'fill in', with neuronal firing, the troughs between actual contact with pheromone filaments.

Conclusions

For some species of moths, pheromone elicits two major responses, one being a self-steered programme of narrow zigzags, the other the optomotor reaction to wind-induced drift which polarizes the zigzags in the upwind direction. The two responses, when integrated, result in optimal pheromone source location. For two species, fluctuating pheromone stimulation is required for upwind zigzagging, whereas continuous uniform stimulation causes the wide casting and eventual downwind displacement characteristic of casting flight in clean air. Pheromone blend ratios and emission rates may modulate movements by producing neuronal fluctuations (flickering) necessary for upwind flight far downwind. Production of flickering may need to be balanced against the blend's tendency to produce a fusion of the fluctuation closer to the source, which may result in arrestment of progress before the source is located. 'Active space' will need to be redefined for those species requiring fluctuating pheromone stimulation for upwind flight.

Future research should be conducted in the field to examine the flight orientation behaviour of males under natural wind and plume-structure conditions. It will be essential to measure, simultaneously with flight track recordings, the wind direction and the plume's physical structure (e.g. with smoke or ions). But in addition the plume's *perceived* structure at the neuronal level will need to be recorded, because the properties of a neuronal system (receptor capacity, latency, etc.) will be what determines the behaviourally relevant type of plume fluctuation.

Such research will bring together for the first time a knowledge of the orientation movements of pheromonally stimulated flying moths *and* neurophysiological recordings obtained under natural conditions at the same time and location as the flight behaviour. Our understanding of how insects perceive and respond to odour should increase with this approach.

Dedication

This chapter is dedicated to Professor J. S. Kennedy for his more than 40 years of contributions to the study of insect behaviour, and for the inspiration he has given to me and to my students.

5 Mechanisms of directional flight in wind

C. T. David

5.1. Introduction

The control of insect flight has been studied for many years in both free-flying and tethered insects with techniques of great ingenuity and precision, but much remains to be discovered. For example, it is still not certain how insects steer to fly into the wind, let alone how they steer at an angle to the wind. These problems are treated in the present chapter.

5.2. The stabilization of height

Flying insects orienting towards an odour source in wind stabilize their height at the level of the plume. If they lose the odour most species fly across the wind in casts of increasing length at the height of the plume, with much smaller vertical than horizontal excursions. Therefore it is not a reaction to the odour alone that keeps them at the height of the plume. Experiments have shown that height is stabilized by visual reactions to surrounding features. This mechanism is best understood in *Drosophila*, where height stabilization forms part of a control system that compensates for errors of both height and groundspeed that are produced by air turbulence and wind changes (Fig. 5.1).

As Fig. 5.1 shows, the image flow in the lateral eye regions of *Drosophila* is resolved into components roughly at right angles to each other, there is electrophysiological evidence for such resolution in moths (Collett and Blest 1966), bees (Fletcher *et al*. 1984), and flies (Eckert 1982; Hausen 1982a,b). The most plausible theories on the steering of up-, down-, or across-wind flight postulate the existence of control systems similar to that stabilizing height, with visual receptors at right angles.

Gypsy moths (*Lymantria dispar* L.) have a system similar to *Drosophila* for controlling their height and groundspeed (Preiss and Kramer 1983 and Chapter 7), except that the ventral region of their eyes is most sensitive to the vertical image flow, whereas in *Drosophila* the equatorial region of the eyes is the most sensitive (David 1979a, b, 1984). If the floor pattern beneath a tethered moth appears to approach or recede, the moth increases or decreases its lift, respectively. Longitudinal movement of the floor pattern in front of the moth makes it change its thrust in a way that could lead to groundspeed control in free flight.

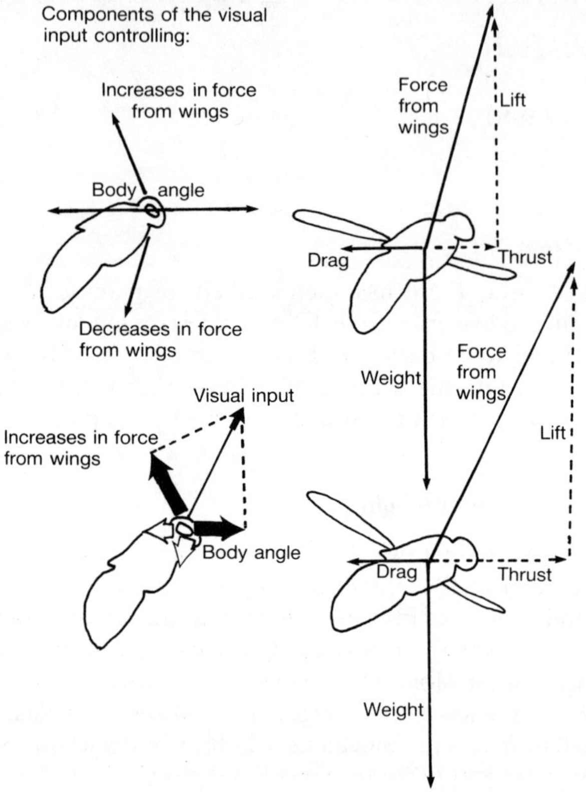

Fig. 5.1. How a fruit fly stabilizes its groundspeed and height. The image flow in *Drosophila* is resolved into four components (top left). The forward horizontal component reduces the body angle (the angle between the body axis and the horizontal) and the backward one increases it if the speed of the image flow is too fast (David 1985). This determines the partition of the force produced by the wings between lift and thrust since in *Drosophila* the direction of this force is always at the same angle to the body axis (David 1978). The upwards component of the image flow increases the force of the wings, while the downwards component decreases it (Götz and Buchner 1978).

In level flight the direction of the image flow is straight backwards. The horizontal component of this produces a particular body angle. The upward component is equal to the downward component, resulting in no change in the force produced by the wings. The result (top right) is that thrust balances the drag and lift balances the weight, hence steady level flight.

If the fly is blown down and backwards by a gust of wind, the direction of the image flow is up and forwards (bottom left). The horizontal component of this is from back to front, reducing the body angle, and so increasing thrust. Also, as there is no longer a component in the downward direction to reduce the force from the wings, this force is increased by the upward component. The result of these two changes (bottom right) is that both thrust and lift are increased and as they are no longer balanced by the drag and weight the fly accelerates up and forwards: the direction which counteracts its deflection by the gust of wind.

As well as the visual feedbacks shown in the figure there are almost certainly mechano- or proprioceptive feedbacks between the body angle, the airspeed, and the force produced by the wings.

A more intricate problem is faced by insects that orient to the source of a vertically moving odour plume. Mosquitoes fly upwards or downwards to reach the source of currents of warm host odour (Daykin 1967). They stabilize their height in still air by visual compensatory reactions to the relative movement in their surroundings. From its image flow alone, the insect cannot distinguish such relative movement due to its own climbing or sinking from that due to vertical movement of the air in which it is flying, and so in which direction the source of an attractant odour lies. Daykin suggested that the mosquito might hover in the vertical air current, stabilizing its height through visual reactions: it could then determine the relative vertical airspeed and direction by some mechanoreceptor, such as its antennae, that protruded outside the air disturbed by its wings, and increase or decrease its lift appropriately to fly against any air current.

5.3. The steering of upwind flight

Insects in continuous flight cannot use mechanoreceptive reactions to steer upwind (Fraenkel 1931; Kennedy 1940) because they are supported by the moving air. Any directional mechanoreceptive stimulation must be due to their own movements through the air and not to the wind direction over the ground. Instead they use visual reactions to the features around them. (But see Gillett (1979), discussed in Kennedy (Chapter 2), who proposes that mosquitoes orient into the wind by responding to the wind shear close to the ground).

Some insects, such as cabbage flies (Hawkes and Coaker 1979) and perhaps tsetse flies (Bursell 1984), orient upwind by flying only in short hops. After they land they turn into the wind using mechanoreceptive cues before taking off again in the upwind direction (see Kennedy (Chapter 2)).

A flying insect could determine the upwind direction by flying in circles and observing the direction in which its groundspeed was the least (Bell and Tobin 1982). However, insects seldom circle, and anyway could not then compensate as quickly as they do for changes in wind direction (David *et al*. 1983), which are very common in nature (David *et al*. 1982). Instead, insects respond to visual drift imposed by the wind to steer upwind (Kennedy 1940).

If an insect is flying due upwind its image flow will be as shown in Fig. 5.2a. Suppose it turns right so that it is flying across the wind. The wind will drift it to the right so that its track over the ground is no longer in line with its body axis (Fig. 5.3). In consequence its image flow will be like that shown in Fig. 5.2c. To return to upwind flight the insect would have to turn to the left in response to such an image flow. There are two ways in which this could be brought about.

The image flow shown in Fig. 5.2b is what would be seen by an insect that was turning right while hovering. In response to such an image flow most insects will turn left. This is the classical optomotor turning reaction

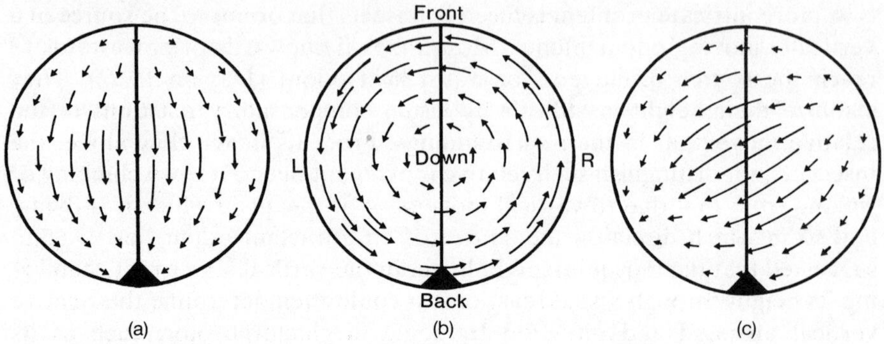

Fig. 5.2. Map of the image flow fields across the downwards-facing hemisphere of an insect's visual field as it flies over flat ground. The length of the arrows is proportional to the speed of image flow at each position: a, straight, level flight; b, turning while hovering; c, either slide-slipping through still air, or flying straight and level through the air but drifted to the right at 45° by a wind from the left.

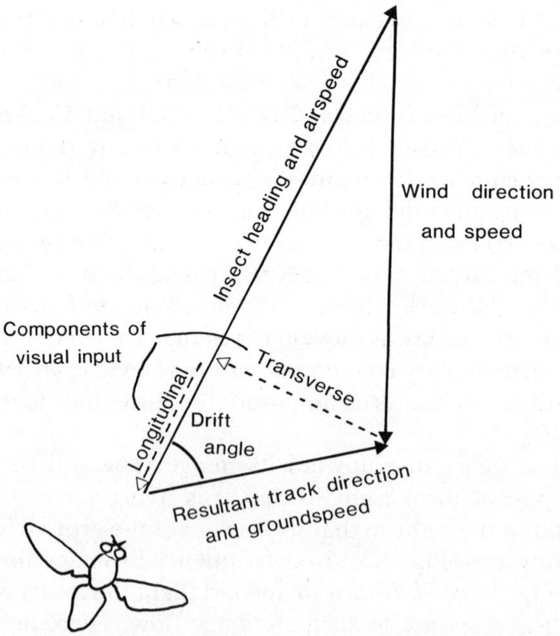

Fig. 5.3. The effect of flying at an angle to the wind on an insect's track, groundspeed and image flow. The image flow, produced by the relative movement between the insect and the ground, is at an angle to the body axis (the drift angle) and is resolved by many insects into its longitudinal and transverse components, shown here as dashed arrows.

(Götz 1972), which serves to keep insects from making 'unintended' turns. The anterior portion of the image flow produced by the wind drift (Fig. 5.2c) shows similar directions of image flow to those in Fig. 5.2b; the posterior region, however, shows directions of image flow that would be produced by a hovering turn in the opposite direction. The upwind turn in response to wind drift could thus be due to the optomotor turning response if the forward region of the insect's eyes was more sensitive than the posterior region; which is certainly true, for many insects that cannot see behind them. Collett (1980) has a similar theory of directional flight control in hoverflies.

The alternative theory is that insects use the downward-facing regions of their eyes for steering upwind. They could distinguish between the image flow in Figs. 5.2a and 5.2c by the direction of the image flow beneath them (Marsh *et al*. 1978), that is, by the 'drift angle' of Fig. 5.3. This theory has more experimental evidence. Insects flying upwind in a wind-tunnel respond to the sideways movement of the floor beneath them, mimicking the sideways drift produced by a wind, by turning to follow the floor movement (Kennedy 1940; David 1982b) Fig. 5.4). They do this even though the equatorial region of their eyes, where the optomotor turning response would be the strongest, is stimulated by the stationary wind-tunnel walls. Indeed flying gypsy moths appear not to respond to rotating movements of the image flow at all (David and Kennedy unpublished).

How could insects be stimulated to turn by the direction of the image flow beneath them? The simplest theory is that they respond to the transverse component of the image flow (Ludlow 1984; Marsh *et al*. 1978), just

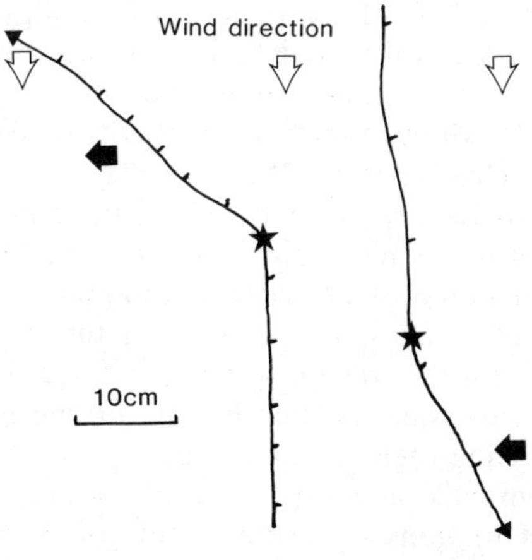

Fig. 5.4. Recorded tracks of *Drosophila* flying upwind (left) and downwind (right). The wind was blowing in the direction of the open arrows. When the fly was at the position marked with a star the floor underneath it started to move in the direction of the solid arrow at 10 cm/s. The upwind flier turned with the direction of the floor movement, that is, in the direction of the transverse component of its image flow. The downwind flier turned against both. (Time marks on tracks are at 0.5 s intervals.)

as *Drosophila* flies respond to the vertical component of the image flow in stabilizing their height. When the insect is flying due upwind there is no transverse component to the image flow. If the insect deviates from the upwind direction a transverse component is produced by the wind drift (Fig. 5.3) and this is the stimulus for the 'compensatory' upwind turn. If an insect is flying downwind there will also be no transverse component to the image flow, but this flight direction will be unstable because any small deviation from the downwind direction will produce a transverse component to the image flow, which will lead to a further turn in the same direction and so on until the insect is facing upwind.

An insect has to fly faster than the wind speed to progress upwind. This is accomplished by keeping the longitudinal component (or possibly the longitudinal plus the transverse component) of the image flow constant (Ludlow 1984). *Drosophila* flies at the same groundspeed over a wide range of wind speeds (David 1979a, 1982a); the system controlling the groundspeed is illustrated in Fig. 5.1. The flies use the longitudinal component of the image flow (Fig. 5.3) to control groundspeed, while they use the transverse component for steering. Many other insects including mosquitoes (Kennedy 1940) and aphids (Kennedy and Thomas 1974) have similar optomotor feedback systems controlling groundspeed.

5.4. The steering of downward flight

Insects often fly downwind. Certain moths (Traynier 1968) and *Drosphila* (personal observation) first cast to and fro across wind, then fly downwind, then cast across wind again when they lose an odour plume. *Spodoptera* (Murlis and Bettany 1977; Murlis *et al*. 1982) and gypsy moths often fly downwind in the field. Downwind flight may be the most energetically efficient way of searching for odour plumes in some circumstances (Sabelis and Schippers 1984). Many migrating insects flying high up, even at night, orient downwind (Schaefer 1976; Waloff 1972).

An insect could fly downwind by reversing the optomotor turning reaction to the apparent rotation of objects in the frontal region of its eyes, or by reversing the reaction to the transverse component of the image flow in the ventral region of its eyes. The optomotor turning reaction seems the less likely candidate for the control system (as it does in the case of upwind steering), because reversing the optomotor turning reactions would presumably lead to circling flight as the normal optomotor reactions serve to keep the insect from making 'unintended' turns. Furthermore, downwind-flying insects react to transverse movements of the floor beneath them (Fig. 5.4). A downwind-steering fly turns against the movement of the floor, whereas an upwind-steering fly turns to follow the movement (Fig. 5.4). This is exactly the response to wind drift which is needed to explain their headings; the upwind-steering fly turns with the transverse component of the image flow, the downwind-steering fly turns against it.

5.5. The steering of cross-wind flight

Flight that is at an angle to the wind is very common, for instance many insects zigzag when flying in odour plumes and also cast to and fro across wind after losing them. The steering of constant cross-wind tracks is more complicated than the steering of flight along the wind-line. This is because a cross-wind-flying insect has to change both its heading and its airspeed in order to keep its ground track the same if the wind speed changes (Fig. 5.5).

Zigzagging *Plodia* moths fly on the same track and at nearly the same groundspeed in different wind speeds (Marsh *et al.* 1978; 1981). *Grapholitha* moths (Baker and Kuenen 1982; Kuenen and Baker 1983) and the gypsy moth (David and Kennedy unpublished) continue on the same track when the wind stops completely, and even switch to a track at 90 ° to the old wind direction when they lose the pheromone after the wind has stopped. These observations are sufficient to eliminate many theories about the steering mechanism. The nature of this has been a particular interest of Ludlow (1984) who computer-simulated many possible mechanisms. He has found one which is simple, produces the sort of tracks observed in experiments and is testable. He supposes that the steering mechanism consists of three negative feedback loops and an internal counterturning generator of the type found by Kennedy *et al*. (1980, 1981) (see also Kennedy (1983) and Baker (Chapter 4)). What the internal

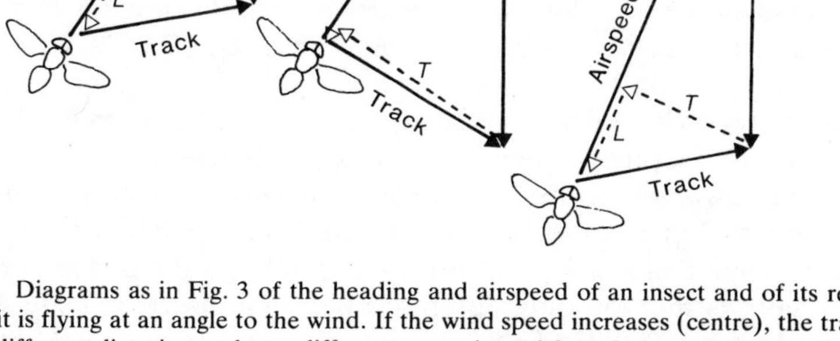

Fig. 5.5. Diagrams as in Fig. 3 of the heading and airspeed of an insect and of its resultant track if it is flying at an angle to the wind. If the wind speed increases (centre), the track will be in a different direction and at a different groundspeed from before, so long as the insect maintains the same heading and airspeed. To maintain a track in the same direction and at the same groundspeed as before the insect has to steer more into the wind and increase its airspeed (right). The longitudinal (L) and transverse (T) components of the insect's image flow are altered by the change in wind speed. These may be the inputs used by an insect to control cross-wind flight (see text).

generator does is to make the moth's wings produce a yawing torque; every so often it reverses the torque so the insect turns the other way, producing the counterturns. The frequency of the counterturns is modulated by, amongst other things, the odour concentration; this produces the different frequencies of counterturning when the moth is casting to and fro in clear air compared with when it is in pheromone. Such a turning mechanism alone would lead the moth to fly in circles, first one way and then the other. Ludlow suggests that what stops this from happening is the first of the feedback loops. This measures the amount by which the moth has turned and decreases the torque produced by the wings as the moth turns further and further. When the moth has turned by a certain amount the torque produced by the wings is zero and the moth flies straight until the torque is reversed by the internal turn generator. The angle between the successive legs of the zigzag track is determined by the balance between the turn generation mechanism and the opposing turn-measuring feedback loop, and this balance point is modulated by the odour stimulation to produce either cross-wind casting or upwind progression.

What is still needed is a mechanism which compensates for the wind, both orienting the zigzag track resultantly into the wind and allowing for its speed. The steering part of this is accomplished by the second feedback loop, the one from the transverse component of the image movement which also steers insects upwind (Section 5.3). As the wind speed increases the transverse component of the image flow also increases leading to a larger turn in the upwind direction, so producing the change in heading shown in Fig. 5.5. This is the upwind polarizing effect of the wind referred to by both Kennedy (Chapter 2) and Baker (Chapter 4). Airspeed is adjusted by the third feedback loop which keeps the image flow speed constant. Since the track is nearly the same at all wind speeds the moths must keep their groundspeed the same. They could do this, if they were resolving the image flow into the longitudinal (L) and transverse (T) components, by keeping ($L^2 + T^2$) constant (Fig. 5.5). Ludlow (1984) has noticed evidence in Marsh *et al*. (1978) that the moths are not as clever as Pythagoras and that they approximate by keeping $L + T$ constant, which is good enough.

Most of Ludlow's theory has yet to be tested. One question is the nature of the turn-measuring system. If this exists, it could work by visual, mechanoreceptive or proprioceptive mechanisms. In the case of gypsy moths the visual mechanism seems to be ruled out as the moths are unaffected by rotating image flow, as mentioned in Section 5.3. The simplicity of the theory is that the only difference between insects that fly straight upwind and those that zigzag is the internal turn-generating mechanism. Even when a moth is zigzagging, the upwind-steering mechanism is in operation (Baker *et al*. 1984).

Acknowledgements

I am grateful to Prof. J. S. Kennedy and Drs A. R. Ludlow, J. E. Moorhouse, and C. Rechten for their helpful comments on the manuscript.

6 Turbulent diffusion and pheromone-triggered anemotaxis

E. Kramer

6.1. Introduction

Until 1981, when Murlis and Jones published their study on the structure of odour plumes originating from point sources, entomologists involved in chemical orientation of insects used to calculate the distribution of odour by means of formulations like that given by Bossert and Wilson (1963). This formulation and some improved derivatives of it still may hold if, for example, active spaces are to be estimated. The stimulus conditions to which insects are exposed when orienting in real odour plumes are, however, completely different from what these formulations describe. If one observes their unsteady and convoluted flight and simultaneously imagines the irregular sequence of odour pulses hitting the insect's antenna it becomes obvious that the insect can hardly obtain information about the position of the odour source by integrating its olfactory input over time or space. We certainly must reject all theories of olfactory orientation towards point sources using such concentration gradients.

In order to understand how the extremely pulsatory structure which Murlis and Jones have demonstrated so impressively may originate we simulated the situation by means of a computer, assuming a small source of odour in a wind with numerous eddies of all sizes (Murlis (Chapter 3)). This simulation taught us that the whole plume consists of only one filament of odorous air which becomes more and more lengthened and convoluted the further it moves downwind from its origin. A few metres from the source the filament becomes a chaotic tangle, some parts of it being stretched only a little, having preserved most of their initial concentration, others having been so stretched as to be extremely diluted. Interestingly, the least diluted parts have preserved best their direction of travel since they left the source (Fig. 6.1). Having been done in ignorance of these circumstances our former experiments with male silkmoths orienting in a wind laden with the pheromone bombykol were necessarily misleading. On the assumption that absolute pheromone concentration and gradients in it would play the most important role in long-range chemical orientation, emphasis was put on making the stimulus as homogeneous as possible at that time. Responses to small and controlled changes in concentration were then observed and quantified and the results put together in order to reconstruct the system used by the insect. The outcome was an 'intercausal

Fig. 6.1. Two-dimensional simulation of the fate of a single filament of air in a turbulent wind. The figure is to demonstrate how the filament is stretched and tangled while travelling downwind. Each small line represents a single molecule and its actual motion. Crowded lines indicate high concentration.

network' (Mittelstaedt 1969) which comprehended all the observations obtained in this way. These data together with the model were published (Kramer 1975). Meanwhile an essential part of the net, namely the flip-flop, had been identified electrophysiologically by Olberg (1983a). Despite some difficulties the system seemed to be well adapted to its task. Nevertheless, what was found was only a small part of the truth, because obviously a kind of stimulation had been applied which under natural conditions is of only little importance.

As a consequence of Murlis and Jones's results, new experiments became inevitable to answer the question of how moths respond to pheromone pulses. Until now this question has been studied very little and in these cases the pulse frequency and pulse duration differed from those found by Murlis and Jones (Bartell and Lawrence, 1977a; Cardé *et al.* 1984). Two questions were therefore of special interest:

(1) How do silkmoths respond to a series of short pheromone pulses with varying pulse frequency?

(2) Are silkmoths capable of detecting small differences in the time of the arrival of a filament at the two antennae, and do they use this ability to obtain information on the direction of the pheromone source?

The above two questions have been studied over the past three years and the results have been integrated into the intercausal net of 1975.

6.2. Methods

All experiments were carried out with male silkmoths between two and four days old. A certain number of individuals stemmed from Silk Products

Taiwan, a greater number from the Instituto Sperimentale per la Zoologia Agraria, Padua, Italy, from which they were sent as pupae. The experiments were performed indoors like our previous ones on a 50 cm servosphere (Kramer 1975). The room was illuminated by fluorescent lamps producing about 500 lx. The temperature was kept between 22 and 25 °C. No care was taken to keep the optical surroundings neutral in respect of visual cues. In many former experiments it had turned out that the visual input had only very little influence on the orienting behaviour of walking silkmoths. Air was blown over the experimental field on top of the sphere with a constant speed of 60 or 20 cm/s. Fresh air was delivered continuously by an array of 105 rectangular (2 × 5 mm) nozzles arranged together with 105 smaller (0.2 × 5 mm) nozzles within a space of 9 × 2.5 cm. Air containing pheromone was blown in pulses through the smaller nozzles. During pulses the speed of the fresh air was reduced so that the net flow over the moth remained unaffected. Pheromone was released in the air by passing it over a 4 cm^2 piece of filter paper soaked with either 1, 0.1 or 0.01 μg, or 1, 0.1 or 0.01 ng of bombykol. Pulse duration was 150 ms throughout the experiments. The pulses were triggered by a minicomputer, which also recorded the moth's path by sampling 10 sets of x, y coordinates every second. All further computation of data is based on these 0.1 s samples, the 'elementary vectors'. Further technical information is given in Section 6.3.

6.3. Results

6.3.1. Response to a pulse series of fixed frequency

During one test of this series a male silkmoth was exposed to five different frequencies of 150 ms pheromone pulses which followed each other in randomized order, each frequency lasting 100 s. The intervals between pulses were 2.4, 1.2, 0.6 and 0.3 s, or pheromone was added continuously over the entire 100 s period. The concentration of pheromone was varied parametrically using × 10 steps as mentioned above. A test was valid if the total path length exceeded 150 cm. If the total path length remained below this value the test was repeated with another moth. 150 valid tests were run using 190 moths. Table 6.1 shows how many trials were needed to obtain 30 valid tests.

A source load of 0.01 ng was not sufficient to arouse the moths enough to fulfill the test conditions. In general the upwind orientation of the moths was best if the frequency of pheromone pulses was set to the maximal value of 3.33 per second. A closer examination of these data showed that the effect of pulsing pheromone on the process of orientation could be best demonstrated by two measures: one is the straightness of the walking direction (*ST*), which is defined as the straight-line distance from the starting point to the end point of a test walk divided by the actual path length between these two points. The other measure (*TQ*) is the time between

TABLE 6.1

Source load	Trials	Valid	Mean walking speed
1 μg	30	30	2.1 cm/s
0.1 μg	32	30	2.3
0.01 μg	30	30	2.1
1 ng	30	30	1.7
0.1 ng	51	30	0.8
0.01 ng	17	0	(<0.3)

two consecutive 'tack changes' from one side of due upwind to the other. The values of these two measures are summarized in Fig. 6.2. *ST* as well as *TQ* show that the accuracy of upwind orientation is greatly improved if the pheromone is pulsed. If the time between pulses exceeds 0.7 s, however, then the accuracy of orientation is dramatically reduced. Furthermore, these figures show a striking parallelism between the curves for different pheromone concentrations, which means that these measures are independent of concentration over a range as large as five orders of magnitude.

6.3.2. *Response to single pulses*

The above experiments were suitable for demonstrating the effect of pulsing the pheromone on the orientation system in general. The data were not

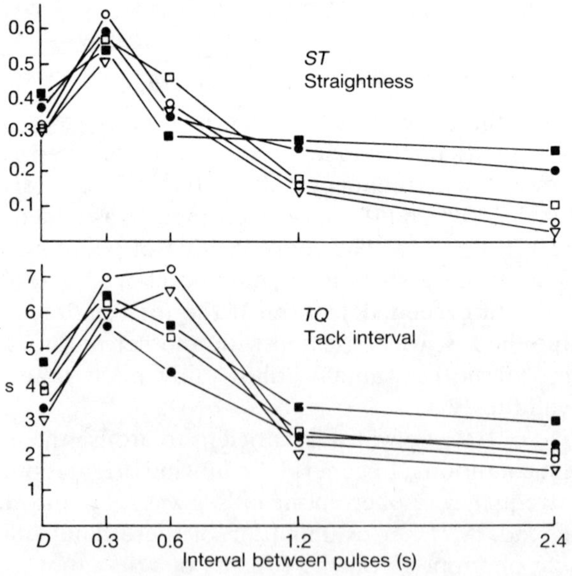

Fig. 6.2. Responses of male silkmoths to a series of pheromone pulses at different intervals, and to constant stimulation (*D*). Pheromone concentration is varied over 5 orders of magnitude: triangle 1 μg, open cirdcle 0.1 μg, open square 0.01 μg, solid circle 1 ng, solid square 0.1 ng of bombykol.

suitable for analysing the effect of individual pulses. This was achieved in a series of experiments in which a special sequence of pulses was precisely timed with respect to the recorded elementary vectors. A run of this series consisted of 10 repetitions of the following sequence: during the first 20 seconds the pulse frequency decreased continuously from 3 to 0.4 per second, then there was a pause of 10 s without any pheromone pulses, this pause was followed by a series of pulses given in the reverse order of the first, and finally the pheromone concentration was kept constant for 10 s at the level of the peak value of the pulses.

These experiments were carried out with two pheromone concentrations (1 and 0.01 μg) and two wind speeds (60 and 20 cm/s). The results, however did not differ significantly with respect to these conditions. The data were therefore pooled. The results of the analysis are summarized in Fig. 6.3. The curves represent the average of 230 periods performed by 23 individuals. The vertical lines mark the onsets of 150 ms pulses within any period.

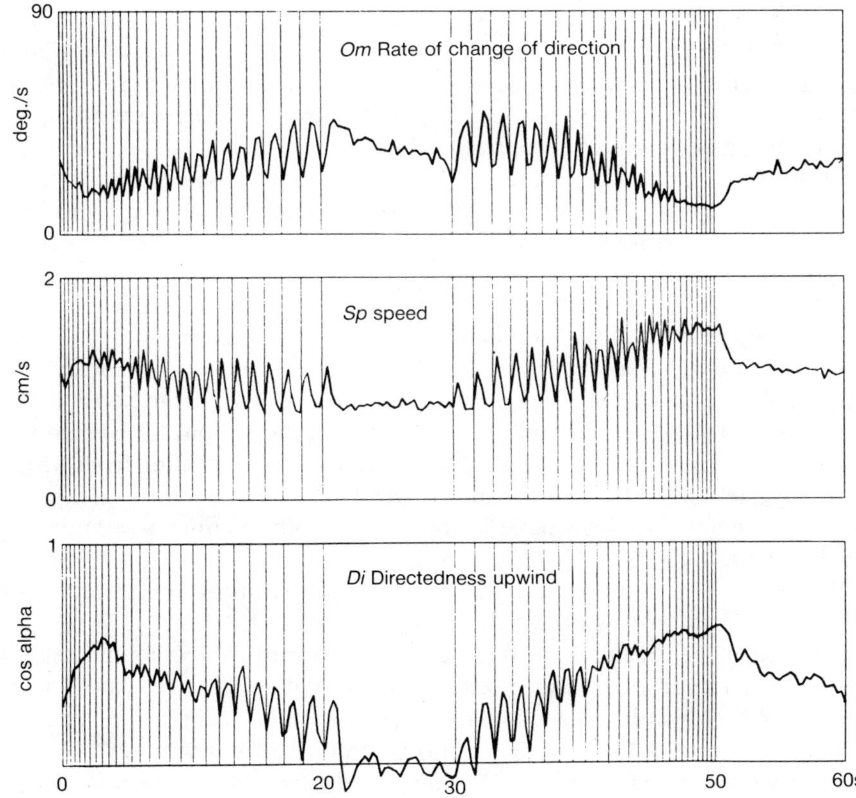

Fig. 6.3. Response of male silkmoths to a complex series of pheromone pulses. A great number of runs have been averaged to show how the moths respond to single pheromone pulses (vertical lines on the abscissa).

Three variables are shown in these diagrams: the walking speed (*Sp*), the rate of change of direction during the walk (*Om*), and the cosine of the walking direction (*Di*) with respect to the wind (alpha). All three qualities show a consistent dependence on the stimulus pattern. Walking speed (*Sp*) and directedness (*Di*) are maximal when the pulse frequency (3.3.I/s) is highest. At least over short spans of time (10 s) the walking speed is about the same when the concentration is permanently high or close to zero. The directedness (*Di*), however, differs in these two cases. While with no stimulus *Di* approaches zero (which means that there is little or no upwind preference), *Di* still reaches 0.5 with a permanently high concentration which corresponds to one pulse every 0.8 s. The rate of change of direction *Om*) is lowest at highest pulse frequencies, which also means that the course is straightest upwind in that case. In the absence of pheromone *Om* is slightly higher than when the pheromone is offered continuously.

These facts serve only to confirm the results of the experiments reported in the last section. The time-course of the different curves is of greater interest here. All three variables follow the underlying time-pattern of pheromone pulses with sharp ups and downs if the pulse frequency is not above 2 to 2.5 pulses per second. About 200 ms after a pulse the walking speed (*Sp*) increases rapidly, after which it falls back to a minimum within the next second if a new pulse does not push it up again. These speed bursts always begin with a sudden change of the walking direction, as can be seen by the two other curves where *Om* always tends to higher values after a pulse. Direction and amount of that change are always such as to reduce the current deviation from due upwind (*Di*), in which ever direction requires the smaller turn.

6.3.3. Response to pulses with delayed arrival at one antenna

Since chemotropo- or chemoklinotactic manoeuvres are obviously not applicable in such odour fields, the only means, apart from sensing the direction of the odourous wind, by which information about the position of the source can possibly be obtained by the insect is by detecting from which side the filaments pass over the insect. In other words, is *Bombyx* capable of detecting and utilizing small differences between the time of stimulation of the two antennae?

The delivery system for the odour was modified in order to study this question. Fresh air was blown through all nozzles of the array. Four wire frames each measuring 1 × 10 cm could be moved by means of a small chain over the front of the nozzle array. A cotton thread soaked with 0.1 μg of bombykol was wound four times around each frame. The air could pass through these frames almost unhindered but was nevertheless loaded with bombykol (Fig. 6.4). The initial filaments had a lateral distance of 7 cm from each other and were moved at speeds of between 1.5 and 15 cm/s through the nozzle array so that the filaments passed over one antenna 0.4 to 0.04 s later than over the other.

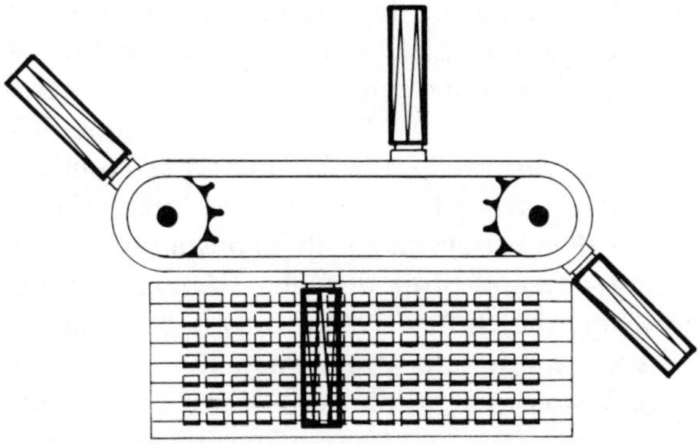

Fig. 6.4. Schematic drawing of the device used to generate pheromone filaments which can be swept over the moth's antennae. Four frames carrying W-shaped threads soaked with bombykol can be moved in front of an array of 2 × 105 rectangular air outlets.

The data obtained with these experiments are not presented here in detail. It suffices to state that they all were negative inasmuch as no significant bias towards one or the other side could be elicited. Besides, the moths behaved exactly as in the above experiments. So it is very unlikely that male silkmoths use information of this type to find the female.

6.4. Discussion

These results appear to be incompatible with former results and the derived intercausal network (Kramer 1975). Where are the features which then seemed so essential: the oblique courses and the switching from one side to the other released by drops in the pheromone concentration?

In the system of 1975 the total turning command was determined by:

$$T_s = f(U) \times g\,(C) \times \sin(\text{alpha}) - \text{sign}(FF) \times T_i$$

With alpha being the actual direction of the moth with respect to the wind, f a function of the windspeed U, g a function of the bombykol concentration, C, sign (FF) the actual sign (status) of a flip-flop element, and T_i an internal turning tendency which is independent of the current chemical stimulation.

According to this formulation a stable course can be only achieved if there is some direction of walk where the part of the equation left of the minus sign exceeds the part to the right. The system will circle if either $f(U)$ or $g(C)$ is zero. If $f(U)$ or $g(C)$ are proportional to U or to C, respectively, then circling will occur if either U, or C, is itself zero because there is no wind or no pheromone, respectively. Such proportionality was a conclusion from the earlier experiments. If we now change $g(C)$ to $g(\mathrm{d}C/\mathrm{d}t)$ then the first term additionally becomes zero if there is no variation in the ambient odour concentration. It becomes greater the stronger these variations are, which means that the course points more upwind.

If this is true, how then, in the earlier experiments, could a constant pheromone concentration cause a rather straight but oblique upwind course? The only explanation we now have is that the pheromone concentration was not as constant as we believed at that time. There must have been unwanted fluctuations of the bombykol concentration in our old delivery system which, for other reasons, was later refined several times. These fluctuations then produced a rather constant signal which was misinterpreted as being somehow propertional to the concentration. At that time there was, however, the difficulty that only a little adaptation could be observed in these experiments even when they were extended beyond 40 min. Electrophysiological studies (Kaissling and Priesner 1970; Zack 1979) had shown that the pheromone receptors adapt very fast so it remained unclear how $g(C)$ could be high for so long a time. Furthermore, nothing special could be seen in the receptor response when the stimulus was turned off, which, however, had a marked effect on behaviour, namely a change in the walking direction. These difficulties are both eliminated if $g(C)$ is substituted for by $g(\mathrm{d}C/t)$ where g represents low-pass filtering with a time-constant of about 1 s (leaky integrator). The term $f(U) \times g(\mathrm{d}C/\mathrm{d}t) \times \sin(\mathrm{alpha})$ is then always high if odour pulses follow each other in intervals smaller than about 0.7 s, and consequently the moth walks more or less upwind (see Baker (Chapter 4)). If however, an interval exceeds that value the moth begins to turn downwind and finally circles without any upwind gain. Kaissling (Chapter 22) has shown that the response dynamics of the pheromone receptors in *Bombyx* fit these behavioural results very well.

These experiments dealt with a moth species domesticated and unable to fly. Do they, nevertheless, teach us something about chemical orientation in flying insects, to which our interest is mainly directed? Firstly, if the temperature is high (>25 °C) a few silkmoths among thousands are able to fly quite normally. So degeneration may not have gone too far. Secondly, one can see that the moths on the sphere really try to fly, but their bodies are obviously too heavy.

These two points encourage us to speculate about how this system might function in flight. Surely $f(U) \times \sin(\mathrm{alpha})$ is no longer available in flight by a direct anemoreceptive system. The direction of wind and its strength are somehow extracted from the optomotor input (David (Chapter 5)). If this signal is combined with that from the anemoreceptive system—the directional output of which is evidently zero in flight—our system is functional. And if we make the one further assumption that while flying in an odour plume it is ‘forbidden’ to drift downwind, then the system will suffice to explain the features which characterize pheromone—regulated anemotaxis (Kennedy 1983 and Chapter 2). The more frequently the pheromone pulses arrive at the moth the more strictly does it fly upwind. If an interval between pulses exceeds a certain value (in our case about 0.7 s) the moth turns more and more downwind and would circle, alternating between left and right. This, however, would lead to a downwind drift which may be not allowed, so it flies more or less transverse to the wind and changes between left and right according to either an internal time-pattern, external stimuli, or both.

7 Pheromone-induced anemotaxis in simulated free flight

R. Preiss and E. Kramer

7.1. Introduction

A male moth, orienting upwind within an odour plume towards a chemically alluring female, detects the wind direction and controls its flight with respect to speed and altitude by watching the movement of ground structures (Kennedy and Marsh 1974; Marsh *et al.* 1978; Miller and Roelofs 1978; Preiss and Kramer 1983; David (Chapter 5); see also Heran 1955; Kennedy 1940; Schneider 1965—experiments with other insect species). Their visual image flow is the sum of self-induced and wind-induced motion. In order to understand how this visual information is processed to result in the observed flight tracks (which under natural conditions are rather erratic with cross-wind excursions in both the horizontal and the vertical plane) two experimental requirements must be met: the optical and chemical stimuli must be known for each instant, and the flight track which reflects the moth's responses must be recorded continuously.

Although the optical stimulus can be quantified in wind-tunnel experiments using free-flying moths, it is difficult to quantify the olfactory stimulus. In addition, reconstruction of the tracks from videotape, for instance, is time consuming. It is therefore, not surprising that in most wind-tunnel studies only the horizontal component of the flight tracks was analysed (e.g. Cardé and Hagaman 1979; Marsh *et al.* 1978; Miller and Roelofs 1978). However, changes in flight altitude strongly affect the apparent movement of the ground pattern and thus the optical signal which the moths perceive. These difficulties can be removed by tethering the moths. Here, not only can the optical and olfactory stimuli be applied in a controlled way, but the moths' responses can also be measured continuously. In the experiments with tethered moths, however, the flight parameters were only measured successively (Preiss and Kramer 1983, 1984). Consequently, closed-loop situations with respect to the visual input were unsatisfying because they were restricted to only one or two degrees of freedom. In order to overcome these restrictions, we developed a 'flight simulator' able to simulate almost completely the visual conditions of free flight.

The moths' responses in this simulator to optically simulated wind, changes in wind direction and changes in speed are compared with results previously obtained, where only single components of flight were accessible (Preiss and Kramer 1983) and with wind-tunnel experiments

(Kennedy and Marsh 1974; Marsh *et al*. 1978; Miller and Roelofs 1978).

7.2. Animals

Experiments were carried out with male gypsy moths, *Lymantria dispar* L., which were kindly supplied as pupae from the USDA. Male pupae and emerged moths were kept in a 16 h/8 h light/dark cycle at 22 °C (room temperature). Experiments were carried out with 24- to 36-hour-old males during the second half of the photophase. Not later than one hour before the experiments, a small metal pin was glued with Pattex® to the moth's thorax.

7.3. 'Flight simulator'

In this flight simulator (Fig. 7.1) the optical situation corresponds to free flight with respect to the three axes of translation and the yaw-axis of rotation. The latter was achieved by just leaving the moth free to turn.

7.3.1. Measuring system

The main part of the measuring system is a horizontal, star-like plate which can be tilted in any direction (Fig. 7.1b, d). This is made possible by suspending it by means of two wires attached to a solid metal ring. These wires are kept taut by a tension spring, adjusted to the size of the insect. The plate acts on the ferrite cores of four friction-free force transducers (linear variable differential transformer displacement transducers—LVDTs), sensing the actual tilt. A small tube is fixed to the centre of the plate and it guides a thin rod, to the end of which the moth is attached by the pin on its thorax. This rod is magnetically suspended within a coil. Its vertical dislocation due to the moth's changes in lift is sensed by a further LVDT. Its ferrite core is mounted at the suspension rod. This signal is amplified and then used to compensate for the dislocation by controlling the current through the suspension coil. Due to vibrations of the rod within its pivots, caused by the flying moth, friction between rod and tube is negligibly small. According to the thrust, which the moth produces, the star-like plate is tilted until the thrust equals the force produced by the tension spring. Due to the orthogonal arrangement of the LVDTs, their signals correspond to $T \times \sin \alpha$ and $T \times \cos \alpha$ respectively, where T = thrust produced by the moth and α = flight direction. The signals of the two opposing LVDTs are added, thus cancelling asymmetries of the measuring system.

To test the measuring system, a known force was applied to the suspension rod from different directions. The measured signal was constant and independent of the direction of force. In addition the signal increased linearly with the applied force, which was also expected from the linear characteristic of the LVDTs and from the small degree of tilt of the star-like plate.

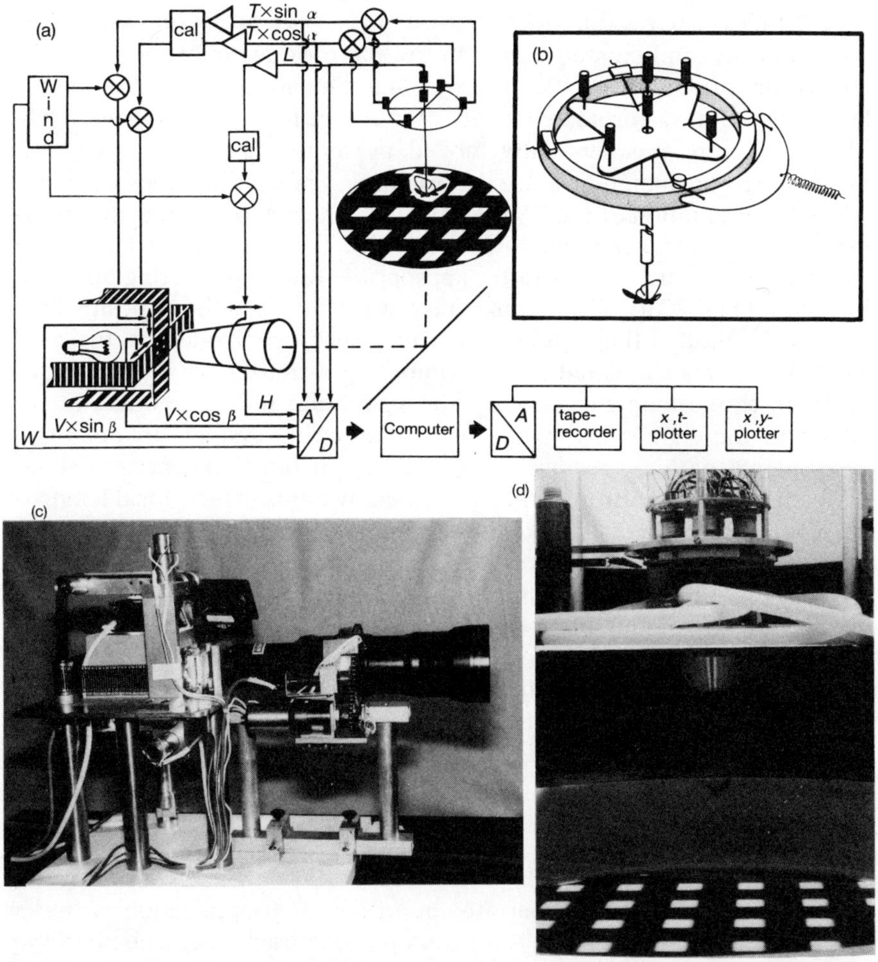

Fig. 7.1. 'Flight simulator'. The optical situation corresponds to free flight with respect to the three axes of translation and the yaw-axis of rotation. The forces produced by the tethered moth are used to control movement and size of a projected ground pattern. If the moth 'flies' in any direction, the ground pattern moves oppositely with a speed proportional to the moth's thrust. Changes in lift are used to simulate falling or rising by modifying the size of the pattern elements. Wind may be simulated by superimposing a translatory pattern movement. Pheromone can be applied from above by means of a circular array of nozzles. (a) Schematic drawing of the 'flight simulator' with signal flow between units. (b) Measuring unit. (c) Pattern-projecting unit. (d) Experimental chamber with projected ground pattern, measuring system and pheromone applicator (for technical details see text).

7.3.2. Visual stimulation

A black/white chequered pattern was projected via a mirror inclined at 45° on to a ground-glass screen (diameter 50 cm) beneath the moth (Fig. 7.1a, d). The pattern was produced by two orthogonally crossed 35 mm film-loops, each bearing equidistant black and clear stripes

(Fig. 7.1a, c). The two film-loops surrounded a slide-projector horizontally and vertically, and crossed each other in the focal plane of the projecting lens (zoom 50–300 mm). When projecting the pattern (500 W projection lamp), the film was protected by an interference heat-reflecting filter. The film-loops were independently moved by motors and their speed was measured by tachometers. Since the orientation of the film-loops was parallel to the orientation of the LVDTs, in closed-loop experiments the signals of the LVDTs corresponding to $T \times \sin \alpha$ and $T \times \cos \alpha$ were used directly to drive the pattern motors (after appropriate low-pass filtering and amplification) (Fig. 7.1a). The pattern thus always moved in the opposite direction to the moth's flight direction, with a speed controlled by the moth itself. Wind was simulated by superimposing a translatory pattern movement. In closed-loop experiments, the signal of a function generator was added to the signals arising from the measuring system. In open-loop experiments, pattern speed was controlled by the function generator alone.

Changes in flight altitude were simulated by adjusting the focal length of the zoom objective lens, and thus the size of the pattern elements, by means of a motor. The actual focal length was measured by a potentiometer. In closed-loop experiments the focal length was controlled by the moth's lift signal, but in open-loop experiments by the output from a function generator.

Except for the ground pattern, the surrounding was homogeneous. A white painted drum (diameter 50 cm) surrounded the moth. A shield above the moth, incorporating the pheromone applicator, gave a homogeneous canopy.

7.3.3. Olfactory stimulation

The pheromone applicator permits the intermittent application of one or two different odour substances, for example an attractant and an inhibitory compound, in addition to clean air. The odour is applied to the moth from above in a radially symmetric fashion (Fig. 7.1d).

The applicator consists of a circular array of 24 vertically oriented nozzles. Each nozzle is connected to one of three chambers, which can contain the odour substances absorbed on filter paper (3.5 × 1 cm). In the radial ring, nozzles are connected successively to different chambers, so that each chamber is connected to eight nozzles. One chamber is left empty to supply clean air. When the application of pheromone is not required, the chambers may be bypassed by using a magnetic valve.

The odour substance was thus blown radially symmetrically in front of the moth and reached (we presume fairly uniformly) the antennae via the airstream produced by the flying moth. In all experiments mentioned here, the attractant pheromone of the gypsy moth '(+)-disparlure' (7*R*,8*S*,-cis-7,8-epoxy-2-methyloctadecan; Bierl *et al*. 1970; Cardé and Hagaman 1979; Cardé *et al*. 1977; Miller and Roelofs 1978; Preiss and Kramer

1983; Vité *et al*. 1976) was continuously applied to the moths from a filter paper source, laden with 0.5 μg (+)-disparlure.

The experimental area was actively ventilated by means of a ring-shaped duct (diameter 50 cm) placed on the pattern-floor. Air was exhausted via a slit within the duct (height 3 mm) and transported by the ventilator to outside the building.

7.3.4. Recording

The signals of the measuring system ($T \times \sin \alpha$, $T \times \cos \alpha$, and L), and the signals of the projecting unit ($V \times \sin \beta$, $V \times \cos \beta$, and H), as well as the optical stimulus forced upon the moth (wind or changes in flight altitude) were applied to an A/D converter whose output was fed into a minicomputer (sampling rate 1 kHz, (Fig. 7.1a). T (thrust), L (lift), α (flight direction), V (pattern speed), β (direction of pattern movement), and H (flight altitude) were calculated. For future processing all data were stored on floppy discs. In addition, the values were stored in analogue form on a seven-channel FM recorder, from which they could be selectively plotted on a four-channel x,t plotter. In some experiments the horizontal component of the flight track was plotted on an x,y plotter.

Answers to questions which would need a statistical treatment of the data will not be given here. In this preliminary publication some results will be presented which can be derived by visual inspection of the x,y and x,t plots.

7.4. Flight behaviour of the moths

In the presence of the attractant pheromone (+)-disparlure, flight periods were usually rather long lasting (up to 2–3 h).

7.4.1. Spontaneous behaviour—without optically simulated wind

Under closed-loop conditions, when pattern movement and size of the pattern elements were controlled by the moths, they flew without directional preference. Straight segments of the tracks were of short duration (several seconds). The moths spontaneously ‘searched around’, apparently in a random manner. Similarly, they ‘spontaneously searched’ with respect to their flight altitude, as apparent from random fluctuations of lift. In several cases, however, lift was not sufficient to compensate for the moths’ body weight.

7.4.2. With optically simulated ‘wind’

(a) Variations in ‘wind direction’ and ‘wind speed’. The biased movement of the ground pattern, optically simulating wind, was detected by the moths. The tortuous flight tracks became stretched and were oriented with the biased pattern movement (upwind) (Fig. 7.2a). Apparently random fluctuations of the course direction within the range of up to 45° resulted

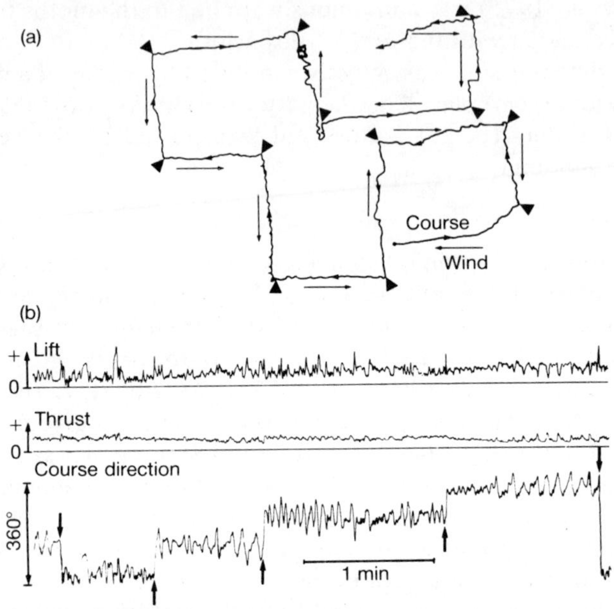

Fig. 7.2. (a) Flight course (top view) of a male gypsy moth in response to optically simulated wind (30 cm/s). Triangles indicate a change in wind direction. (b) An abrupt change of the wind direction by 90° (arrows) is immediately answered by an upwind turn.

in a noisy but otherwise straight upwind-oriented flight track (Fig. 7.2b). An abrupt change in the wind direction of 90° was usually answered within less than 1 s by an upwind turn (Fig. 7.2b). When modulating the wind speed in a triangular manner, combined with a periodic change of the wind direction of 180°, a speed of 15.0 ± 10.8 cm/s in the opposite direction was needed to elicit the moths' U-turn (calculated from 40 periods). This corresponds to an angular velocity of the pattern elements of about 25°/s (flight altitude 15 cm). This mean value, however, does not represent the moths' threshold for detection of the upwind direction. In some cases, the U-turn was elicited at a wind speed of about 4 cm/s (7°/s), which corresponds to a wind speed of about 30 cm/s with respect to a more natural flight altitude of 1 m. In several cases, the 180° change in wind direction was obviously misinterpreted by the moths as a change in wind speed and, therefore, answered by an alteration in thrust instead of turning, which under the closed-loop conditions resulted in an ongoing progression with respect to the ground pattern movement.

The moths responded to an alteration of wind speed by changing the thrust alone (Fig. 7.3a; compare thrust and direction of flight with 'wind speed'), or both thrust and lift (Fig. 7.3b). The polarity of the responses was such as to counteract the changes in pattern speed. With increasing wind speed (headwind) and thus decreasing groundspeed, thrust was increased, and vice versa. Groundspeed was thus kept rather constant.

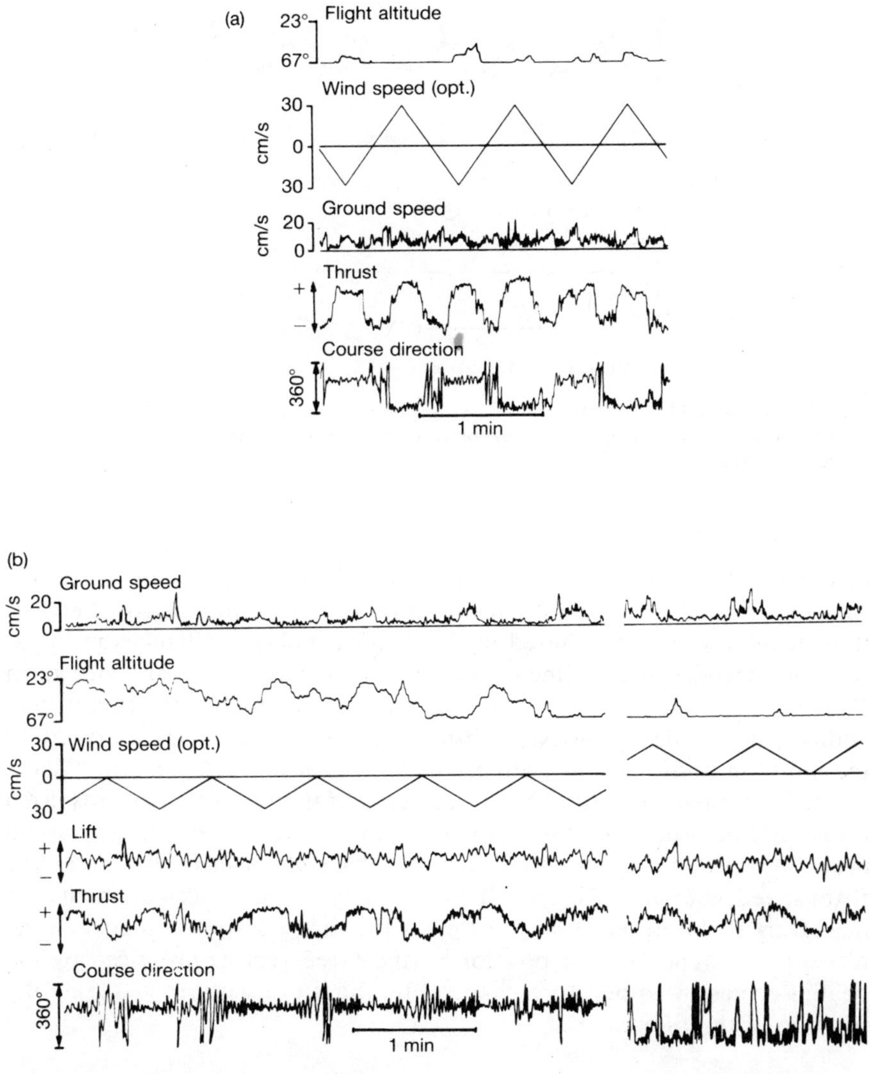

Fig. 7.3. Flight parameters in response to changes in wind speed and wind direction. (a) Changes in wind speed are answered by compensatory changes of thrust alone, or (b) by a combination of thrust and lift, as indicated by the induced changes in flight altitude (defined by the angular width of one black/white pattern element). In both cases groundspeed is kept rather constant.

Though in many cases the lift response itself was hardly detectable, inspection of the induced changes in flight altitude (recorded as the focal length of the zoom objective lens) easily revealed its correlation with the changes in wind speed (Fig.7.3b). This is not surprising since the moth's flight altitude results from the integration of its lift, whereas flight speed is prop-

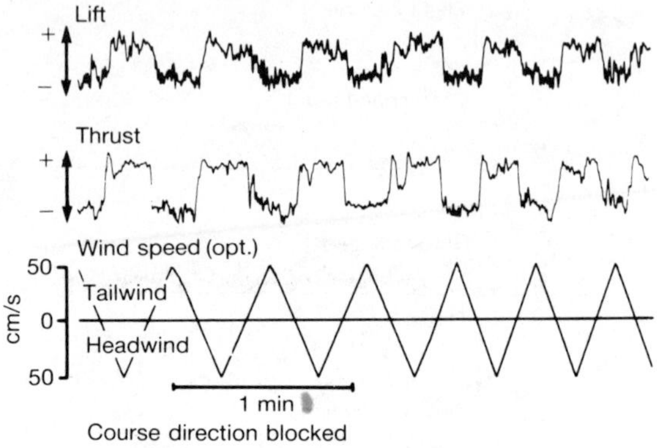

Fig. 7.4. Adjustments of thrust and lift in response to changes in wind speed and wind direction (open-loop experiment). In several cases the thrust response obviously precedes the lift response (arrows).

ortional to the moth's thrust when measured over similar time intervals. This experiment also shows that lift and thrust were not strictly coupled, though the moth was tethered in this flight simulator. Changes in thrust were not necessarily combined with a parallel change in lift, and vice versa (Fig. 7.4). In this open-loop situation (the moth's flight direction was in addition mechanically blocked) thrust and lift were strongly adjusted by the moth, in both cases to counteract the optically simulated wind. When the pattern movement simulated regression of the moth (backwards drift) thrust and lift were increased, which under closed-loop conditions would probably result in an overcoming of the backwards drift (by increasing flight speed) and in a reduced angular velocity of the pattern elements (by increasing flight altitude). In most cases the thrust response preceded the lift response. The angular position of the force vector, produced by the beating wings, was therefore changed by the moth without changing the position of the thorax.

(b) Upward and downward 'wind' movements. Variations in flight altitude forced upon the moth by upward and downward air movements were optically simulated by periodically increasing and then decreasing the size of the ground pattern elements in a triangular way. Independent of whether the pattern was otherwise stationary (open-loop experiment, Fig. 7.5) or moving according to the moth's thrust signal (closed-loop experiment), lift and thurst were controlled with a strong negative correlation with the simulated changes of flight altitude. Lift and thrust decreased as the pattern size decreased (increasing flight altitude) and vice versa. The lift response thus counteracted the optically simulated impression of changes in flight altitude.

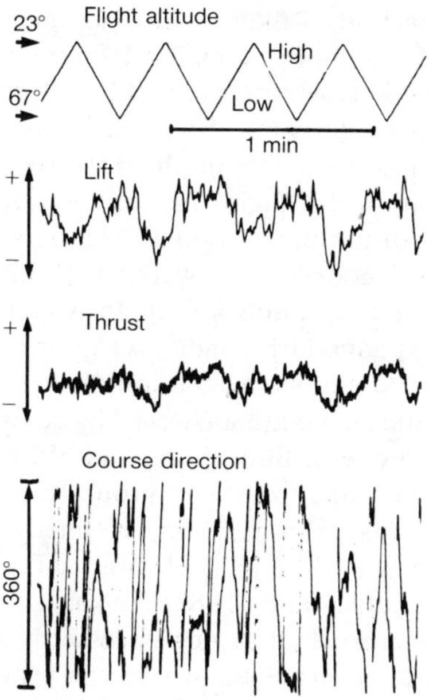

Fig. 7.5. Variations in the size of the pattern elements, optically simulating falling and rising, are answered by compensatory modulations of lift and thrust (flight altitude defined by the angular width of one black/white pattern element).

7.5. Concluding remarks

The above experiments have shown that the 'flight simulator' provides an experimental method which permits the simulation of the optical situation corresponding to free flight of a tethered flying moth. From the six degrees of freedom of a free-flying animal (rotation and translation in each of the three coordinates of space), sideslip, roll, and pitch are not accessible by this method. However, the measurement of thrust, lift, and flight direction, and the simulation of the optical stimuli caused by these, seem to be sufficient for an adequate simulation of free flight. Wind-tunnel experiments indicate that flight direction coincides fairly well with the longitudinal body axis, which means that sideslip is negligible (Preiss unpublished). We therefore restricted our system to those flight parameters though a system measuring forces in all six degrees of freedom had already been developed (Blondeau 1981).

The results obtained with the 'flight simulator' can be compared with those from wind-tunnel experiments, previous tethered flight experiments and experiments with walking males:

(1) Optically simulated wind is detected by the gypsy moth. The flight courses are oriented straight upwind, without alternating menotactical orientation with respect to the wind direction. This was

already found in both walking and flying gypsy moth males (Preiss 1980; Preiss and Kramer, submitted for publication). In all cases cross-wind excursions were explained by internal noise functions.

(2) Abrupt changes in the wind direction by 90° are answered by an immediate upwind turn. The moth's high sensitivity in detection of the wind direction in flight in the presence of the attractant pheromone (+)-disparlure is comparable to its very low threshold for an upwind orientation when walking (Preiss, 1980; Preiss and Kramer, submitted for publication). In walking moths this lowered threshold was explained by a higher weighting of the anemoreceptive signal by the simultaneous perception of (+)-disparlure.

(3) Variations in wind speed are answered by compensatory adjustments of either thrust alone or both thrust and lift. Compensatory changes in thrust result in a stabilization of groundspeed, also found in wind-tunnel experiments (Heran 1955; Kennedy 1940; Kennedy and Marsh 1974; Marsh *et al*. 1978) and tethered flight experiments (Preiss *et al*. 1980; Preiss and Kramer 1983). The observed, combined response of both lift and thrust does not necessarily contradict the results from wind-tunnel experiments, where wind speed is compensated by an adjusted modulation of flight speed alone, since flight altitude is restricted due to the limited diameter of the odour plume and its preselected height. In the field, flight altitude can vary with the wind speed (Cameron personal communication; see also von Frisch and Lindauer 1955; Heran and Lindauer 1963—field experiments with honeybees, when no odour plume was present).

(4) Compensatory changes of lift upon variations in the size of the pattern elements are responsible for an additional stabilization of flight altitude. As shown by Preiss and Kramer (1983), the radial movement of pattern elements either towards (centripetal) or from (centrifugal) the point directly beneath the moth is the effective stimulus for the moth's recognition of changes in flight altitude, with the two transverse sectors laterally beneath the moth being sufficient.

(5) Although the moths are tethered in the 'flight simulator' (fixed inclination) thrust and lift are not strictly coupled, but can be varied simultaneously by the moth. This has also to be postulated from wind-tunnel experiments, where wind speed is compensated by changes in flight speed but without changes in flight altitude.

The 'flight simulator' allows for a controlled optical and olfactory stimulus application and a continuous registration of the flight tracks. The experimental requirements for an analysis of the mechanism underlying the optical detection of the wind direction, and for further analyses of the optical control of flight speed and altitude are met by this experimental arrangement. This apparatus also allows further examination of the modulatory influence of the attractant pheromone as well as of inhibitory compounds on the flight control system.

Acknowledgements

This work was supported by the Deutsche Forschungsgemeinschaft (Schn. 14/61-1).

8 Visual and olfactory stimulus interaction in resource finding by insects

R. J. Prokopy

8.1. Introduction

To quote J. S. Kennedy (1978) 'All behaviour is the product of central nervous integration of different stimuli from both outside and inside the insect and likewise of different responses'. It would require far more space than is permissible here to apply the elements of this concept to detailed analysis of how insects integrate and respond to visual and olfactory information from the environment when attempting to find resources such as food, mates, egg-laying sites, and refuges. Hence, this paper can be only an abbreviated account of several aspects one might consider when investigating insect responses to visual and olfactory stimuli. The focus will be on the resource-finding behaviour of airborne herbivorous insects up to the point of arrival at a resource. A principal theme will be that, although the olfactory ecology of insects has received the most attention by researchers, the visual ecology may be no less important in locating resources. Near the end, examples will be given to illustrate the interplay of visual and olfactory stimuli during insect search for mates or egg-laying sites.

8.2. Stimuli external to the insect

8.2.1. Habitat structure

An insect foraging for a resource may perceive the environment several hierarchical levels: the habitat, the patch, and the resource item itself (Hassel and Southwood 1978). A habitat (e.g. a woodland) can be considered a collection of patches (e.g. groups of trees). A patch may be identified as an aggregation of resource items (e.g. individual trees, limbs, leaves). It is important to recognize that insect habitats considered by humans as having discrete levels of resource distribution might be viewed by foraging insects as having 'blurred' levels. Hence, as Hassel and Southwood (1978) suggest, we should look to changes in the forager's behaviour to identify resource level boundaries. This is a particularly weak area of understanding, especially in regard to insect response to plant stimuli.

The texture of the habitat with respect to abundance, quality, and distribution of resource items, as well as the pattern of mixture with non-resource items, can all profoundly affect resource-foraging behaviour

(Kareiva 1983; Stanton 1983). In some habitats, visual or olfactory stimuli from non-resource items may mask or otherwise interfere with ready detectability of resources, or even elicit repulsion from resources. In other habitats (e.g. agricultural croplands), non-resource items may be less abundant or absent, thereby facilitating detectability of resource stimuli. Whichever the case may be, one should be aware that over evolutionary time, the structure of the resource habitat has undoubtedly played a major role in shaping the foraging behaviour of the insect and the manner in which the insect responds to visual, olfactory, and other environmental stimuli. At the same time one should be cognizant that all animals are constrained by their phylogenetic history, and thus may not be able to perceive and utilize the entire wealth of visual and olfactory information emanating from a habitat, patch, or resource item.

8.2.2. Visual stimuli

The detectability of a resource item at any distance by an insect's visual system, as by any visual system, is a function of the item's dimensions, pattern, contrast against background, optical properties of the medium, and intensity of illumination (Lythgoe 1979; Prokopy and Owens 1983). Contrast may arise from intensity or spectral differences between resource items and the background, dimension or pattern differences, or motion differences, all of which may constitute stimuli to the insect. The nature of photons renders their distribution relatively independent of air movement. Therefore, visual stimuli are effectively omnidirectional (provided there exist no physical obstructions) and are not acutely variable with small changes in distance from the source (Miller and Strickler 1984). Even so, it is paramount to recognize that human scales of distance and the viewing angle from which humans approach a resource are unlikely to be applicable to animals as small and as mobile as airborne insects. Furthermore, subjective guesses as to the reflectance properties of resource items are not an adequate substitute for quantitative analysis using photographic, spectrophotometric, or spectral radiometric techniques (Prokopy and Owens 1983). Such analysis can provide accurate characterization of brightness (intensity of reflected light), hue (dominant wavelength), and saturation (spectral purity).

8.2.3. Olfactory stimuli

The olfactory signal from a resource under conditions of moving air (rarely is there ever total calm in nature) is a filamentous plume having a comparatively flat concentration gradient along its windward axis but sharper boundaries laterally (Elkinton and Cardé 1984). The distance over which an active odour space extends from the source depends upon the number of odour molecules released from the source, height of source above ground, wind speed, turbulent diffusion, and degree of interference by vegetation or other objects. Thresholds of perception and response may change with time, owing to the discontinuous (filamentous) nature of an odour plume. An insect downwind of an odour source will be likely to experience odour in bursts (Elkinton and Cardé 1984). Rate of emission may also change with time. Together, these properties suggest that the active space of olfactory signals is far more limited in supplying information on direction of source than is true for visual signals, but under certain environmental conditions it may extend further from the source than that for visual signals. For higher-flying insects, visual stimuli might be the more readily discerned, because odour plumes tend to become increasingly diffuse with increasing altitude (Miller and Strickler 1984).

8.3. Stimuli internal to the insect

Stimuli internal to the insect are no less important in affecting the insect's response pattern than are external stimuli. Relevant internal stimuli include information from the insect's peripheral visual and olfactory sensory systems, information from central neural processing of peripheral events, and information from developmental, hormonal, neural, and prior experience events that influence interpretation of incoming peripheral signals by the central nervous system (Dethier 1982). With regard to the insect's ability to perceive visual stimuli, one should appreciate that the insect's visual system is a fixed-focus type, that visual acuity (ability to resolve patterns) is maximum at very close range (centimetres or decimetres) and decreases rapidly with distance, and that the larger the compound eye the greater the acuity (Wehner 1981). Also, the properties of most (but not all) insect photoreceptor pigments constrain wavelength discrimination ability to a zone between about 350 and 600 nm (ultraviolet to yellow).

The nature of an insect's response to environmental stimuli is very likely to be a product of combined positive internal and external stimuli and combined negative internal and external stimuli (Dethier 1982). Miller and Strickler (1984) present a highly appealing graphic mechanical 'rolling fulcrum' model (an analogue of Dethier's verbal model) that vividly illustrates how internal behavioural thresholds of resource specialist versus resource generalist insects might affect the strength of potential response to external stimuli. The model highlights the importance, when analysing reactions of an insect to visual or olfactory stimuli, of being as aware as possible of the insect's physiological state with regard to threshold levels of responsiveness.

8.4. Response pattern of the insect

8.4.1. Optomotor reactions

One of the principal factors that has shaped visual pattern perception capability in flying insects is the necessity of maintaining visual contact with the ground during flight so that flight speed and direction can be monitored. How insects respond to visual information from the ground beneath them while orienting to wind direction, olfactory plumes, and other stimuli will not be addressed here, but is discussed by Cardé (1984).

8.4.2. Resource finding

Prior to arriving at a resource item, an insect may engage in movements that are random (i.e. unaffected in any way by resource stimuli) or whose speed or direction are in fact influenced one way or another by resource stimuli. As Kennedy (1978) argues convincingly, little insight into the nature of insect response to environmental stimuli is gained from viewing the response in such general terms as 'attraction' or 'arrestment'. Rather, detailed tracking and analysis of the slowing or speeding up of movement and the increase or decrease in turning rate and angle are far more useful parameters for response measurement. Ideally, one would like to know the speed and direction of every movement for individuals in a setting devoid of resource stimuli, compared with these parameters for individuals in settings that provide synthetic olfactory stimuli alone, synthetic visual stimuli alone, the combination of these synthetic stimuli, and finally the combined stimuli of the natural resource. One would also like to know how movement speed and direction change according to changes in level of resource distribution. One might expect greater specificity in response pattern as the resource distribution level narrows from the habitat to the patch to the resource item itself.

Obtaining this sort of information under the heterogeneous conditions typical of totally natural settings would be very difficult at best. However, much more than has been done to date could be done toward this end through creating, under semi-natural conditions, prescribed resource-containing patches that mimic natural patches in structure and quality, and through tracking the movements of released individuals of prescribed (so far as possible) physiological state. Till now, the comparatively little detailed information available on analysis of insect movement under non-stimulus versus resource-stimulus conditions stems almost exclusively from studies under highly artificial conditions in the laboratory (conditions which do not include resource distribution in patches and, with a few exceptions, involve insect response to olfactory stimuli only) (Bell 1984; Cardé 1984).

Two crucial factors remain to be considered in analysing response patterns to visual and olfactory stimuli in any environmental setting. These involve (a) presenting exact synthetic equivalents of stimulus components to permit delineation of cues according to their relative importance (this involves mimicking the natural stimulus in both quality and quantity of emitted information), and (b) determining the distance over which a resource stimulus is detectable by the insect (this is almost certain to vary according to the quantity and quality of stimulus). As regards pheromonal stimuli from potential mates, there exist a few cases in which the natural emission rate of the pheromonal blend has been closely approximated in quality and quantity via artificial dispensers of synthetic material, and a few cases in which distance of response to pheromone source has been analysed (e.g. Elkinton *et al*., 1984). There is a near-total lack of information along these lines about plant olfactory stimuli (Finch 1980). This is due partly to the difficulty inherent in identifying all volatile emission components of a plant stimulating an insect and in creating a synthetic equivalent of the natural blend. It is due also to the difficulty up till now of discerning consistent measurable change in an individual's behaviour that signals to the observer that recognition of plant olfactory components has occurred at a given distance from the source under identified environmental conditions. One might guess that much progress along these lines has been made in regard to insect response to visual resource stimuli. This is not so, however, because little attention has been paid to quantifying precisely the spectral quality characters of a resource and creating a synthetic mimic of equivalent natural spectral quality and quantity, together with the natural equivalent of stimulus pattern and dimensions (Prokopy *et al*. 1983a).

The hierarchical levels of resource distribution (habitat, patch, resource item) may have, in many cases, selected for a hierarchical organization of the insect's response pattern (Markl 1974) when it is seeking a resource. It is important to recognize, therefore, that at the habitat level of resource distribution the response to an olfactory (or visual) stimulus may be predominant, whereas, at the patch level the response to a visual (or olfactory) stimulus may dominate, and at the resource item level an olfactory (or visual) stimulus may again dominate. In addition to interacting in this sort of sequential fashion (where one stimulus type may or may not release response to another type, according to behavioural threshold level changes), olfactory and visual resource stimuli may also interact simultaneously. Furthermore, if one sensory system suffers injury or is otherwise impaired (e.g. through adaptation, central habituation, or decline with age) the insect may still locate a resource through exclusive use of another sensory system. As shown by Dyer and Gould (1983), possession of redundancy of sensory mechanisms organized in a hierarchical fashion can be of great selective advantage to an insect. But it also can create difficulty in analysing response patterns, because an individual's ability to locate a resource without employing a particular sensory modality does not necessarily mean that it or its conspecifics would not under other circumstances use information from that modality.

8.5. Examples of visual and olfactory stimulus interaction

8.5.1. Rhagoletis pomonella, the apple maggot fly (AMF)

In AMF, mating occurs exclusively on host plants bearing fruits that serve as oviposition and larval development sites. When approaching sexual maturity and searching for fruiting host trees, both sexes respond positively from a distance (yet undetermined) to host fruit olfactory and tree visual stimuli. Details on this and other aspects of AMF response to mate-finding cues are summarized in Prokopy (1982) and Prokopy and Roitberg (1984). The blend of seven esters which comprises the olfactory fruit stimulus elicits, over time, accumulation of both sexes at an upwind source in laboratory wind-tunnels (Fein *et al*. 1982). The tree visual stimuli consist of green leaf colour (or its supernormal equivalent, yellow colour), silhouette of tree against the background, tree size (the larger, the more detectable), and tree shape (compact). Unlike the fruit volatile blend, none of these visual characters is specific to host trees. Under field conditions, natural fruit volatiles alone (release rate unknown) do not elicit as many alightments by either sex as do tree-mimicking visual stimuli (large yellow rectangles) alone. The combination elicits the most alightments. We remain essentially ignorant of the course of inter-plant movement of either sex in nature in the absence of host tree stimuli or upon first encountering host tree visual or olfactory stimuli.

After arrival on a fruiting host tree, males that have reached sexual maturity search first for leaves, where they emit a volatile sex pheromone (chemistry unknown) that is 'attractive' or 'arresting' over a short distance (probably no more than a metre or two) to virgin females. Individual leaves are located exclusively by vision during short lateral or upward flights. Larger leaves elicit more landings than smaller ones. Leaf shape is not important. Leaf colour (particularly the characteristic green hue peaking in reflectance at 550 nm) is very important as a stimulus, as revealed by experiments in which leaf hue, saturation, and brightness were mimicked closely using artist pigment combinations (Owens 1982). Virgin females arriving within about 40 cm of pheromone-releasing males on leaves provide a suite of visual cues that stimulate males to fly, hop, or crawl directly to females and attempt copulation. When females are mature enough to oviposit (but not before) they as well as males seek out individual fruits, which are detected at short range only (1 m or less). There, the males establish territories from which they launch force-mating activities against already-mated, unreceptive females that arrive on the same or nearby fruit (sperm mixing in the females confers a fitness advantage to males engaging in this behaviour). Under conditions of abundant fruit (e.g. several per cubic metre of tree space) both sexes locate individual fruits during short lateral or upward flights and do so solely on the basis of fruit visual properties (spherical shape, characteristic size, and contrast of colour against the background). Under conditions where fruit is scarce, AMF may respond to fruit olfactory as well as visual stimuli to locate individual fruits within trees.

Although much more needs to be learned about the movement pattern and distance of response of AMF to plant stimuli under varying habitat structure conditions, this example illustrates the interaction of one set of olfactory and visual stimuli at the level of foraging for host trees within a habitat or patch, and the interaction of a different set of visual and olfactory stimuli at the level of foraging for mating sites, mating territories, or mates within host trees.

8.5.2. Lymantria dispar, the gypsy moth (GM)

In GM, mating occurs almost exclusively on tree trunks, where flightless virgin females emit a monocomponent sex pheromone (the synthetic equivalent is (+)-disparlure) that is attractive to males (Cardé 1981). Males seeking trees harbouring virgin females respond positively both to female sex pheromone and to the visual character of vertically-oriented tree trunks. After arrival on a tree trunk, a male finds a female either by walking upwind to the female or, in situations where the male lands outside the female's pheromone plume, by a 'preprogrammed' walking pattern that results in tactile contact.

The GM is noteworthy here particularly on account of the highly detailed analysis that has been made of movement patterns of individual males in the absence of female stimuli, as well as at varying distances from real or synthetic female sex pheromone sources, and from real or synthetic tree trunks under natural conditions and laboratory wind-tunnel conditions (see Cardé 1981; Cardé and Hagaman 1984; David *et al*. 1983; Elkinton and Cardé 1984; Elkinton *et al*. 1984; Schal *et al*. unpublished). Males can detect the pheromone of a single female at least 80 m away. They orient to the pheromone source at long distances such as this through odour-modulated anemotactic processes. At close range, they may engage in aerial odour-trail following as well. Males may also engage in visually-stimulated tree-oriented flights during which they fly up and down along the tree trunk, evidently in 'search' of females. Orientation to tree trunk visual stimuli occurs (a) in the absence of female pheromone above the threshold eliciting upwind flight along a plume, (b) in high-density female populations, where tracking the pheromone trail of an individual female may pose special difficulties, and (c) under conditions of rather turbulent air.

Although a visual stimulus (yet unquantified) appears operative at only one level of the mate-foraging behaviour of GM, and although habitat structure parameters have not yet been altered in a systematic experimental fashion, studies to date on mate finding in GM are exemplary of the sort of analysis of movement speed and direction in relation to behavioral thresholds that is needed to quantify insect response patterns under varying habitat conditions.

8.5.3. Delia radicum, the cabbage root fly (CRF)

In CRF, egg laying takes place in soil at the base of host plants, the roots of which are fed upon by the larvae. The host-finding process (summarized in Finch 1980; Hawkes and Coaker 1979; and Prokopy *et al*. 1983b) commences with the upwind dispersal of gravid females. On arriving within 5–15 m downwind of a patch of host plants, at least a proportion of females is able to perceive host olfactory stimuli (a complex of hydrolysed glucosinolates, many of which have been identified and synthesized). Females then become increasingly active and move upwind toward the patch in short flights or hops averaging about 50 cm in length. A zigzagging flight path typical of GM and other airborne insects responding to sex pheromone appears not to be characteristic of CRF when seeking host plants. When the release rate of synthetic host volatiles is at a greatly elevated level, CRF may be stimulated further downwind than normal to orient toward the source, but may not approach the source as close as normal owing to the high release rate.

After arrival within a host patch, CRF detect individual plants largely or exclusively on the basis of visual stimuli when plants are spaced close together (e.g. 25 cm apart). Plant to plant flights in such patches are primarily straight-line, with CRF seemingly being able to detect a plant by vision at about 40–60 cm or more. To find individual plants spaced further apart (e.g. 1 m) in a patch, CRF appear to respond to both plant olfactory and visual stimuli, with inter-plant flights more curved than direct. The most important visual stimulus is plant colour, particularly the peak reflectance at 550 nm characteristic of green leaves. This was determined through use of artist's pigments to mimic closely CRF host plant reflectance. The larger the plant, the more CRF alight. Plant pattern appears rather unimportant. Unlike host olfactory stimuli, plant visual stimuli are not specific to CRF hosts. After arrival on a host plant leaf, CRF crawl to the base of the plant to lay eggs.

Although we remain relatively ignorant of the exact nature of the movement patterns of CRF females in the absence versus the presence of plant olfactory or visual stimuli, the CRF represents a good example wherein both host olfactory and visual stimuli eliciting positive responses have been rather well defined and mimicked qualitatively and quantitatively and wherein the nature of the interaction between these stimuli has been studied up to the point of arrival on a host plant.

8.5.4. Delia antiqua, the onion fly (OF)

In OF, as in CRF, egg laying takes place in soil at the base of host plants. The response pattern of female OF to plant stimuli (summarized in Dindonis and Miller 1980; Harris and Miller 1983, 1984) commences when females enter plumes of host volatiles several metres downwind of the source (maximum distance unknown). They then alight on the ground, turn into the wind, and move upwind toward the source in a series of short non-zigzagging flights or hops, much as CRF do. Several alkyl sulphides have been identified as comprising a part of the host olfactory stimulus blend. After arriving close by a host plant, OF were found in greenhouse cage studies to respond positively to plant visual stimuli, alighting most frequently on plant mimics having yellow colour (= supernormal equivalent of foliar hue) and narrow upright stems. Following alightment, the colour, pattern, and dimensions of the plant, as well as volatiles emanating from the base of the plant if a host, were shown to interact in simultaneous or sequential fashion to elicit characteristic pre-oviposition behaviour and egg laying.

It is the elucidation of interactions between host visual and olfactory stimuli during the examining phase of oviposition site-foraging behaviour (after alightment on a host but prior to arrival at the oviposition site itself) that is unique to this work on OF compared with studies on CRF and other herbivorous insects.

8.6. Conclusions

The concepts presented here, though abbreviated, and the examples, though far from reflecting adequate understanding, none the less are highly suggestive that insect responses to olfactory signals should not, indeed cannot, be investigated meaningfully apart from the total sensory ecology of the insect. There may in fact exist few, if any, cases wherein an insect uses only a single sensory modality to find a resource. Typical experimental settings in which insect responses to resource stimuli are investigated do not usually resemble, either qualitatively or quantitatively, the natural structure of the resource habitat sufficiently well to merit firm conclusions as to the precise character of all relevant responses. One recognizes that a certain amount of simplicity in environmental setting is required to record and analyse in detail exact movement patterns of individual insects. Nevertheless, the foregoing offers ample indication that movement patterns as well as the nature of the stimuli eliciting movement (olfactory or visual) can change significantly not only from one resource level to another (habitat to patch to resource item) but also according to resource abundance and distribution within each level. Fuller attention to conducting analyses of insect stimulus response patterns under conditions more closely approximating the texture of natural resource habitats, together with consideration of several other internal and external elements discussed that affect insect response patterns, should give rise in the future to a more realistic appraisal of the true nature of interactions between visual and olfactory stimuli in the resource-finding behaviour of insects.

[illegible] on plant [illegible] [illegible] [illegible] [illegible] [illegible] of [illegible] [illegible] [illegible] [illegible] [illegible] [illegible] the [illegible] [illegible] of the plant, as well as [illegible] from the [illegible] of the plant [illegible] [illegible] [illegible] [illegible] [illegible] [illegible] [illegible] [illegible] [illegible] [illegible] [illegible] [illegible] and [illegible].

[illegible] [illegible] [illegible] [illegible] [illegible] [illegible] [illegible] [illegible] visual and olfactory [illegible].

[illegible] Conclusion

The [illegible] [illegible] [illegible] [illegible] [illegible] [illegible] [illegible] [illegible] [illegible] [illegible] [illegible] [illegible] [illegible] [illegible] [illegible] the [illegible] [illegible] [illegible] [illegible] [illegible] [illegible] [illegible] [illegible] [illegible] [illegible] [illegible] of the [illegible] [illegible] [illegible] [illegible] [illegible] [illegible] one sensory modality to the [illegible]. [illegible] [illegible] [illegible] [illegible] [illegible] to resource [illegible] [illegible] [illegible] [illegible] [illegible] [illegible] qualitatively or quantitatively [illegible] [illegible] [illegible] [illegible] [illegible] [illegible] [illegible] [illegible] [illegible] [illegible] to the [illegible] [illegible] [illegible] responses [illegible] [illegible] [illegible] [illegible] [illegible] [illegible] [illegible] and analyses [illegible] analyses of insect [illegible] of the true nature of [illegible] [illegible] visual and olfactory stimuli in resource [illegible] [illegible] [illegible].

9 The effect of habitat on the flight of moths orienting to pheromone sources

J. N. Perry and C. Wall

9.1. Introduction

This paper synthesizes recent work on the flight of the pea moth, *Cydia nigricana* F., orienting to pheromone sources in a wheat crop. We contrast its behaviour in this habitat with that of other moths not in crops. Adult males fly on warm, summer afternoons, but observation of moths responding in the field is difficult, because flight is usually within the crop or at crop height, so much data has been obtained by trapping. The traps in this study contained 100 μg of the natural pheromone (*E,E*)-8,10-dodecadien-1-yl acetate (Greenway 1984). Investigation of the effect of habitat on trap catch, and therefore on flight behaviour, has led us to seek a model to explain these data.

9.2. Review

In 1977 experiments to investigate the effect of the presence of one trap on the catch in another showed that moths were attracted over at least 100 m (Perry *et al*. 1980; Wall and Perry 1978). In lines of three interacting traps along the wind, most moths were caught in the upwind trap. An initial model postulated that responding moths flew in the mean upwind direction. Those that passed close enough to the centre or downwind traps might be caught (e.g. moth A, Fig. 9.1). Moths further cross-wind of the line would fly on upwind, parallel to the line, past the downwind and centre traps (e.g. moth B, Fig. 9.1), and eventually encounter the 'edge' of the 'trapping zone'. Moths would follow a path along this 'diffuse edge' by an unknown mechanism and be caught in the upwind trap. Further experiments during 1978 and 1979 suggested that the range of attraction exceeds 400 m (Wall and Perry 1980); this was confirmed later in a mark and recapture experiment (Wall and Perry unpublished). The initial model predicted that, given a large range of attraction, most of those moths not caught in the upwind trap should be caught in the extreme downwind trap, which they would encounter first. This was demonstrated in the field for lines of two and five traps, and later confirmed for lines of four, nine (Perry and Wall 1982), and three (Perry and Wall 1984a).
Wall 1984a).

Meanwhile, discussions with Drs F. B. Smith and N. Thompson of the Meteorological Office during 1978 revealed that, over open ground, the

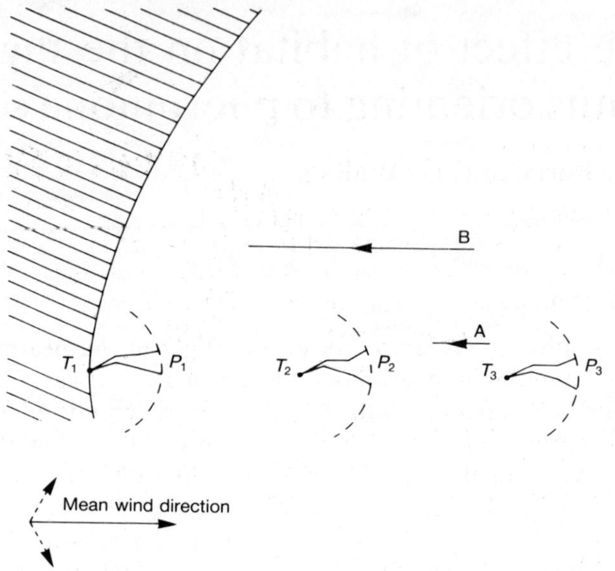

Fig. 9.1. Schematic plan view of three interacting pheromone traps (T_1, T_2, T_3) for *Cydia nigricana* F. aligned along the wind in a wheat crop. The arc around each trap indicates the assumed limit of the well-defined plume (P_1, P_2, P_3). Moths in the shaded area are assumed not to receive sufficient pheromone to respond. Moth A flies up (mean) wind, subsequently encounters plume P_2 and is caught in trap T_2. Moth B flies up (mean) wind, past trap T_2, and is subsequently caught in trap T_1 (after Perry and Wall 1984b).

wind passing through any source continues in a straight line for hundreds of metres, even though subsequent changes in wind direction at the source give rise to a snaking plume. Therefore, over open ground, an insect would detect pheromone only when the wind was blowing directly from the source; if responding with upwind flight it would move directly towards the source. This straight-line hypothesis was confirmed for smoke over short grass during 1981 by David *et al*. (1982). They suggested that a strategy of flying upwind on detecting pheromone, but holding station against the wind (with or without casting across it) on losing the odour, would aid source location by flying moths over open ground. This odour-modulated anemotactic mechanism was compared with an inferior strategy of odour-trail following. David *et al*. (1983) demonstrated elegantly that the proposed anemotactic strategy is used over short grass by *Lymantria dispar* L.

In the initial pea moth model the assumed average resultant ground track—up (mean) wind, not directly towards the source—is explicable only if the moths detect pheromone continuously. Wall and Perry (1980) therefore qualified the model by hypothesizing that the discrete plume breaks down in a wheat crop at relatively small distances from the trap, to give a continuous spatial field of pheromone further downwind (Fig. 9.1).

The results of two experiments in 1979 showed why a crop should have such an effect. In the first experiment, vegetation in the immediate vicinity of a trap was shown to have absorbed sufficient pheromone to be detectable later by electroantennographic and gas chromatography techniques. After trap removal, the vegetation released sufficient pheromone to attract males for a considerable period (Wall *et al*. 1981). Thus, a molecule of pheromone may be successively absorbed and released by many plants before being perceived by a moth. In the second experiment, Legg *et al*. (1980) released nitrous oxide into a barley crop monitored by a downwind sampler. The plume became entrained by the crop at distances no greater than 10 m from the source; material velocities within the crop were far smaller than had been expected from the wind velocities above it.

The substantial effects of contaminated vegetation were quantified during 1981. Wall and Perry (1983) reported that wheat exposed to pheromone for only 30 min attracted males for longer than 180 min, and that while the clump of wheat plants that had been immediately downwind of the trap remained attractive after trap removal, the removal of this clump did not at all affect the attractiveness of the neighbouring plants. It was also noted that changes in responsiveness could be independent of changes in random flight activity, and that moth responsiveness often increased when a trap was removed.

A wider variation than had previously been reported in the proportional catch in each of a line of three traps along the wind was found during 1981 (Perry and Wall 1984a). In the previous experiments, traps had been left to acquire a catch over an entire trapping period; over consecutive short intervals within a trapping period the proportional catch in the centre trap was constant and that in the upwind trap showed trends in time. Two lines of traps that were operated simultaneously gave similar results, emphasizing that systematic processes were involved and the need to explain such results with a quantitative model. During these experiments observers frequently reported moths flying upwind parallel to and up to 20 m crosswind of the line of traps, past downwind and centre traps and not directly towards any of the sources. These observations, although not quantitative, reinforced the initial model, and confirmed previous isolated and similar observations (Fig. 9.1).

In a discussion paper Wall and Perry (1982) presented preliminary field results from 1981, and a speculative explanation of short-term variations in proportional trap catch through changing response thresholds and random flight, which is now thought to be unlikely. However, their discussion of variations in response thresholds between and within individuals (see also Baker and Roelofs 1981), and associated difficulties with Wall and Perry's (1978) definition of a 'trapping zone', demonstrated the need for spatial quantitative models either to include stochastic elements or to refer to the average behaviour of a population of moths. The effects of a crop on pheromone dispersal were reiterated and contrasted with the model of David *et al*. (1982).

The drawback to Wall and Perry's (1978, 1980) initial model was the lack of a mechanism for the moth to reach the upwind trap, once it has flow upwind and encountered the 'diffuse edge of the trapping zone' where pheromone concentration is relatively low. However, following David *et al.* (1982), Perry and Wall (1984b) assumed that such a moth, on encountering a region where the concentration of pheromone is too small to continue responding, moves in a cross-wind direction (e.g. moth A, Fig. 9.2). If this movement brings it closer to the line of traps we assume it resumes upwind flight; in this way its net motion is diagonal to the line of traps (e.g. moth B, Fig. 9.2) and the upwind trap is approached. This modification forms the basis of a spatial mathematical model (Perry and Wall 1984b) which simulates the average behaviour of a population of moths:

> Downwind of a trap, where pheromone is assumed to exist at all positions, it is suggested that a 'steady-state' occurs, so that the pheromone concentration at a given point changes little with time. Time-averaged

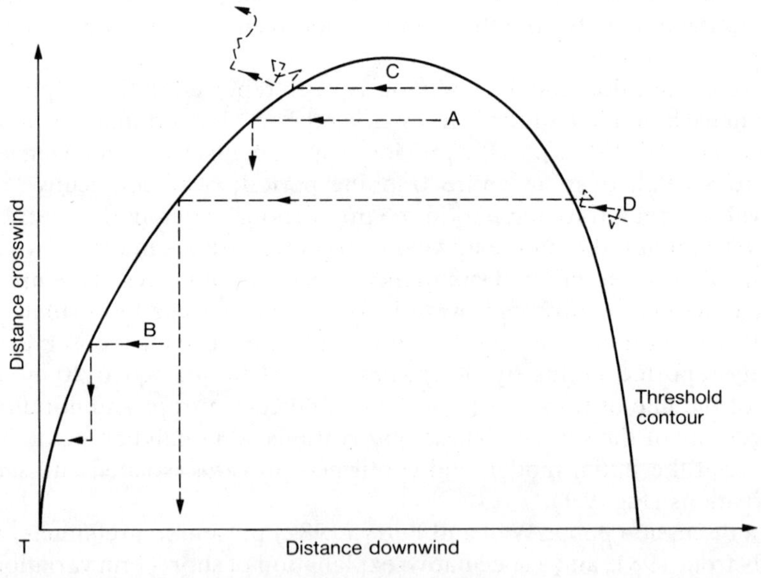

Fig. 9.2. Schematic plan view of assumed net movement of simulated *Cydia nigricana* F., responding some distance from a pheromone trap, T, in a wheat crop (only half the symmetrical area bounded by the threshold contour is shown). Diffusion equations are used to estimate the position of the threshold contour, which bounds the region within which pheromone concentration, at all positions, is sufficiently large for up (mean) wind flight to continue. A moth encountering the threshold contour moves cross (mean) wind, either (i) a net distance inside the contour (moth A) with subsequent resumption of upwind flight and net motion towards the trap (moth B), or (ii) a net distance outside the contour (moth C), where pheromone concentration is below the response threshold, with subsequent flight in random directions. A moth flying randomly outside the contour may cross inside it by chance (moth D) and subsequently respond by flying upwind (after Perry and Wall 1984b).

diffusion equations, considerably simpler than those employed by Bossert and Wilson (1963), are derived to approximate pheromone concentrations. These equations are used only to estimate the threshold contour, the boundary of the region where concentrations are sufficiently large for upwind flight to continue. Random flight is allowed: moths which encounter the threshold contour and move cross-wind away from the line of traps are assumed to fly in random directions thereafter (e.g. moth C, Fig. 9.2), some moths are assumed to cross the threshold contour after flying randomly outside it and subsequently to detect pheromone and respond (e.g. moth D, Fig. 9.2). Moths passing close enough to a trap are assumed to be caught, except a proportion that fly upwind to the next trap. The effects of trap efficiency or different doses and attractants (Wall and Perry 1981) may be incorporated.

Since the model is based on the average behaviour of a population of moths it is compatible with both the anemotactic theory and a system of

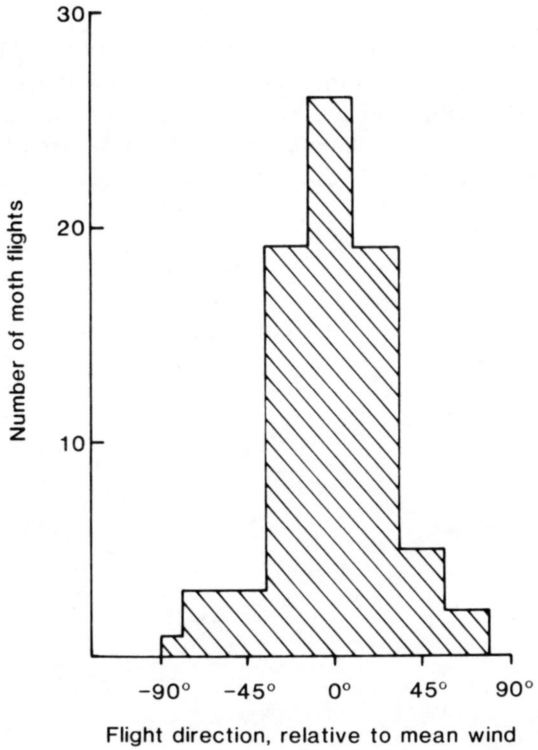

Fig. 9.3. The ground track of responding *Cydia nigricana* F. and the wind direction were observed in 5 m × 5 m areas, cross-wind of a line of interacting pheromone traps aligned along the wind. Flight directions were expressed relative to the mean wind direction (up mean wind = 0°). The frequency distribution shown (n = 78) refers to an area 17.7 m from the nearest trap; the sample mean was 1.78° ± 4.1° (after Perry and Wall 1985).

integrated anemotaxis and longitudinal chemo-klinotaxis (Kuenen and Baker 1983). Seven parameters are involved, four of which are concerned solely with aspects of flight behaviour. The model has provided excellent fits to field data concerning trap interactions, reproducing quantitative and qualitative results previously discussed from different experiments with very moderate changes in parameter values, set within realistic ranges.

During 1983, experiments to quantify previous observations (Fig. 9.1) of pea moth ground tracks (Perry and Wall, 1985) confirmed that the average direction of moths through 5 m × 5 m areas whose centres were 12.5 m cross-wind of a line of three traps was up (mean) wind (Fig. 9.3). There is as yet no observational evidence for zigzagging upwind flight in the horizontal plane; casting to and fro across the wind and hovering were noted only occasionally. Closer to a trap moths appeared to respond as if to a discrete plume; the distance downwind of the source that the plume broke up was estimated to be between 12.5 m and 17.7 m.

9.3. Conclusions

The flight of *C. nigricana* orienting to pheromone sources in a wheat crop differs strikingly from that of other moths so far studied over open ground. We believe similar habitat effects will be found in other species. Further work is required to elucidate how moths might make use of the attractant released from vegetation and other substrates in addition to that released from the source.

10 Orientation in forest Coleoptera: an uncertain outcome of responses by individual beetles to variable stimuli

J. H. Borden, D. W. A. Hunt, D. R. Miller, and K. N. Slessor

10.1. Introduction

A great deal of research has been done on the effects of olfactory stimuli on forest Coleoptera. However, in contrast to the intricate and often elegant experiments on moths, very little of this research has investigated orientation by flying beetles, nor has it addressed the intricacies of orientation behaviour.

Research on various components of the chemical ecology of forest Coleoptera, particularly bark beetles (Scolytidae), has allowed syntheses to be made which describe orientation as a complex of sequential events (Birch 1984; Borden 1974; Renwick and Vité 1970; Rudinsky and Ryker 1977; Vité and Bakke 1979; Vité and Francke 1976). In the Scolytidae this series of events leads to the aggregation of populations that mass-attack a host tree or log. Pioneer insects orient to an uninfested host; all of the others orient to a host with insects already on or in it. There is a complex of visual stimuli and olfactory signals released by both the host and other insects, including both conspecifics and competitors, to which orienting beetles must respond.

Inevitably, an understanding of such complex systems of orientation requires generalizations about orientation behaviour. Lost in the generalization is the individuality of the insects involved. One tends to regard all insects as equal, disregarding Atkins's (1966a) and Wellington's (1977) findings and advice that they are not. This individuality in orientation responses is superimposed upon variability in the stimuli associated with potential hosts and other insects, as well as on diverse environmental factors. Thus, the textbook orientation systems, which appear so definitive in print, are so variable in nature that successful orientation may always be in doubt.

With emphasis on the Scolytidae, and utilizing our recent results on the mountain pine beetle, *Dendroctonus ponderosae* Hopkins, and the pine engraver, *Ips pini* (Say), this chapter approaches two aspects of such variation: individuality in behavioural responses by orienting beetles, and variability in olfactory and visual stimuli to which the beetles are exposed.

10.2. Individuality in behavioural responses

10.2.1. Responsiveness to abiotic stimuli

It is of obvious selective advantage for an insect to subject itself to favourable environmental conditions which permit essential types of behaviour to occur. Thus scolytid beetles in the processes of taking flight, dispersing, and orienting to olfactory stimuli are sensitive to changing temperature, light and humidity (Atkins 1959; Borden 1967; Henson 1962; McCambridge 1971; Perttunen and Boman 1965), barometric flux (Lanier and Burns 1978), and wind and rain (Coster *et al*. 1978; Moser and Dell 1979). They must respond rapidly to changes in their environment. If they make a mistake, as did those that Furniss and Furniss (1972) found on high-altitude snowfields, they are likely to die. Beetles that can make precise meteorological observations and change their behaviour accordingly are the survivors that orient successfully to olfactory stimuli.

10.2.2. Variation in physiological readiness

Orientation to olfactory stimuli by insects is in part governed by their physiological state. Unless the correct internal signals are struck, some insects simply won't respond to the most attractive olfactory stimulus. The evidence for individual variability in responsiveness by bark beetles remained unnoticed for years. For example, many bark beetles emerge from one tree and respond immediately to another. But what about those that establish a new centre of attack, far removed from the brood trees from which they emerged? Why don't they attack the nearest available trees?

Investigations on both aphids (Kennedy and Booth 1963a,b) and bark beetles (Table 1) independently discovered the existence of physiological control of orientation behaviour. Flight exercise was found to release a host-positive response in the striped ambrosia beetle, *Trypodendron lineatum* (Olivier) (Graham 1959, 1962). Atkins (1966b, 1969) found that Douglas-fir beetles, *Dendroctonus pseudotsugae* Hopkins, with a high lipid content were unresponsive to host odours, while those that were poor in lipid were immediately responsive to olfactory stimuli. On flight mills, *T. lineatum* and *D. pseudotsugae* required an average of 30 and 90 min of flight exercise, respectively, before they were arrested by the odour of pheromone-laden frass (Bennett and Borden 1971). Choudhury and Kennedy (1980) found that, after 6 h of flight activity, male smaller European elm bark beetles, *Scolytus multistriatus* (Marsham), responded upwind in pheromone-permeated air, whereas unflown beetles did not.

Here was a remarkable mechanism by which bark and ambrosia beetles optimized their opportunities according to their metabolic fitness. Those that were least fit (poor in lipid reserves) were capable of responding to the first appropriate olfactory signal, but, at least in Douglas-fir beetles, such individuals were still capable of dispersal flights (Atkins 1966b, 1969).

TABLE 10.1

Examples of variability in and physiological control of responsiveness by bark beetles to olfactory stimuli

Species	Example
Dendroctonus frontalis Zimmerman	Significant proportion of emergent beetles positively responsive to attractive semiochemicals in laboratory bioassay (Payne *et al.* 1976), but flight exercise on flight mill for mean durations of 43 and 55 min for males and females, respectively, resulted in 79 and 50% orientation responses to attractive semiochemicals in laboratory bioassays, as compared with 62 and 25% for unflown males and females, respectively (Andryszak *et al.* 1982).
Dendroctonus pseudotsugae Hopkins	Variable proportions of beetles in different populations either photopositive and host-negative, or photonegative and host-positive. Host-negative beetles contain >20% lipid and are flight-positive. Host-positive beetles contain <20% lipid, and are capable of flight if lipid content >10% (Atkins 1966b, 1969). Beetles flown on flight mill >60 min contained a mean of 37% lipid versus a mean of 45.6% in unflown beetles (Atkins 1969). Flight exercise of >90 min on flight mill released arrestment response to the odour of female frass (males) or host phloem tissue (both sexes) (Bennett and Borden 1971). Certain fatty acids were selectively oxidized during flight (Thompson and Bennett 1971).
Ips paraconfusus Lanier	In laboratory olfactometers, overwintering beetles were much less responsive to the odour of male frass than beetles from spring or summer generations (Borden 1967; Wood and Bushing 1963). Beetles of both sexes that were responsive to the odour of male frass contained significantly less lipid than non-responsive beetles. Lipid contents for overwintered beetles were 14.0 and 21.9% for responders and non-responders, respectively; equivalent values for spring generation beetles were 14.0 and 19.7%. Overwintering non-responders contained significantly more lipid than spring generation non-responders (Hagan and Atkins 1975).
Ips pini (Say)	In laboratory olfactometers, overwintering females were significantly less responsive to the odour of male frass than beetles that emerged in the spring or summer (Birch 1974b).

Table 10.1 (*cont.*)

Table 10.1 cont.

Species	Example
Ips typographus L.	Laboratory experiments demonstrated that only 4% of emergent beetles did not take off if placed on a Norway spruce log. Beetles of both sexes stayed on the log after a mean weight loss during succeeding days that was equivalent to weight lost during a 7 km flight. Similarly, 26% of emergent beetles stayed on a log baited with pheromone, but the remainder stayed on the pheromone-baited log following weight loss in succeeding days that was equivalent to weight lost during a 2–3 km flight (Gries 1984).
Pityogenes chalcographus L.	Emergent males were neither host-positive or host-negative. Starvation or walking activity released host-positive response by host-negative beetles (Sanders 1983).
Scolytus multistriatus Marsham	Newly emerged male beetles in a wind-tunnel flew predominantly downwind in pheromone-permeated, moving air, but after 6 h of pre-test flight activity, about 70% of the responses were upwind (Choudhury and Kennedy, 1980). Beetles in a field experiment allowed to emerge from host logs surrounded by traps in five concentric rings responded in greatest number to traps in nearest ring (20 m) and most distant ring after a flight of at least 600 m (Lanier *et al*. 1976) (See Fig. 10.1).
Trypodendron lineatum Olivier	Emergent beetles were primarily host-negative and photopositive, but became responsive to host odours after up to 6 h of sustained flight exercise (Graham 1959, 1962). There was an arrestment response by both sexes to odour of attractive female frass released after 30 min of flight on laboratory flight mill (Bennett and Borden 1971).

Those that were more fit (lipid rich) could disperse widely until some internal feedback mechanism, possibly the accumulation of metabolic by-products during flight (Atkins 1969), released their responsiveness to olfactory stimuli. These beetles sacrificed the opportunity to attack immediately-available hosts for the chance to establish themselves in more distant, and perhaps more suitable hosts, and to interbreed with beetles from other populations.

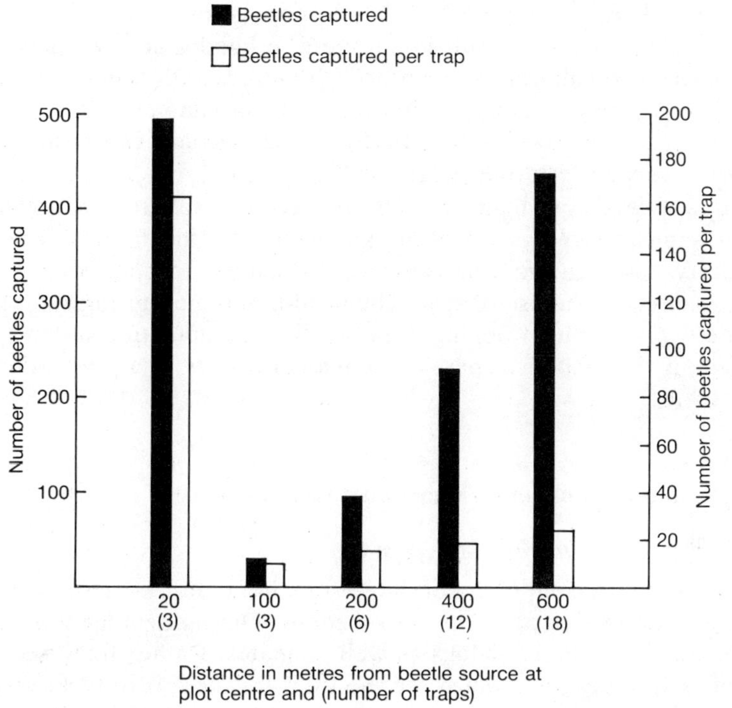

Fig. 10.1. Capture of *S. multistriatus* on multilure-baited traps placed in concentric circles around a pile of infested elm logs from which brood beetles were emerging (data from Lanier *et al*. 1976).

If lipid-dependent, flight-driven behavioural control mechanisms underlie orientation responses of bark beetles to olfactory stimuli, then one would expect a bimodality in dispersal ranges prior to responses to olfactory stimuli. In a rare example of how a natural population responds Lanier *et al*. (1976) (Fig. 10.1), found just such a bimodality. Some *S. multistriatus* emerging from a point source of infestation (a pile of infested elm logs) responded to nearby semiochemical-baited traps, whereas the majority were caught in much more distant traps, probably after considerably more time in flight. If Atkins's (1966b, 1969) hypothesis is valid, those beetles caught on more distant traps should contain more lipid than those caught nearer to the logs from which they emerged.

The physiological character of a population might also determine how semiochemicals should be used in nature. If it were possible to determine the lipid content profile of a population prior to emergence and dispersal, traps might be placed optimally around an infestation so as to catch both the lipid-poor, immediate responders as well as those that require flight activity prior to becoming responsive to olfactory stimuli. And what of the physiological character of a bark beetle population during the course of an infestation? If fewer lipid-rich, dispersive beetles are produced as the beetles attack smaller and smaller, residual trees with thinner, less nutritious phloem tissue, semiochemical-based containment of infestations (Borden *et al.* 1983b; Pitman 1973) might become increasingly more successful as an infestation progressed.

One wonders how many insects in diverse taxa are not affected by semiochemicals deployed without knowledge of the insects' interacting obligatory dispersal mechanisms and olfactory responsiveness. Great improvements in the use of semiochemical-based monitoring systems, and in the efficacy of mass trapping or population manipulation systems, might be made if the responsiveness of individual insects in a population were predictable.

10.3. Variation in olfactory and visual stimuli

10.3.1. Visual stimuli from the host tree

Vision is a critical component of the sensory input utilized in orientation by forest Coleoptera, particularly bark beetles. During aggregation, scolytid beetles must find trees or logs as well as mates. Rather than 'seeing the forest for the trees', they must see the trees for the forest, picking out the best ones from a vast array of possibilities (Fig. 10.2). This is an exacting task considering the diversity of acceptable and unacceptable host sizes and the changing perception of a silhouette at different distances.

It is possible that there is a visual cut-off point at which beetles reject a tree or log that is too small or too large, too upright or too close to horizontal. In laboratory experiments, the white pine weevil, *Pissodes strobi* (Peck), has shown a marked preference for wide, upright silhouettes

Fig. 10.2. Simulated distribution of 25 tree silhouettes faced by the mountain pine beetle on the Diamond S Ranch in the interior of British Columbia (Borden *et al.* 1983a). Smallest diameter silhouettes depict trees 11 cm in diameter at breast height (dph) (1.3 m), and largest silhouette represents a tree 38 cm dbh. The two silhouettes at the far right represent trees 22 cm dbh, approximately the minimum diameter tree from which significant brood production can be expected, and the silhouette seventh from the right depicts a tree 30 cm dbh, from which there would be >1 emergent brood beetle per attacking parent (Cole and Amman 1980).

over narrow or off-vertical ones, corresponding with its preference in the field for thick, upright leaders of Sitka spruce (VanderSar and Borden 1977). Similarly, *D. ponderosae* responded more readily in the laboratory from a distance of 23 cm to wide black cards than to narrow ones (Shepherd 1966).

For the mountain pine beetle, the evidence for visual discrimination of optimal silhouette size (Fig. 10.2) does not seem to hold up in the field. The beetle does show a fourfold preference for upright over horizontal cylinders (Billings *et al.* 1976). However, landing rates by *D. ponderosae* were found to be independent of tree size (Hynum and Berryman 1980), and the beetles responded in very large numbers to traps as small as 6.4 × 13.6 cm (Pitman *et al.* 1978), far below the minimum diameter for suitable hosts (Cole and Amman 1980). When conventional multiple funnel traps (Lindgren 1983) were doubled in height and/or width (by lateral panels) there was no effect of increased width on trap catch, but a doubling of beetles caught with double-height traps, were they wide or narrow (L. C. Ryker, L. J. Chong and J. H. Borden, unpublished data). Thus for *D. ponderosae* the height of the upright silhouette is more important than the width.

10.3.2. The complex of host-associated visual and chemical stimuli

How then does a beetle like *D. ponderosae* sense trees of acceptable or larger size and attack them at higher densities than other trees (Borden *et al.* 1983a)? Firstly, there may be a greater length of unbroken, vertical silhouette, which is visible from a greater distance. Secondly, there is more surface area, and thicker phloem tissue (Cole and Amman 1980) in such trees, and thus more beetles can produce more aggregation pheromone, attracting in turn more beetles.

In addition, the tree under attack may produce kairomones which synergize the bark beetle pheromones. Pines under attack physically react to tunnelling mountain pine beetles by producing copious amounts of monoterpene-containing resin. Two of these monoterpenes, α-pinene and myrcene, act as synergists for beetle-produced pheromones in white pine (Pitman 1971) and lodgepole line, respectively (Billings *et al.* 1976; Borden *et al.* 1983a; Conn *et al.* 1983).

However, α-pinene lacks integrity (Table 10.2). When exposed to air, it autoxidizes (Moore *et al.* 1956) to a variety of compounds including *cis*-verbenol and *trans*-verbenol, the latter being an aggregation pheromone of *D. ponderosae* (Pitman 1971; Pitman *et al.* 1968). These two compounds are then apparently converted in turn to verbenone, an anti-aggregation pheromone of the mountain pine beetle (Ryker and Yandell 1983) and numerous other bark beetles (Borden 1982a, b). Verbenone becomes the predominant oxidation product after 24 h (Table 10.2).

TABLE 10.2

Cumulative amounts of α-pinene oxidation products detected by gas chromatographic analysis in a 0.2 ml sample of α-pinene exposed to air at room temperature in a 1.86 ml vial; prior to the experiment, the α-pinene was purified to 99% by distillation and contained no detectable oxidation products before being stored at −9 °C for 41 days*

Time after exposure to air (h)	Amount of oxidation products (ng/mg of α-pinene)		
	cis-verbenol	*trans*-verbenol	verbenone
0.0	<1.0	14.0	5.1
3.5	4.6	21.7	13.7
9.5	8.1	28.1	21.0
20.0	15.9	46.4	44.1
43.5	51.0	121.4	145.2
92.0	143.8	277.2	377.7

* Hewlett Packard 5880A in the direct injection mode on a glass capillary column (30 m × 0.66 mm ID) coated with SP-1000 (Supelco Inc., Bellafonte, PA).

Discounting the 14.0 ng/mg of *trans*-verbenol present in our α-pinene after its storage at −9 °C (Table 10.2), and ignoring loss by evaporation and conversion to verbenone, the production rate of *trans*-verbenol by autoxidation in the first 20 h is 1.6 ng/mg of α-pinene/h. If an average female beetle were to produce 256 ng of *trans*-verbenol in 24 h (Fig. 10.3), the hourly production rate of *trans*-verbenol by a beetle (based conservatively on content, not release) would be 10.7 ng/h. Therefore, 6.7 mg of α-pinene exudation/h would be required to produce *trans*-verbenol at the rate at which one female would produce it. At a content of 5.9 per cent α-pinene in lodgepole pine xylem volatiles (Shrimpton 1972), 114 mg/h of exuded resin would be required to produce the female equivalent of 10.7 ng/h of *trans*-verbenol. Lodgepole pines that sustained low, average and high attack densities exhibited pre-attack resin flows of 0.124, 0.425 and 0.719 ml/h, respectively, from a single 11 mm diameter hole driven into the cambium layer (Raffa and Berryman 1983). Therefore, the value of 114.3 mg/h appears well within the range at which a tree attacked at a density of about 40 galleries/m, the threshold for successful attack (Raffa and Berryman 1983), would exude resin yielding many female equivalents of *trans*-verbenol.

Obviously the dynamics of autoxidation must be studied closely in the first few minutes and hours after pure α-pinene or resin is exposed to air. Only then will it be possible to determine when verbenone exceeds the level of approximately 15 per cent of the level of *trans*-verbenol, at which it should exert its antiaggregative properties (Ryker and Yandell 1983).

The autoxidation of α-pinene to *trans*-verbenol indicates that fresh resin flowing from wounds made in a tree by attacking bark beetles may actually

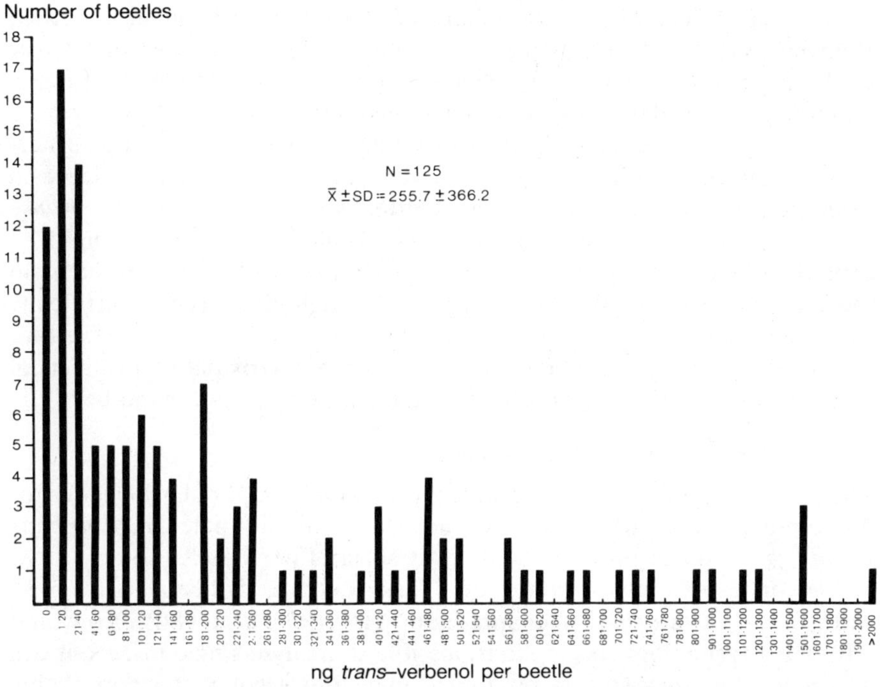

Fig. 10.3. Content of *trans*-verbenol in individual female *C. ponderosae* extracted aftwr 24 h exposure to air saturated with α-pinene vapour and subjected to gas chromatographic analysis (see Table 10.2) (data combined from several experiments, all using beetles from the interior of British Columbia and subjected to identical conditions).

supplement the pheromone produced by the beetles. If the density of initially-attacking beetles were low, resistance by heavy resin producers could be expressed in part by the build up in verbenone to levels that exceeded the ability of the beetles to counteract it by their own production of *trans*-verbenol. These hypotheses are consistent with Raffa and Berryman's (1983) observation that if initial attack occurred on 'resistant' trees that were heavy resin producers, the mass attack was first delayed (due to the physical action of exuded resin on the attackers plus possible high levels of verbenone produced by autoxidation). If mass attack occurred on these trees, they were subjected to higher attack rates and remained attractive for longer periods of time than light resin producers (possibly due to high levels of *trans*-verbenol produced by autoxidation as fresh resin flow increased).

The autoxidation of α-pinene to *cis*- and *trans*-verbenol and then to verbenone suggests that experimental findings involving α-pinene (Borden 1982a, b) should be re-examined. The finding that for the southern pine beetle, *Dendroctonus frontalis* Zimm, *trans*-verbenol can substitute for α-pinene as a synergist for the pheromone frontalin (Payne *et al*. 1978; Renwick and Vité 1969, 1970) could be explained by the fact that *trans*-verbenol is produced when α-pinene is exposed to air. Similarly, although α-pinene is considered to be a syngergist for lineatin, an aggregation pheromone of *T. lineatum*, both *cis*- and *trans*-verbenol also act as synergists (Klimetzek 1984). And the observation that day-old xylem oleoresin from ponderosa pine had become ineffective as a synergist for *trans*-verbenol in attraction of *D. ponderosae* (Billings *et al*. 1976) might be explained by a build-up of verbenone through autoxidation of α-pinene in the exposed resin. Finally, the vulnerability of lightning-struck or artifically wounded trees to aggregation and attack by the southern pine beetle (see Payne (Chapter 7), could be explained in part if the oxidation products of α-pinene were to act as primary attractants or arrestants for the beetle.

10.3.3. Variable stimuli from conspecific beetles

The use of pooled extracts of volatiles from large groups of bark beetles for pheromone analysis has resulted in such statements as 'male *Gnathotrichus sulcatus* (LeConte) produce 14.4 ng of sulcatol per beetle' (Byrne *et al*. 1974), or '*I. pini* males from New York produce ipsdienol in a 65:35 ratio of *S*-(+) and *R*-(−) enantiomers' (Lanier *et al*. 1980). Now that advancing technology has made it possible to analyse single insects, it can be shown that such statements fail to take into account the remarkable complexity and variability of biological systems.

10.3.3.1. Variation in the amount of pheromone produced. The amount of pheromone produced by an individual beetle may vary greatly, for example from 0 to 2204 ng of *trans*-verbenol per female *D. ponderosae* exposed to α-pinene vapours for 24 h (Fig. 10.3), and from 0 to 2638 ng of ipsdienol per male *I. pini* fed on lodgepole pine for 24 h (Fig. 10.4).

The implications of these individual variations in nature are profound. For example the question of 'pioneer' beetles might be re-examined with regard to their pheromone-producing capacity. Are the pioneers only those few beetles with a capacity for producing very large quantities of pheromone? Or do most beetles given the opportunity attempt to become pioneers, with only those few that can produce large amounts of pheromone succeeding? The latter hypothesis would explain in part why many trees in nature sustain a few attacks but are not mass-attacked (personal observations), and why 32 of 47 trees entered by at least one of three beetles in a field experiment failed to be mass-attacked (Raffa and Berryman 1983).

Alcock (1982) contended that in the process of mass attack, tree-killing scolytids conform to conventional evolutionary theory and are in competition with each other rather than in harmonious cooperation. Thus, as proposed for *D. ponderosae* (Borden 1984), those that produce more pheromone will be more successful than poor pheromone producers in

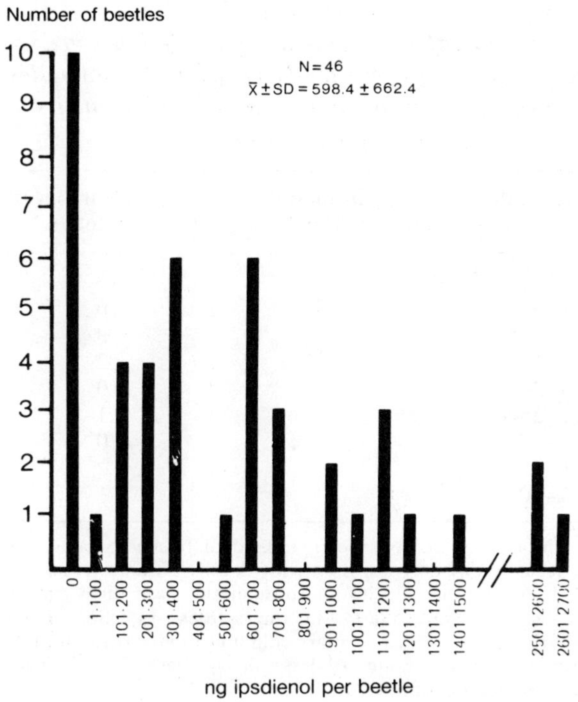

Fig. 10.4. Content of ipsdienol in individual *Ips pini* males extracted after 24–48 h feeding in lodgepole pine bolts and subjected to gas chromatographic analysis (see Table 10.3.)

inducing other beetles to 'work' for them in mass-attacking and killing a tree, so that it becomes a suitable host. As they are the most attractive, they may also be the first to attract mates, and subsequently to produce antiaggregation pheromones (Borden 1982a, b), reserving suitable real estate in which their brood will succeed. Conversely, poor pheromone producers might compete with the successful pioneers by establishing relatively unfragrant and unremarkable galleries that infringed on the optimal gallery space for their more attractive competitors, and would be less likely to attract kairomone-responding predators and parasites.

10.3.3.2. Variation in pheromone chirality. Superimposed upon the individual variation in total amount of pheromone produced is the potential for individuality in the enantiomeric ratio of chiral pheromones (Silverstein 1979) within a population. In a laboratory population of 41 *I. pini* obtained from the Kootenay region of British Columbia, 41 per cent produced greater than 99 per cent pure *R*-(−)-ipsdienol (Table 10.3). The other 69 per cent contained varying amounts of *S*-(+)-ipsdienol, and one beetle contained ipsdienol in a 60 : 40 ratio of *R*-(−) to *S*-(+) enantiomers.

TABLE 10.3

Production of enantiomers of ipsdienol detected by gas chromatographic analysis of 41* I. pini *males allowed to bore in lodgepole pine phloem tissue for 24–48 h*

Percentage of *S*-(+)-ipsdienol in extracts of individual male beetles	Number of beetles
0% [>99% *R*-(−)]	17
1–5%	10
6–10%	8
11–15%	2
16–20%	0
21–25%	1
26–30%	0
31–35%	2
36–40%	1

* Chirality of pheromone in individual beetle extracts was determined by derivatization with chirally pure acetyl lactate and separation of the diastereomeric derivatives on a Hewlett Packard 5880A in the splitless injection mode utilizing a fused silica capillary column (30 m × 0.25 mm ID) with a bonded 0.25 μm DB-1 stationary phase (J & W Scientific, Rancho Cordova, CA).

S-(+)-ipsdienol is inhibitory to *I. pini* from California (Birch *et al.* 1980b) and Idaho (Lanier *et al.* 1980). However, there is very little known about how individuals in a northwestern population respond to individual enantiomers. Are they all repelled by *S*-(+)-ipsdienol, effectively selecting against and weeding out producers of this enantiomer each generation? Or do some respond preferentially (possibly in an obligatory manner) to producers of both enantiomers, as in New York populations (Lanier *et al.* 1980)? If the latter case were true, the population would be a mosaic of variable producers and responders, with segments of the mosaic segregated by their capacity to produce and respond to overlapping ranges in enantiomeric ratios of their pheromones.

Such a population might have enough evolutionary plasticity to impede pest management programs. For example, what would happen if a population of *I. pini* in British Columbia were subjected to a mass trapping program employing *R*-(−)-ipsdienol. Lanier *et al.* (1972) hypothesized that resistance to one pheromone stimulus could result from selection of individual *I. pini* 'capable of using other compounds or combinations in a pheromone system'. Our results suggest that in British Columbia, resistance might well be expressed through the selection of individuals that could produce and respond to the *S*-(+) enantiomer or mixtures of the two.

10.3.3.3. Variation caused by anti-aggregation pheromones. Three compounds have been shown to exert anti-aggregation activity on mountain pine beetles. *Exo*-brevicomin is a male-produced multifunctional pheromone that at low concentrations attracts conspecifics, primarily females (Borden *et al*. 1983a; Conn *et al*. 1983), while at high concentrations it has anti-aggregative properties (Ryker and Rudinsky 1982). After joining females, males also produce the anti-aggregation pheromone frontalin (Ryker and Libbey 1982), and both sexes produce verbenone after feeding (Ryker and Yandell 1983). In addition, our discovery (unpublished) that male *D. ponderosae* produce ipsidenol, an anti-aggregation pheromone of the western pine beetle, *Dendroctonus brevicomis* (LeConte) (Byers 1982), suggests that ipsdienol may serve as a fourth anti-aggregation pheromone in this species. These compounds would act in conjunction with verbenone produced through autoxidation of α-pinene (Table 10.2) by trees under attack.

Although the dynamics of the production of and responses to anti-aggregation pheromones have not been studied in detail, their effect must surely be influenced by such factors as the degree of resistance of the host tree (Raffa and Berryman 1983), the rate at which a new host is attacked, the rapidity with which the sexes are united, and the total population of beetles in a given host. Anti-aggregation pheromones add dramatically to the complexity of semiochemicals to which mountain pine beetles must respond and to the uncertainty that positive orientation responses will occur.

10.4. Conclusion

We have addressed two phenomena that govern the successful orientation by forest Coleoptera to olfactory stimuli. Firstly, the individual beetles are particular in selecting the stimuli to which they will respond, and often change in responsiveness, orienting positively only when they are physiologically ready. Secondly, the orienting beetle is faced with an interacting and highly variable complex of visual and olfactory stimuli emanating from both host trees and the beetles themselves. Although these stimuli promote such events as successful selection and mass attack of host trees by bark beetles in nature, their variability and complexity suggest that the probability of an individual beetle in a responsive state encountering an optimal blend of attractive stimuli is highly uncertain. Moreover, we have not considered the multitudes of interspecific signals (Birch 1984; Borden 1982a) which add further complexity to the system.

Scientists have achieved considerable satisfaction from the study of olfactory stimuli, and have even begun to exploit semiochemicals in manipulating natural populations. However, a more complete understanding of both the stimuli and the responsiveness of individual beetles may lead to a greatly enhanced appreciation of nature and to significant improvements in the applied use of semiochemicals.

11 Olfaction and vision in host finding by a bark beetle

T. L. Payne

The processes at work in host finding by the southern pine beetle, *Dendroctonus frontalis*, and under the influence of pheromones are well documented (Payne 1980 and others). Brood adults on emerging from their host tree generally fly to the nearest vertical object. When the object is an unattacked host tree the beetles bore through the bark and start constructing galleries in the phloem–cambium tissues. Just prior to, or at the onset of the boring activity, the adults release pheromones (secondary attractants). Perception of the pheromones, as well as host odours released from the freshly wounded tree, stimulate aggregation on the tree by other *D. frontalis* in the immediate area. As a result of this aggregation behaviour, the tree is successfully attacked, broods develop, and adults emerge to attack adjacent unattacked host trees. But how *D. frontalis* locates its host in the absence of pheromones, outside of existing infestations, has not been experimentally defined.

Conceptually for *D. frontalis*, as well as other bark beetle species, host finding has been attributed to the efforts of those beetles that initially land upon and attack host trees. Such beetles, commonly referred to as 'pioneer beetles' (Borden 1974), are the beetles which establish a focal point of development for the next generation. Pioneer beetles must locate suitable host trees without the aid of secondary attractants.

It is generally recognized that for many bark beetle species, including *D. frontalis*, the initiation of infestations is associated with attack on stressed trees (Blanche *et al*. 1983; Moeck *et al*. 1981; Wood 1982). Of the many stress-causing agents, lightning has been long recognized as having a profound effect in making pines susceptible to attack by *D. frontalis* (Blanche *et al*. 1983; Coulson *et al*. 1983; Hopkins 1909). From surveys carried out during the past 20 years, a range of 23 per cent to 77 per cent of over 2000 infestations surveyed were associated with lightning-struck trees (Hicks 1980; Hodges and Pickard 1971; Lorio and Bennett 1974).

Stress on the lightning-struck tree is visibly obvious. In most cases a strip of bark and phloem is torn away from the bole from the crown to the base of the tree. In some cases the tree may be literally blown to pieces or set on fire. It has been suggested that trees surrounding the struck tree become stressed, although to a much less obvious degree (Komarek 1964; Schmitz and Taylor 1969; Sharples 1933). It is conceivable that the peak temperatures which result from a lightning discharge at ground level could cause damage to root systems of the surrounding trees. As a result, the lightning-struck tree can serve as a focal point of initiation for the beetle infestation and the surrounding stressed trees provide susceptible hosts for growth of the infestation (Coulson *et al*. 1983).

Coulson *et al*. (1983) recently considered the role of lightning in the epidemiology of *D. frontalis*. From historical data, they calculated that the occurrence of lightning strikes in the forest is at a frequency of between 45 and 180 per square mile during the course of a year. The temporal distribution of the lightning strike varies seasonally and follows the pattern of occurrence of thunderstorms, which is greatest from June through August. This correlates well with the findings of Hodges and Pickard (1971) and Lorio and Bennett (1974) that over 70 per cent of the *D. frontalis* infestations reported in August were associated with the presence of lightning-struck trees. Coulson *et al*. (1983) concluded that lightning is the only stress-causing disturbance as yet reported to be associated with the population dynamics of *D. frontalis* that occurs with sufficient consistency, including number of strikes and spatial and temporal distribution, to be considered as a primary component of the life history of the beetle. Even though lightning has been shown to be highly correlated with the initiation of *D. frontalis* infestations, how the beetles find the stressed trees is still unknown.

There are two main hypotheses for how bark beetles, including *D. frontalis*, locate and select their hosts. Some investigators have proposed that primary attraction in response to olfactory stimuli from the stressed host tree is the means by which the beetle accomplishes host finding and selection (see extensive review by Moeck *et al*. 1981; Person 1931). It has been further hypothesized that pioneer beetles are attracted to and find susceptible hosts by changes in the volatile compounds resulting from deterioration of host tissue (e.g. *Heikkenen* 1977). Random landing by pioneer beetles is the second main hypothesis proposed (Hynum and Berryman 1980; Moeck *et al*. 1981; Wood 1972, 1982). Hypothetically, beetles land at random on host trees, both healthy and stressed, and on non-host trees alike. Once on a tree, the beetles go through further stages of host selection in response to the presence or absence of various gustatory stimuli after the initiation of biting behaviour.

Primary attraction has been demonstrated in several ambrosia beetles in experiments where cut host material was used (Moeck *et al*., 1981), and also recently in *Tomicus* bark beetle species (Byers *et al*. 1986). And the ability to perceive host odours has been demonstrated in *D. frontalis* and other bark beetle species (Dickens and Payne 1977; Dickens *et al*. 1984; Payne 1970, 1975). However, in all studies in which standing tress were predisposed to bark beetle attack, primary attraction could not be demonstrated (Moeck *et al*. 1981; Wood 1972, 1976).

Experiments with stressed and healthy standing, living trees (the host tree condition and profile with which *D. frontalis* and some other bark beetles interact) have demonstrated no differences in landing rates as well as no differences in the landing rates on host and non-host tree species (Berryman and Ashraf 1970; Hynum and Berryman 1980; Wood *et al*. 1974). The most definitive experiments were carried out by Moeck *et al*. (1981) on ponderosa pines. They found that on standing, living trees, both healthy and unhealthy, the landing rates of bark beetles were not significantly different. Their results strongly suggest that the initial landing on host trees by beetles is not governed by the health of the tree or primary attraction.

There have been no reports of experiments on random landing by *D. frontalis* in the absence of sources of secondary attraction. However, Gara *et al*. (1965) found that host pines and hardwoods in the immediate vicinity of an infestation centre were landed upon by similar numbers of *D. frontalis*. They also found that tree profile was a significant factor, with virtually no *D. frontalis* attacking felled host trees in the horizontal plane.

Gara *et al*. (1965) hypothesized that, in the absence of secondary attractants, host finding in *D. frontalis* might occur through response of the beetles to trees struck by lightning, damaged by logging or infested with other bark beetle species. Tests carried out to date provide little evidence of primary attraction of *D. frontalis* or response by the beetles to phermones of other species (Birch *et al*. 1980a; Dixon and Payne 1980; Payne *et al*. 1978; Renwick and Vité 1969; Vité *et al*. 1964).

Although there is no substantial evidence to show that primary attraction occurs in *D. frontalis*, it has been reported that host tree odours appear to function in host selection by arresting the flight of the beetles (Payne *et al*. 1879; Renwick and Vité 1969). That is, *D. frontalis* may not be attracted to host odours, but in close proximity its flight may be arrested. This would explain how host odours may function to enhance the catch of *D. frontalis* on traps baited with pheromone (Kinzer *et al*. 1969; Payne *et al*. 1978). Beetles attracted to a trap by pheromone would be arrested by host odours. In the absence of host odours, beetles were found to resume flight eventually where non-sticky traps were used (J. A. A. Renwick and J. P. Vité personal communication). Those findings are indirectly supported by electrophysiological studies which showed that significantly smaller and fewer muscle potentials were elicited by host odour compared with pheromone (Payne 1973). Furthermore, added indirect support is found in electrophysiological studies in which the terpene hydrocarbon *trans*-verbenol was found to suppress pheromone-elicited muscle potentials (Dickens and Payne 1978). This is particularly interesting since the host volatile α-pinene has been reported to auto-oxidize to *trans*-verbenol when exposed to the air, and may indeed represent a 'host' volatile to which the beetles respond [Borden *et al*. (Chapter 10)].

In an effort to gain some insight into the host-finding phenomenon, we monitored trees which had been artificially stressed using a modification (pheromones were not used) of the 'simulated lightning' technique (Miller

1983) and others). Successful attack by *D. frontalis* was induced on over 50 host trees which had been physically scarred by activated detonating cord over the bole of each tree. The detonating cord caused physical damage to the tree bole resembling that of a natural lightning strike. Other than the physical injury to the bole, it is unlikely that the detonating cord duplicated the entire effect of a lightning strike. However, this is not known. In every case the detonated standing trees were successfully attacked by *D. frontalis*. Adjacent control trees were not attacked until after the detonated trees were colonized and the attack had spread to the adjacent trees. Prior to creating the simulated lightning-strike-stressed trees, trees attacked by *D. frontalis* could not be found in the area and beetles were not caught in survey traps baited with pheromone. These observations led to the consideration that the presence of a vertical object (standing host tree) together with a large volume of host volatiles (vertically-scarred bole) play a significant role in host finding by *D. frontalis* and in the initiation of a new infestation.

In a preliminary experiment five host trees, each separated by a few hundred metres, were scarred to determine whether or not vertically positioned physical scarring (mass and vertical release of host odours) would elicit landing and/or attack by *D. frontalis*. Trees attacked by *D. frontalis* were not in the immediate vicinity. Three trees received a 15 cm wide scar from the base of the crown to ground level cut to the phloem tissue by use of a sickle. The other two trees each received a similar scar. However, one was scarred from the base up to 3 m and the other was scarred over a 3 m distance at midbole. All three trees with full bole scars were attacked by *D. frontalis* within three to five days after the scars were made. Neither of the trees with partial bole scars was attacked by the beetle.

Subsequently five stands, each separated by about 3 km and with no *D. frontalis* infestations, were selected for added experimentation. Each month, beginning January 1984, two host trees separated by about 20 m were selected in each stand. One tree received a 15 cm wide scar over the length of the bole as described earlier, while the other acted as a control. Landing of *D. frontalis* on the scarred and control trees was monitored by use of two 100 cm^2 screen wire sticky traps positioned at about 3.5 m and 6.5 m on the bole of the tree. This is the range of height on the host tree where the majority of initial landing and attacks takes place by the beetles (Coster *et al*. 1977; Fargo *et al*. 1978). At the time of writing (September 1984) six scar/control tree pairs had been established in each of the five stands. Of the 40 scarred trees, seven became mass-attacked by *D. frontalis*. Only one control tree was mass-attacked, which occurred because the tree was in the direct line of spread of a *D. frontalis* infestation initiated from one of the scarred trees. The relative number of beetles caught on the landing traps was low for those trees not mass-attacked and subsequently not under the influence of secondary attractants. A total of 67 beetles were caught on the 43 non-attacked, scarred trees, and three beetles were caught on the control trees.

These results could be interpreted to support the hypothesis of primary attraction of *D. frontalis* to olfactory signals produced by the host tree. However, since the trees were not screened to prevent any level of attack by the beetles and because the potential for production of secondary attractants, the possibility of primary attraction can be justifiably questioned. The results could also be interpreted to be due to random landing by *D. frontalis* on suitable (stressed) hosts with the subsequent release of secondary attractants upon successful attack by the beetles.

If the release of host volatiles by scarring represented a source of primary attractant, it might logically be expected that more beetles would have been trapped on the scarred trees. This is especially so when it is recognized that the populations of flying beetles in at least three of the stands were large enough to have facilitated mass attack on six host trees. However, though host volatiles may not operate as primary attractants in *D. frontalis*, they may operate as primary olfactory stimuli which facilitate host habitat location and host finding.

Based upon the previously reported finding cited above, it can be hypothesized that through random landing on vertical objects *D. frontalis* eventually finds itself on a host tree. In the absence of a sufficient concentration of host-released volatiles, the beetle resumes flight and continues to land at random on both host and non-host trees. In stands with high concentrations of released host volatiles the probability is increased that larger numbers of beetles will be arrested in the area. Within this host habitat those trees, for whatever reason, emitting a relatively high level of host volatiles would be likely to be stressed and more susceptible to attack by the beetles. The fact that *D. frontalis* is generally not found attacking trees damaged through logging, wind storms and otherwise, respresenting sources of host volatile release, is probably due to the fact that the source and concentration of the volatiles (the location of the damage) is not within the general 3 to 10 m vertical flight, landing, and attack range of the beetle. Over 90 per cent of the flying population occurs within that vertical range (Coster *et al*. 1977; Fargo *et al*. 1978; Moore 1979). Logging damage generally occurs below 3 m. However, *D. frontalis* does occur in logging sites, as indicated by catches of beetles in pheromone-baited traps (R. F. Billings personal communication). The presence of a relatively large amount of host volatiles apparently arrests beetles in the area, and subsequently they aggregate to the source of pheromone. This same phenomenon was found in tornado-damaged stands (R. F. Billings personal communication). Generally, however, sufficient numbers of beetles are not arrested on standing trees to accomplish successful attack and initiate secondary attraction. The lightning-struck tree, for which there is a high correlation with successful attack by *D. frontalis*, appears to provide the key olfactory and visual stimuli required for host finding by the beetle. That is, the lightning-struck tree generally provides a larger volume of released host volatiles over the entire vertical flight and landing range of the beetle than do trees damaged by other means. It also represents a verticle visual profile. Consequently, there is a higher probability that the randomly landing beetles will be arrested in sufficient numbers on the tree to initiate successful attack. This behaviour of *D. frontalis* would increase the probability that when sufficient host odour is being released the tree is stressed and likely to be susceptible to attack by the beetle.

This interpretation of the role of visual and olfactory stimuli in host finding by *D. frontalis* requires further research before it can be sufficiently justified. The needed definitive studies involving stressed trees screened against beetle attack are currently under way to determine whether or not primary attraction occurs.

Acknowledgements

I thank B. O'Donnell, K. Walsh, T. Potter, J. Delorme, C. Russell and D. Klimetzek for assistance with the observations and studies originally reported here.

Endnotes

Texas Agricultural Experiment Station paper no. TA–19720. The work was funded in part by McIntire–Stennis project 1525 and the USDA–Forest Service IPM Bark Beetle Program cooperative agreement No. 19–83–109, and developed in part while Alexander von Humbolt Senior Scientist and Visiting Professor at Forstzoologisches Institut Universität, Freiburg, West Germany. The findings, opinions, and recommendations reported herein are those of the authors and not necessarily those of the United States Department of Agriculture.

12 The role of pheromone concentration in male moth flight behaviour

C. J. Sanders

12.1. Introduction

The potential application of research on lepidopteran sex pheromones is twofold: (a) luring moths to traps to detect and monitor populations, and (b) regulating populations by disrupting mating processes.

Luring males in sufficient numbers to trap is not a critical problem. Changes in blend and in release rate certainly affect the numbers, but, even in the absence of chemical components which are critical for courtship and copulation, males may still approach the lure close enough to be captured. Allowances for these factors can be made and catches will still give a relative indication of the presence and abundance of the target insect.

The process of disruption, however, is a more complex problem. In his recent review, Bartell (1982) lists five possible mechanisms: (i) adaptation/habituation, (ii) false-trail following, (iii) masking of odour plumes, (iv) imbalance in sensory input by wrong blends of chemical components, and (v) modification of responses by 'antipheromones.'

So far, for the spruce budworm (*Choristoneura fumiferana* (Clem.)) we know of no effective antipheromones, and available evidence indicates that the most effective disruptant is a pheromone blend close to the natural pheromone (Sanders 1981), indicating that imbalance in pheromone blend is not an effective way to achieve disruption.

Recent work has therefore focused on habituation (adaptation), false-trail following and masking.

12.2. Habituation

Sanders (1982a) showed that leaving male spruce budworm moths in a pheromone-permeated atmosphere in a wind-tunnel for up to 24 h had no effect on the numbers of males responding to calling females. But this was based on evidence from only one concentration of background pheromone. More recently this work has been extended to cover a wider range of pheromone concentrations. For simplification, one duration of exposure was selected, 3 h. The synthetic pheromone was incorporated into poly-vinyl chloride (PVC) pellets at various concentrations and these pellets were placed behind baffles in the wind-tunnel to permeate the air with the pheromone. Male moths were divided into two groups, one was placed in the wind-tunnel and left for 3 h to adapt to conditions in the tunnel: the other group was kept in charcoal-filtered air. Lures of various strength were placed one at a time in front of the baffles to provide discrete pheromone plumes, and the males were exposed to these in the presence of the background pheromone. Responses of the males were then recorded. If an exposure of the males to pheromone for 3 h was causing habituation

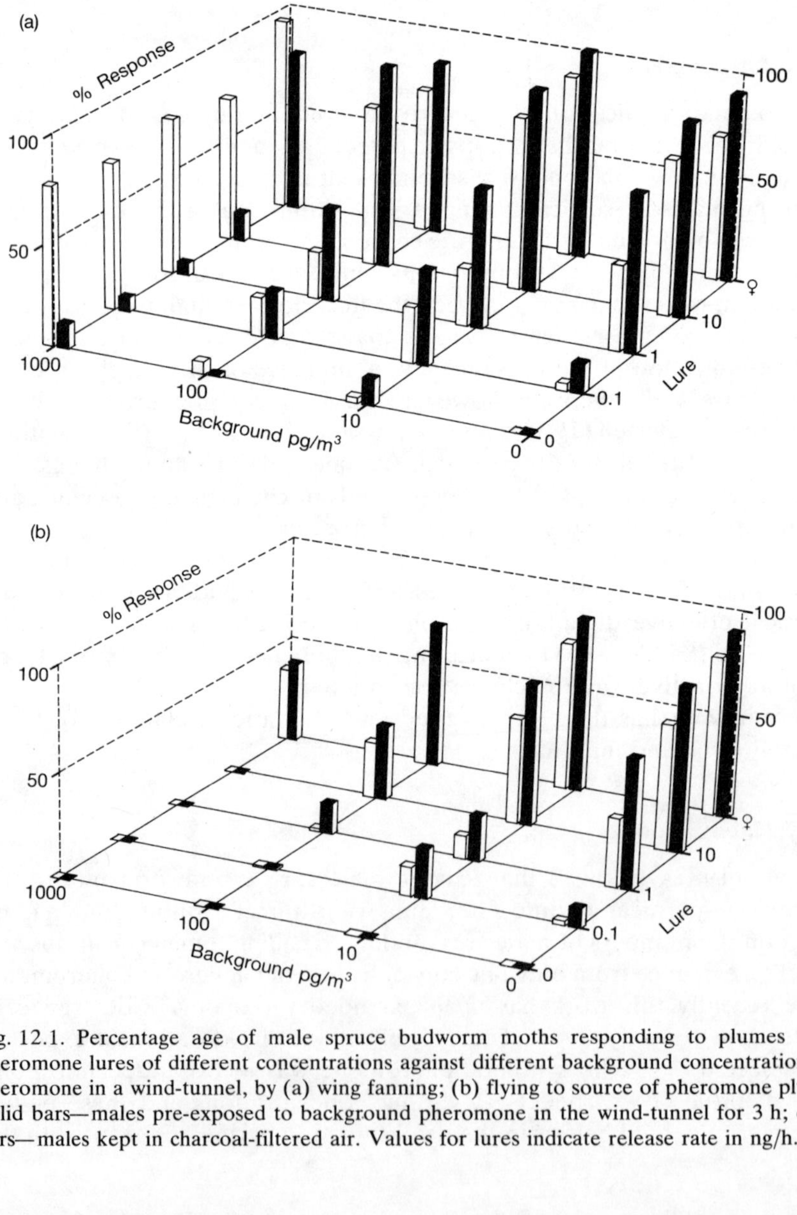

Fig. 12.1. Percentage age of male spruce budworm moths responding to plumes from pheromone lures of different concentrations against different background concentrations of pheromone in a wind-tunnel, by (a) wing fanning; (b) flying to source of pheromone plume. Solid bars—males pre-exposed to background pheromone in the wind-tunnel for 3 h; open bars—males kept in charcoal-filtered air. Values for lures indicate release rate in ng/h.

and/or adaptation then the responses of the exposed males should have been lower than those of the naive males (those kept in charcoal-filtered air). The results (Fig. 12.1a, b) show that in 'clean air' (i.e. no background pheromone) and at lower concentrations of background pheromone, responses were actually higher among the exposed group. Presumably the additional handling in transferring the naive males to the wind-tunnel depressed their responsiveness slightly. Only at a background concentration estimated at 1000 pg/m^3 was an effect seen, and this was only in the number of males wing fanning. At this level of background pheromone, orientation is already disrupted by the masking of the pheromone plumes (Fig. 12.1b; Sanders 1982b). In background concentrations of less than 1000 pg/m^3 male spruce budworm can perceive and will follow superimposed pheromone plumes of higher concentration; similar observations were reported by Kennedy *et al*. (1981) on *Adoxophyes orana*. These more recent observations show that even prolonged exposure to similar background concentrations will not prevent male moths responding to plumes of higher concentrations, and the conclusion is that habituation and adaptation are not important factors in disrupting orientation.

12.3. False trails and masking of plumes

Earlier work in a wind-tunnel (Sanders 1982a, b) led to the conclusion that a formulation in which each particle emits pheromone at a higher rate than a female moth is more effective in disrupting male orientation than one in which the same total amount of pheromone is released from more numerous but less potent particles. In other words, it appears that the creation of false trails is a better strategy than masking the plumes emitted by calling females.

Although several field experiments have been carried out with various Lepidoptera to evaluate the effectiveness of different spacing and different concentrations of pheromone emitters, so far none have been designed to test the theory of false trails versus masking. Part of the problem is logistical. To out-compete a female a lure should emit pheromone in the order of ten times the rate of a female. To permeate the atmosphere without creating distinct plumes, the pheromone emitters should be at least one-tenth the concentration of a female, that is, 100 more emitters are required to emit the same total amount of pheromone. In an experimental design with one lure of high concentration at each corner of a cubic metre, 800 of the lower-concentration lures spaced out evenly would be required per cubic metre, a difficult task to achieve by hand.

If in fact the best strategy is the creation of false trails, then it becomes extremely important to ensure that the emitters are creating the most attractive plumes possible, plumes which are at least competitive with virgin females.

Although it has been shown for spruce budworm in the field that traps baited with lures containing the known pheromone (a 95:5 binary blend of (*E*:*Z*)-11-tetradecanal) capture as many males as traps baited with virgin females during the females' calling period, the same lures do not evoke the same duration of flight in a sustained-flight wind-tunnel (Sanders 1984). Evidently, obtaining the males' attention for two to three minutes is sufficient to lure them to a trap. However, for maximum disruption it is important that the males remain occupied by the synthetic pheromone source as long as possible, and considerable research is now going into upgrading the potency of the synthetic pheromone. Addition of the known secondary components, tetradecanal and *E*-11-tetradecenyl acetate, makes only slight improvements to the duration of sustained flight. The search for additional components continues. Research is also being conducted on male flight behaviour under different conditions.

12.4. Pheromone concentration and ground speed of moth

Kuenen and Baker (1982a), Marsh *et al*. (1978), and Preiss and Kramer (1983) showed that male moths following pheromone plumes compensate for different wind speeds by adjusting their airspeed and their turning angles, so maintaining a 'preferred' constant upwind groundspeed (net upwind velocity). Similar results have been found in the spruce budworm (Table 12.1). But estimates of pheromone concentration, made by measuring plume dimensions at the different wind speeds, show that by adjusting their airspeed the rate of flow of pheromone that the moths experience also remains constant. A logical question then is whether the adjustment in airspeed is due to the males selecting a preferred rate of flow of pheromone

TABLE 12.1

Net upwind groundspeed of male spruce budworm moths responding to either PVC lure or virgin females in three different wind speeds (each speed based on flight times of 16–25 moths)

	Wind speed (cm/s)		
	7.5	22.5	32.5
Moth speed (cm/s)*			
—to PVC lure	19.5[a]	19.9[a]	18.7[a]
—to virgin females	26.3[b]	28.1[b]	25.8[b]
Plume cross-section at 1.5 m (cm^2)	1065	390	245
Rate of flow of pheromone (pg/l/s)†	0.35	0.31	0.35

* Numbers followed by same letter not significantly different ($P = 0.01$).
† Calculation based on release rate of 10 ng/h.

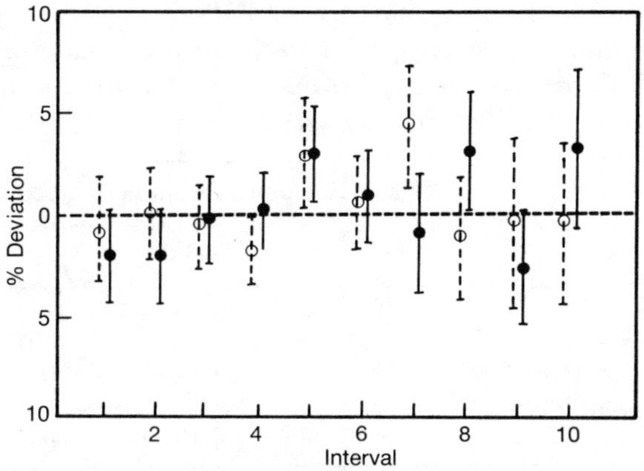

Fig. 12.2. Deviations from average speed (±1 SE) over successive 0.5 min intervals, of male spruce budworm moths responding in a wind tunnel to pheromone plumes from PVC lures (open circles) or virgin females (solid circles).

over their receptors, rather than to the selection of a preferred groundspeed.

In recent experiments we have also found that male spruce budworm held in flight in the centre section of a wind-tunnel by moving optomotor cues (and therefore also held in a constant time-averaged concentration of pheromone) do not slow down, as one might presume they would do as they tire. Instead they maintain a constant groundspeed, and consequently a constant rate of flow of pheromone, until they leave the pheromone plume (Fig. 12.2). This evidence supports both theories, that is, that males have a preferred groundspeed or a preferred rate of pheromone flow.

Against the theory of a preferred groundspeed is the evidence that male moths slow down their upwind progress as pheromone concentration increases (Cardé and Hagaman 1979; Kuenen and Baker 1982b; Sanders *et al*. 1981). Presumably this is the reason why males slow down as they approach a pheromone source, where the narrowing plume leads to higher concentrations of pheromone (Farkas *et al*. 1974; Marsh *et al*. 1978; Murlis *et al*. 1982).

Recently, we have carried out experiments on spruce budworm in a tunnel consisting of a tube 7 m long and 75 cm in diameter, modelled after that of Jones *et al*. (1981), which is kept inflated by the airflow. Males responding to a pheromone source were timed over successive 1.5 m sections of the tunnel. As anticipated, speeds decreased over successive sections (Table 12.2). Again it is temping to theorize that this is an attempt by the males to maintain a constant rate of flow of pheromone over their olfactory receptors.

Certainly the idea that male moths adjust their speed to keep the flow of pheromone over their receptors constant is appealing in its simplicity. Max-

TABLE 12.2

Net upwind groundspeeds (cm/s) of male spruce budworm moths over successive 1.5 m sections of a pheromone plume produced by either a PVC lure or virgin females

Pheromone source	Distance from pheromone source (m)			
	6–4.5	4.5–3.0	3.0–1.5	1.5–0
PVC (N = 30)	48.1	32.5	23.8	18.9
Virgin females (n = 42)	54.3	44.6	41.0	39.1
Plume cross-section (cm^2)*	2320	1245	670	390
Rate of flow of pheromone (pg/∝/s)†	0.05	0.10	0.18	0.31

* Measured at upwind end of 1.5 m section.
† Calculations based on release rate of 10 ng/h.

imum speed will occur in the absence of pheromone, and the jerky upwind motion of males in a plume of pheromone can be accounted for by responses to changes in pheromone concentration due to turbulence.

The strongest evidence against the theory is that in wind-tunnels where the airflow, and hence the pheromone concentration, is constant, the speed of flying moths can be regulated independently of the pheromone concentration by moving optomotor cues. Nonetheless, it is clear that concentration plays an important role in the flight behaviour of male moths. Resolution of its role will depend upon more accurate measurements of the actual pheromone concentrations experienced by flying moths, and far more detailed analysis of flight tracks than has been attempted here, and which possibly can be achieved only by the analysis of tracks in three dimensions.

13 Species specificity in response to pheromone substances in diprionid sawflies

J. Löfqvist

13.1. Introduction

Sex attraction in diprionid sawflies was reported by Coppel *et al.* in 1960, only a year after the term pheromone was coined and the first insect pheromone was chemically characterized. They studied male attraction to virgin females of *Diprion similis* and wrote: 'A virgin female was set out at 11 a.m. and the first males flew into the trap within 30 seconds. Activity continued until 4 p.m. at which time over 7000 males had been attracted. This female continued attracting males at approximately 1000 per day until she died on the fifth day. Even after death there was sufficient attractant left to lure males in dwindling numbers for the next three days. In all, over 11,000 males were removed from the trap and the ground below was littered with many not included in the count.'

Diprion similis is not unique. Attraction to sex pheromones has been studied in eight diprionid sawfly species and very potent substances have been found for all of them. Even though the number of diprionid species studied is only about 10 per cent of the total, the studies most probably reflect the role of sex pheromones in the whole family. There are about 85 diprionid species and races recognized today, with *Diprion* and *Neodiprion* as the two dominating genera (Knerer and Atwood 1973; Ross 1955). *Diprion* is a Eurasian genus with some species introduced into North America. *Neodiprion*, on the other hand, is an American genus with only one representative in Eurasia, *N. sertifer*, which is thought to have crossed the Bering Strait and then expanded throughout the Palaearctic region. *N. sertifer* became extinct in North America and was then accidentally reintroduced from Europe about a century ago.

Pheromone communication in sawflies as discussed here is based primarily on results from *N. sertifer*. A comparison of pheromone communication in closely related species will elucidate the species specificity in an insect family with many sympatric species.

13.2. Male antennae

The biological role of pheromone communication in diprionid sawflies is suggested by the large plumed antennae of males in contrast to the slender ones of females. The sensilla on the antennae of *N. sertifer* have been described by Hallberg (1979) who found five different types, four of which were common to males and females. The unique type of male sensillum is long, single-walled, with 8–12 neurones and numerous pores in the sensillum wall. These sensilla are similar to the pheromone sensilla in moth species and are the most common type in male European sawflies. The rather high number of neurones per sensillum may indicate a complex pheromone consisting of several components. Different sensory cells in morphologically similar sensilla have been shown in moth species to respond to different pheromone components (Kaissling *et al*. 1978; Löfstedt *et al*. 1982; Van Der Pers and Löfstedt (Chapter 27).

13.3. Female sex pheromone

Acetates and propionates of 3,7-dimethylpentadecan-2-ol (Fig. 13.1) have been identified as attractants for diprionid sawflies (Jewett *et al*. 1976). The alcohol, diprionol, was isolated from a hydrolysed extract of 27 000 females of *N. lecontei,* and has also been isolated by a similar procedure from female extracts of *N. sertifer* and *D. similis*. However, diprionol was inactive both in electroantennogram and field experiments. The assays indicated that the diprionol acetate was a major pheromone component for the two *Neodiprion* species, while *D. similis* used the corresponding propionate.

To date, no other compounds have been identified as sex pheromones from diprionid sawflies. Female extracts from *N. sertifer* and *D. similis* have been fractioned with thin-layer and gas chromatography, but the only fraction attracting males contained derivatives of diprionol (Kikukawa *et al*. 1982, 1983). In addition, no fraction was synergistic to the active one.

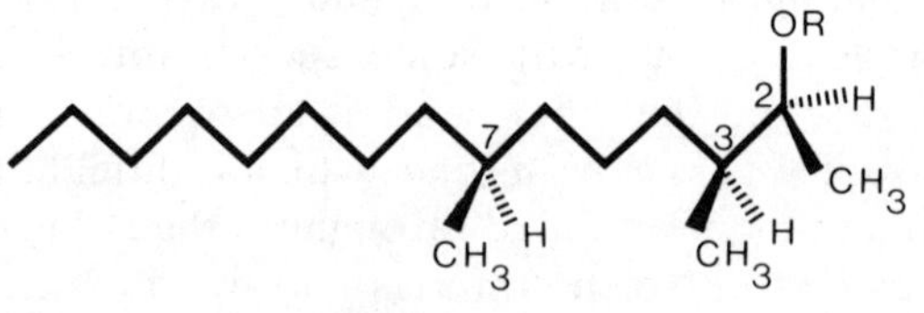

I R = H

II R = CH_3CO-

III R = CH_3CH_2CO-

I 3,7-Dimethylpentadecan-2-ol

Fig. 13.1. Chirality in positions 2, 3 and 7 results in eight (2^3) isomers of diprionol(I) identified from female diprionid sawflies. Chiral acetate(II) or propionate(III) isomers of diprionol are potent sex pheromones for several species of diprionid sawflies.

13.4. Species specificity

Jewett *et al*. (1976) suggested that species specificity in diprionid sawflies is based on different chiral isomers combined in different ways. This has been confirmed in all later studies. It is well known that chirality can cause dramatic differences in the response to pheromones in many insects. The chiral pheromones of bark beetles are among the best examples of this (Birch 1984), but there are examples of chirally sensitive species in almost all studied taxa (Silverstein 1979).

Diprionol has asymmetric carbon atoms in positions 2, 3, and 7, which means that it can be present in $2^3 = 8$ different chiral isomers. These are abbreviated below as, for example, 2*S*,3*S*,7*R*, which means that substituents in the 2 and 3 positions are arranged in the *S*-form, but in the 7 position they are in the *R*-form. These eight chiral isomers can potentially occur in 254 ways in combinations with one to eight of them. The number of combinations increases to 65,492 if the possibility of an acetate or a propionate is also considered. This high number of possible blends is a significant contrast to the approximately 85 species or races known in Diprionidae (Knerer and Atwood 1973) and can most probably account for the species-specific pheromones even if biosynthetic and sensory limitations are considered.

13.4.1. Acetate or propionate

Specificity of the acetate and propionate groups was studied in a field experiment with *N. sertifer* in which we compared the 2*S*,3*S*,7*S* acetate with the corresponding propionate as attractants in traps at various release rates. The substances were synthesized by Byström *et al*. (1981). Both compounds were potent attractants for males of *N. sertifer* (Fig. 13.2). The acetate was somewhat more attractive than the propionate, a finding also reported by Kikukawa *et al*. (1983) and by Kraemer *et al*. (1983). The most probable explanation for this is that both the acetate and the propionate are components in the pheromone of *N. sertifer*. We did not find any synergism between the two at any ratio. The lack of specificity in *N. sertifer* for either the acetate or the propionate, however, is not typical for diprionid sawflies. Most species respond to only one of the esters, with a high selectivity (Table 13.1).

13.4.2. Chirality in position 7

Attractiveness of the two 7-position chiral isomers has been studied in field experiments with many diprionid species (Table 13.2). Males of all but one of the seven species studied were attracted to only one of the isomers.

All *Neodiprion* species studied responded to the *S* isomer in the 7 position as a sex attractant. The only exception was *N. nanulus nanulus*, which responded to both the *S* and the *R* isomer. Apparently both compounds are components in the sex pheromone of the species. However, chiral composi-

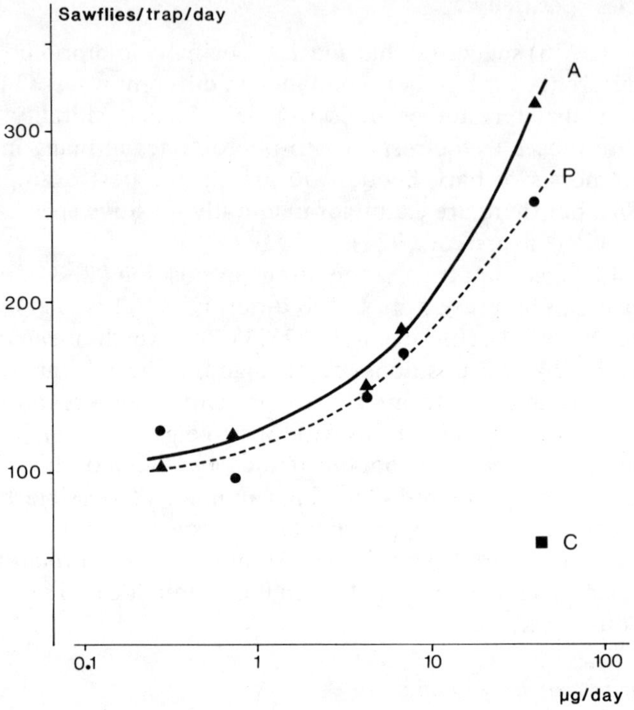

Fig. 13.2. Both (2*S*, 3*S*, 7*S*) diprionyl acetate(A) and the corresponding propionate(P) are attractants for males of *N. sertifer*, as shown in a field trap dose–response experiment. *C* represents the mean number of males trapped in unbaited control traps.

Points in the diagram are means for 21 trap-days. Glass capillaries of different diameters were used to provide release rates of steadied precise magnitude. The amounts released were determined after the experiment by migration of the meniscus in the capillary using a microscope.

tion of sex pheromone compounds has not yet been chemically analysed for any diprionid sawfly species.

13.4.3. Chirality in positions 2 and 3

All potent attractants identified so far in *Neodiprion* have the 2*S*,3*S* configuration. An *erythro* arrangement (2*S*,3*S* or 2*R*,3*R*) was also found in the original chemical identification of diprionid sawfly pheromones (Jewett *et al*. 1976). The only *threo* isomer (2*S*,3*R* or 2*R*,3*S*) yet found is 2*S*,3*R*,7*R* propionate, which is an attractant for *D. similis* (Kikukawa *et al*. 1982). Many more *erythro* and *threo* isomers have been tested in field experiments as attractants. All were essentially or entirely unattractive. This indicates that the chiral arrangements around positions 2 and 3 are of great importance, as has been shown for *N. sertifer* (Kikukawa *et al*. 1983) and *N. nanulus nanulus* (Kraemer *et al*. 1983).

TABLE 13.1
Response in field tests of diprionid sawflies to acetates (A) and propionates (P) of diprionol

Species	Substance	A	P	References
Neodiprion lecontei	2*S*,3*S*,7*S*	A	—	Kraemer *et al*. (1981)
Neodiprion pinetum	2*S*,3*S*,7*S*	A	—	Kraemer *et al*. (1979)
Neodiprion taedae	2*S*,3*S*,7*S*	A	P	Kraemer *et al*. (1983)
Neodiprion pratti	2*S*,3*S*,7*S*	A	—	Kraemer *et al*. (1983)
	2*S*,3*R*,7*R*	—	P	Kraemer *et al*. (1983)
Neodiprion sertifer	2*S*,3*S*,7*S*	A	P	Löfqvist *et al*. unpubl., Kikukawa *et al*. (1983), Kraemer *et al*. (1983)
Neodiprion nanulus nanulus	2*S*,3*S*,7*S*	A	P	Kraemer *et al*. (1983)
	2*S*,3*S*,7*R*	A	—	Kraemer *et al*. (1983)
Diprion similis	2*S*,3*R*,7*R*	—	P	Kikukawa *et al*. (1982)
	2*S*,3*S*,7*S*	—	P	Longhurst *et al*. (1980)
	2*R*,3*R*,7*R*	—	P	Longhurst *et al*. (1980)
Gilpinia frutetorum	2*S*,3*R*,7*R*	A	—	Kikukawa *et al*. (1982)

A = acetate; P = propionate.

TABLE 13.2
Response in field tests of diprionid sawflies to chirality in position 7 of diprionol

Species	Substance	7*S*	7*R*	References
Neodiprion lecontei	2*S*,3*S*,7xA	7*S*	—	Kraemer *et al*. (1981)
Neodiprion pinetum	2*S*,3*S*,7xA	7*S*	—	Kraemer *et al*. (1979)
Neodiprion taedae	2*S*,3*S*,7xA	7*S*	—	Kraemer *et al*. (1983)
	2*S*,3*S*,7xP	7*S*	—	Kraemer *et al*. (1983)
Neodiprion pratti	2*S*,3*S*,7xA	7*S*	—	Kraemer *et al*. (1983)
Neodiprion sertifer	2*S*,3*S*,7xA	7*S*	—	Kraemer *et al*. (1983)
	2*S*,3*S*,7xP	7*S*	—	Kikukawa *et al*. (1983)
Neodiprion nanulus nanulus	2*S*,3*S*,7xA	7*S*	7*R*	Kraemer *et al*. (1983)
	2*S*,3*S*,7xP	7*S*	—	Kraemer *et al*. (1983)
Diprion similis	2*S*,3*S*,7xP	7*S*	—	Longhurst *et al*. (1980)
	2*R*,3*R*,7xP	—	7*R*	Longhurst *et al*. (1980)
	2*S*,3*R*,7xP	—	7*R*	Kikukawa *et al*. (1982)

A = acetate; P = propionate.

13.5. Inhibition and synergism

It seems apparent that sex pheromones in diprionid sawflies consist of different combinations of chiral isomers, acetates and propionates of one single substance, diprionol. In such a narrow system, inhibition and synergism between these derivatives can be expected to maintain species specificity. This was studied in a field experiment with *N. sertifer* where the attractant, 2*S*,3*S*,7*S* acetate, (Byström *et al*. 1981) combined with the acetate of the racemic *threo* form (diastereomeric mixture of 7*R* and 7*S* isomers), 2*S*,3*R*,7*R*/*S* + 2*R*,3*S*,7*R*/*S* acetate, (Magnusson 1977, 1978), were released from separate dispensers in traps (Fig. 13.3). The attractant was released at a constant rate of 10 μg/d and the *threo* acetate was released in amounts ranging from 1 ng/d to 100 μg/d in half decadic steps. The *threo* acetate was found to essentially eliminate the attraction of the 2*S*,3*S*,7*S* acetate, even at amounts as low as 3 per cent of the attractant. The effect decreased with amounts below 3 per cent and was not present below 0.3

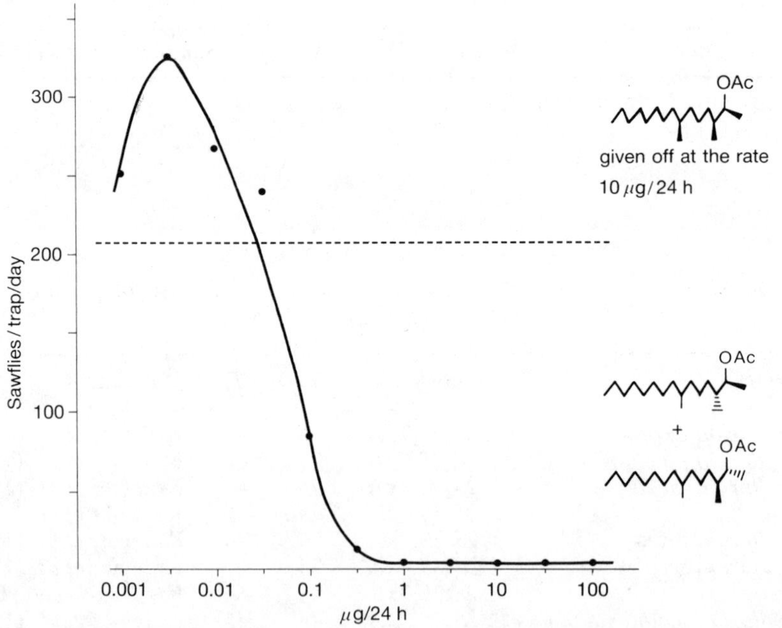

Fig. 13.3. Inhibition and synergism of attraction of males of *N. sertifer* by (2*S*, 3*R*, 7*R*/*S*) diprionyl acetate to traps baited with (2*S*, 3*S*, 7*S*) diprionyl acetate as an attractant in a field experiment. The dashed line shows mean trap catches in traps baited with the attractant only at a release rate of 10 μg/d Attraction is inhibited by as little as about 0.03 μg of (2*S*, 3*R*, 7*R*/*S*) diprionyl acetate released per day. At still lower proportions, this substance is a synergist with the attractant, increasing attraction more than 50 per cent. The mean number of males trapped in unbaited control traps was 0.8/d/trap. The substances were released from capillaries as described for the experiment shown in Fig. 13.2. Points are means for nine trap-days.

per cent. However, the *threo* acetate acts as synergist to the attractant at proportions below 0.3 per cent, with an optimal effect at 0.03 per cent. The synergistic effect, measured as increased trap catches, was about 56 per cent at the optimum value. This same concentration dependent inhibition–synergism has been reported in Lepidoptera (Priesner (Chapter 26)).

Similar results were obtained by Kikukawa *et al*. (1983). They found that male attraction to the 2*S*,3*S*,7*S* acetate in *N. sertifer* was reduced by 2*R*,3*R*,7*R* propionate presented in equal amounts with the attractant. However, the 2*R*,3*R*,7*S* propionate was not effective, indicating that the inhibitory effect depends on the 7*R* molecular arrangement. The propionate of 2*S*,3*R*,7*R*/*S*, however, was effective in reducing attraction. Based on these results it is plausible that the most effective inhibitor is a propionate with the structure 2*S*,3*R*,7*R*. This is the same structure that is an effective attractant for *D. similis*, a sympatric species. The same pattern of inter-species specificity—0.5—found in sympatric *Ips* bark beetles (Birch 1984) and some moth species may also occur in sawflies.

13.6. Conclusions

All information to date indicates that the very potent sex pheromones in all diprionid sawflies are based only on derivatives of 2,7-dimethyl-pentadecan-2-ol, diprionol. Derivatives of diprionol are the only sex pheromone compounds identified from females of diprionid sawflies. Only gas chromatography (GC) fractions containing diprionol derivatives show any attraction of males. Acetates and/or propionates of diprionol are very potent sex attractants for all diprionid species studied.

Species specificity of diprionid sex pheromones can be expected to be based on specific combinations of different chiral isomers of diprionol. The very similar chemical structure of these isomers means that all sex pheromone compounds have almost the same volatility. Proportions between two or more compounds can be expected to be essential in such a system. The number of species-specific combinations will then be almost infinite.

14 Orientation along terrestrial trails by ants

J. M. Pasteels J.-L. Deneubourg, J.-C. Verhaeghe, J.-L. Boevé, and Y. Quinet

14.1. Introduction

Many ant species lay olfactory trails to recruit sisters to more or less definite locations, for example new food sources or nest sites. The mechanism of trail orientation seems well understood. It involves osmotropotaxis and osmoklinotaxis (Hantgartner 1967) in a narrow and elongated active space (Bossert and Wilson 1963). The answers to questions such as how and why ants follow trails appear thus rather trivial.

In identical conditions different species vary in their efficiency, and theory predicts that the most accurate species is not necessarily the most efficient. In complex environments, where several food sources are present and aggregated, a certain amount of 'noise' optimizes food collection (Deneubourg *et al.* 1983). In this context how, when, and why ants lose their trail are highly relevant questions; new quantitative experiments were designed to study the subject.

In this paper we will discuss the following:

(1) How ants orient themselves on the trail when they cross it or when they reach a trail bifurcation. This will allow us to test the tropotaxis hypothesis.
(2) How trail pheromone concentration influences trail-following behaviour.
(3) How 'lost ants' explore the environment.

Finally, we will discuss how these behavioural mechanisms are integrated to contribute towards efficient food collection, in the light of food recruitment experiments.

14.2. Methods

Trail-following behaviour in *Myrmica rubra* was observed:

(a) on artificial circular trails (diameter 10 cm) drawn with pure pheromone (3-ethyl-2,5-dimethylpyrazine) (Evershed *et al.* 1982) dissolved in hexane, and

(b) on Y-shaped, artificial trails, each 5 cm branch separated from the others by an angle of 120°. One branch received a smaller amount of pheromone than the two others, which received the same quantity. The trail was rotated between each of the three replications.

Both series of experiments lasted 10 min, during which the behaviour of all ants reaching the trail was recorded. The experiments on circular trails were repeated two or three times until a sample of around 200 ants was observed. All quantifications were from video-recordings.

Food recruitment experiments were performed with *Tetramorium impurum*. Food sources (0.1 ml of sucrose solution) were placed at 10 or 40 cm from the nest entrance, symmetrically when two sources were offered. The distribution of recruits was recorded by time-lapse photography (3 min interval) and population growth at the food source was calculated from video-recording.

Mathematical models were constructed either to predict or to interpret the ants' behaviour. Only their principles will be briefly described here. Equations as well as more details of the methods will be reported elsewhere.

14.3. Why did some ants fail to respond to the pheromone?

A certain proportion of ants crossed circular trails without following them. Even at the most efficient concentration of pheromone, this proportion amounted to nearly 20 per cent, which is far too high to conclude that all these foragers did not perceive the pheromone. In fact, the same individuals followed the trail later in the experiment.

We hypothesized that ants failed to follow the trail because they crossed it at an angle which did not allow tropotactic orientation.

To test this hypothesis, a mathematical model was constructed based on the following: (1) the pheromone propagation mechanism in still air is Fickian diffusion (Bossert and Wilson 1963), (2) the difference of concentrations (ΔC) perceived between the two antennae was calculated as the ant crosses the active space at various angles, and (3) ΔC had to reach a threshold level below which no tropotactic orientation was possible.

Two predictions of the model were experimentally tested:

(1) Success in orientation depends on the angles at which the ants reach the trail (leading angle). Acute angles facilitate tropotactic orientation in contrast to more perpendicular angles. This dependence on the leading angle is stronger when the concentration of pheromone is lower, and success in orientation is then expected to be lower even for acute angles.

All leading angles were measured for two concentrations in trail pheromone (1.2×10^{-11} and 4×10^{-13} g/cm). As expected, amongst the ants which reached the trail at acute angles (<60°) a much larger proportion followed the trail than amongst those which reached it at an angle greater than 60° (respectively 78 and 24 per cent at 1.2×10^{-11} g/cm, 65 and 24 per cent at 4×10^{-13} g/cm, $P < 0.001$, χ^2). Additionally, at 1.2×10^{-11} g/cm, only 27 per cent of the ants which failed to orient themselves reached the trail at an angle smaller than 60°, in contrast to the 40 per cent at 4×10^{-13}g/cm. This difference however, was not highly significant ($0.1 > P > 0.05$, χ^2).

The mathematical model is deterministic and predicts a clear-cut dependence of success in orientation on leading angles. The experimental data were not so clear-cut, probably because of the klinotactic component and random events.

(2) Another prediction of the model was that the proportion of no-orientations to the trail should increase non-linearly with lower amounts of pheromone (Fig. 14.1a). Experimental data showed this increase (Fig. 14.1b). The proportion of ants not following the trail remained, however, unexpectedly high with high amounts of pheromones. This seems to be a consequence of the particular geometry of the experimental set-up, in which the ants tended to reach the trail at right angles. This constraint was not included in the model. The proportions of no-orientations at low concentrations (close to threshold) or with pure hexane were unexpectedly low. This could indicate that the ants respond weakly to hexane and/or that our criteria for accepting an orientation was too broad (e.g. random orientation).

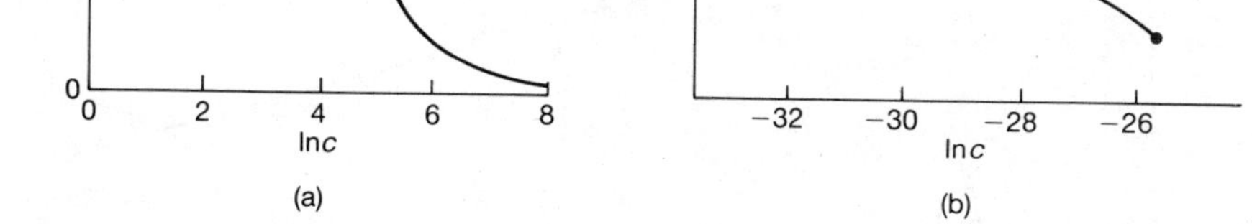

Fig. 14.1. Proportion of ants which fail to respond to the trail pheromone according to pheromone concentrations (c): (a) theoretical curve: diffusion coefficient 0.6 cm^2/min, distance between antennae 0.4 cm; (b) experimental curve.

14.4. The distribution of ants at a trail bifurcation

The same tropotactic mathematical model was used to describe the behaviour of ants reaching a bifurcation on the trail. The ants enter the double active space at various angles, due to their sinusoidal walk when following a trail. As in the previous model, the ants orient themselves towards the highest concentration perceived if ΔC (see above) reaches a threshold level. The model predicts that discrimination between two trails for the same ratio of pheromone concentration should be easier when the absolute concentrations in pheromone are high. Ants were able to discriminate between two trails, one having twice the amount of pheromone as the other, when a large quantity of pheromone was laid on the strongest trail (1.2×10^{-11}–4×10^{-12} g/cm), but not with a smaller quantity.

14.5. The probability of leaving a trail

For all the pheromone concentrations used, the number of ants following at least a certain length on the circular trail is a negative exponential function of the length they followed. The natural logarithm of the number of ants still following the trail was linearly proportional to the length of trail followed ($P \ll 0.005$, t-test) (Fig. 14.2). Holt (1955) made similar observations with wood-ant foragers following a natural track.

The exponential distribution means that the probability of an ant losing a

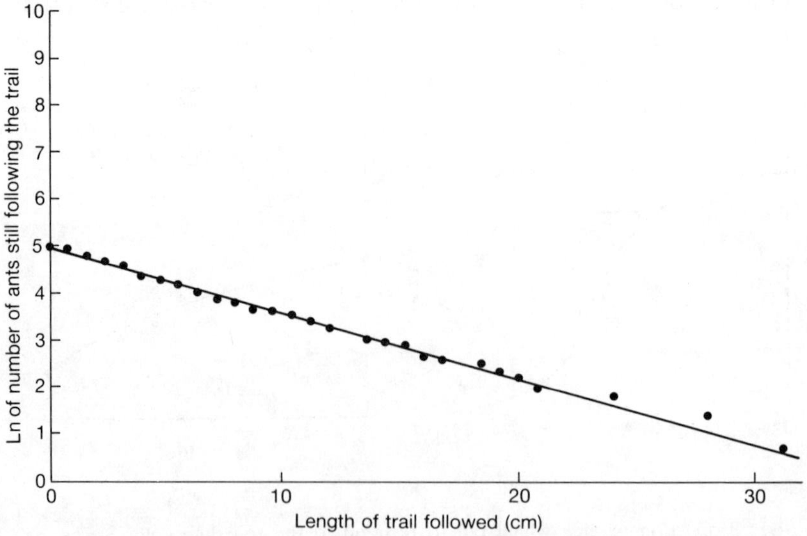

Fig. 14.2. Linear regression between the natural logarithm of the number of ants still following the trail (4×10^{-12} g/cm) and the length of trail followed (number of ants observed = 155).

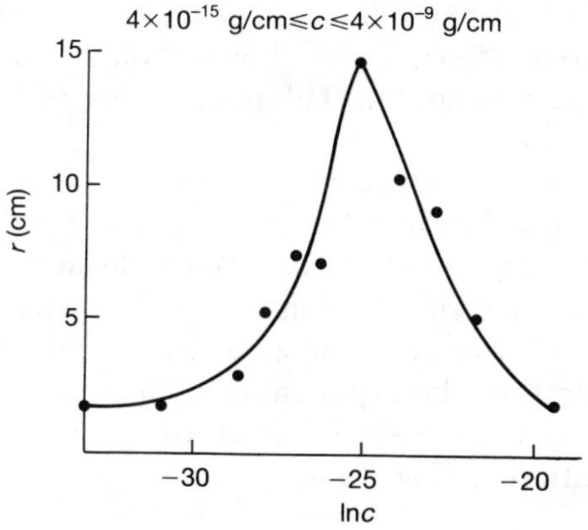

Fig. 14.3. Mean length (r) of trail followed by the ants as a function of the natural logarithm of the amount of pheromone (c) on trail.

trail remains constant with length, independent of learning or fatigue. This could be due in part to the tropotactic mechanism itself. Tropotactic orientation induces a sinusoidal walk (Hantgartner 1967) during which the ΔC between the two antennae is periodically inversed. No saturation of sense organs can occur, and loss of the trail is probably due only to casual movements of the ants.

As shown by Evershed *et al*. (1982), the probability of leaving a trail is a complex function of pheromone concentration. Between 4×10^{-14} and 1.2×10^{-11} g/cm this probability decreases, but it increases again for higher concentrations. As a result, the mean distance followed by the ants on the trail increases first and then decreases (Fig. 14.3).

This effect is not yet fully understood. The ants did not seem to be

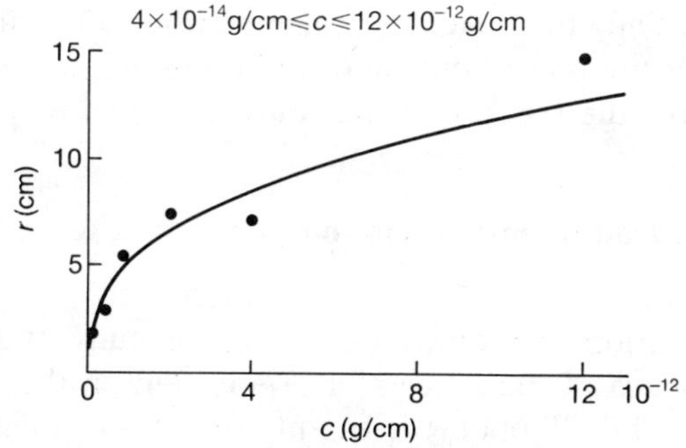

Fig. 14.4. Mean length (r) of the trail followed by the ants as a function of the amount of pheromone (c) on trail ($4 \times 10^{-14} \leq$ pheromone concentration $\leq 1.2 \times 10^{-11}$ g/cm) and its fitting by the law: $r = 1096$ (pheromone concentration)$^{0.362}$; r is in cm and the pheromone concentration is in g/cm.

repelled by high levels of pheromone. It could be that with large amounts the active space becomes too large to allow tropotactic orientation and that the olfactory sensilla are saturated. This phenomena probably never occurs in nature.

The increase in performance between 4×10^{-14} g/cm and 1.2×10^{-11} g/cm is non-linear. The best linear regression was obtained between the natural logarithm of the mean distance and the natural logarithm of pheromone concentration ($P < 0.005$, t-test). Thus the mean distance followed is best approximated by a power dependence of pheromone concentration, the exponent being less than 1. In other words, efficiency of orientation increases faster at low pheromone concentration than at high concentration (Fig. 14.4).

14.6. The distribution of ants during food recruitment

During food recruitment, ants observed along a perpendicular to the trail showed a bell-shaped distribution centred on the trail. The distribution became narrower with time as the amount of pheromone increased on the trail (Fig. 14.5). This agrees with the fact that the probability of leaving the trail decreases with pheromone concentration.

A similar distribution can be obtained on the assumption that the movement of lost ants results from two main influences: a lateral diffusion from the trail, and a tendency to keep the previous direction (Deneubourg *et al*. 1982). Such a distribution can be approximated by a Gaussian curve.

14.7. Trail orientation to two food sources

During trail orientation towards two food sources, one source was always preferred to the other, even when the two sources were identical. A less rewarding source (0.5 M sugar solution) was only fully exploited when the most rewarding source (1 M sugar solution) dried up. Ants always exploited preferentially the richer source, even when it was introduced into the arena 1 h after the poorer one, and trail recruitment had already been established towards the less rewarding source.

14.8. Conclusion: trail orientation to food sources, a self-regulating mechanism

Tropotactic orientation was confirmed during the study. It agrees with the observed dependence of the success of orientation on the angle at which the ants reach the trail. Tropotactic orientation also explains the distribution of ants at a bifurcation, and the constancy of the probability of leaving a trail.

Trail orientation to food sources is a highly sophisticated mechanism which allows the society to exploit the most rewarding sources.

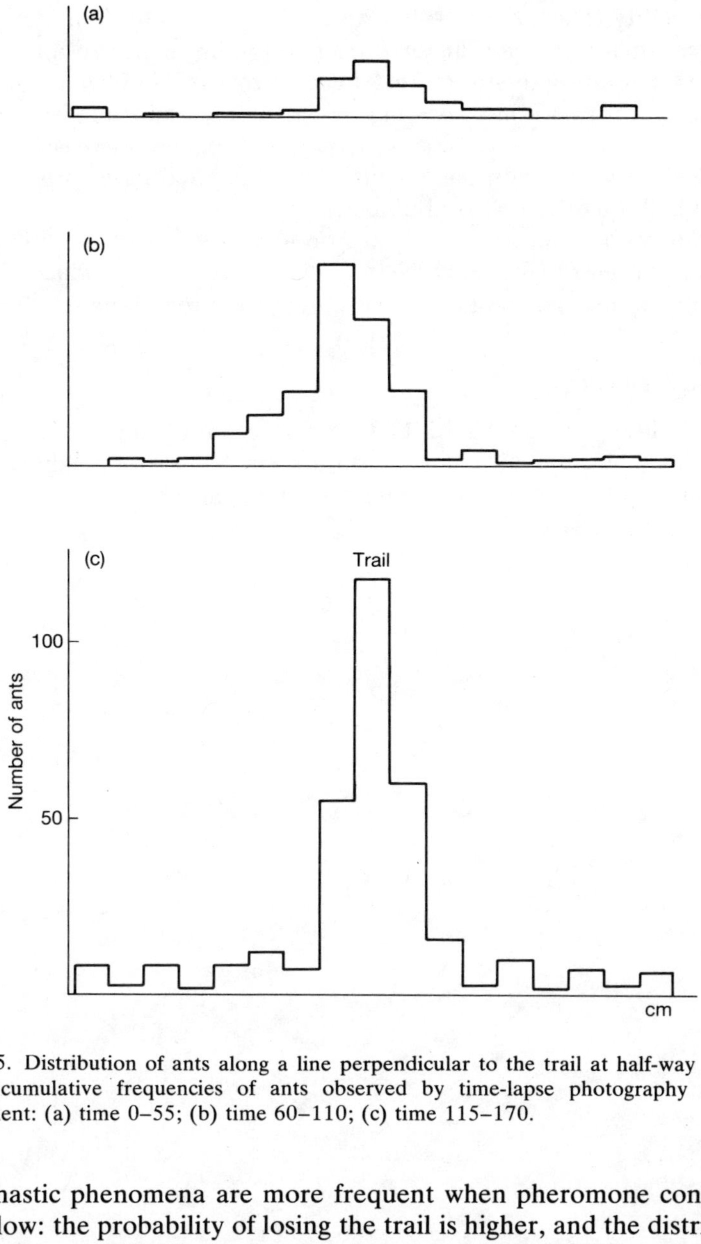

Fig. 14.5. Distribution of ants along a line perpendicular to the trail at half-way along its length—cumulative frequencies of ants observed by time-lapse photography during a recruitment: (a) time 0–55; (b) time 60–110; (c) time 115–170.

Stochastic phenomena are more frequent when pheromone concentration is low: the probability of losing the trail is higher, and the distribution of the ants at a trail bifurcation remains undetermined. This occurs at the beginning of a recruitment, or when the recruitment is towards a poor source. In both circumstances, lost ants can be valuable for the society through exploring the environment and discovering rich sources.

Adaptive reactions of the society are possible because, firstly, each recruiter seems able to control the amount of pheromone it lays according to food quality (Hangartner 1969) and, secondly, trail orientation during food recruitment is a non-linear process. Non-linearities result not only from the fact that successful recruits become recruiters (Wilson 1962), but also from the rapid increase in efficiency of trail following at low pheromone concentration. Trail orientation becomes increasingly more deterministic when conditions for high levels of pheromone are met, that is, when high-quality food is abundant.

A mathematical model based on non-linear equations describing the behaviour of ants will be presented elsewhere. It reproduces all the behaviours of a society so far known during food recruitment.

Acknowledgements

We would like to thank Dr M. C. Cammaerts who furnished us with the solution of pheromone. Thanks also to Dr M. Rowell-Rahier, H. J. Jacobson, and S. Goss for their help with the English.

15 Responses of arthropods to temporal chemical stimulus changes: simulation of a humidity differential and a pheromone plume

W. J. Bell

15.1. Introduction

Turning manoeuvres in response to temporal stimulus changes have been described as klinokinesis and klinotaxis (Fraenkel and Gunn 1961). In longitudinal klinotaxis an animal turns through a definite angle relative to its immediate previous direction (Kennedy 1978), as in *Phormia regina* remaining within a 'patch' of sucrose (Nelson 1977) and the parasitoid *Nemeritis canescens* remaining within a 'patch' of host odour (Waage 1978) by executing large-dimension turns immediately upon leaving the 'patch'. This strategy of changing the mean turn angle velocity following an abrupt stimulus change increases the chances of returning to the 'patch', and, if performed successively, of remaining inside the 'patch'. Adaptation can be important in longitudinal klinotaxis, as for example when the parasitoid *N. canescens* adapts to host odour and thereby leaves the 'patch' (Waage 1978).

Experimental protocol can restrict the nature of a stimulus change so that an experimenter can accurately correlate a stimulus change and the response. The manipulations of Kramer (1978), Havukkala (1980), and Tobin and Bell (1985) restrict the stimulus differential available to an arthropod, allowing it to perceive only a temporal comparison. In contrast to spatial (= instantaneous) comparisons, temporal comparisons lead to turns which are not necessarily correlated with any directional information supplied by a stimulus differential. Turning is regulated by internally stored information, permitting an animal to turn relative to its own track in some defined pattern (Bell and Tobin 1982).

The objective of the present study is to determine if arthropods can 'recognize' borders defined by temporal chemical concentration changes, and use this information to remain within a favourable 'zone' or to orient upwind along the edge of an odour plume.

15.2. Methods and materials

The data in this analysis were collected from a locomotion compensator (servosphere) apparatus at a rate of 10 data points per second (Bell *et al.* 1983; Sorensen and Bell 1985; Tobin and Bell 1985).

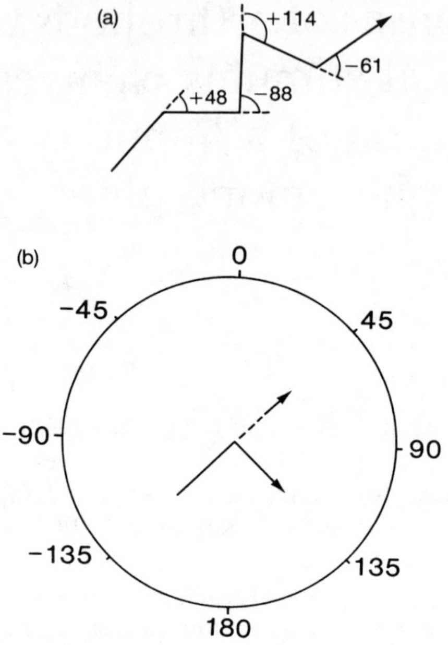

Fig. 15.1. (a) Turn angle velocity (TAV) relative to an animal's immediate track. Distances between turns are equal, representing the equal time periods in the data. TAV values with signs are relative TAVs; without the signs the values are absolute TAVs. (b) Compass directions measured for two legs. +450° before turning and +135° after turning.

Frequency distributions were calculated for relative turn angle velocities (± degrees/0.1 or 0.2 s) (see Fig. 15.1a), including both left (−) and right (+) turns, the relative mean TAV (theta) and its circular variation, and the mean vector length. Relative TAV is also referred to in the text as 'turn bias', as the magnitude of its value indicates the extent to which a given sign of turning is maintained. Turn angle velocity values were then folded 180°, combining left and right turning values, to calculate frequency distributions of absolute turn angle velocities, mean absolute TAV and its circular variation.

Frequency distributions of compass directions (see Fig. 15.1b) for each step were also calculated relative to arbitrary compass directions in the experimental apparatus or to a directional cue. Course direction is defined as the compass azimuth, and is used only to indicate the overall mean direction of movement.

15.3. Results

5.3.1. Orientation responses of an isopod, Armadillidum nasatum, *in a temporal simulation of a circular zone of high humidity*

Two levels of humidity were generated by switching on a computer-controlled valve (for 90 per cent RH) while the isopod was inside a simu-

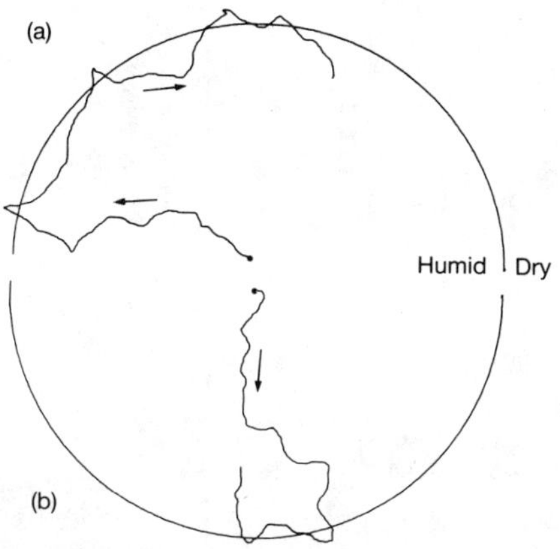

Fig 15.2. (a) and (b) Path of *A. nasatum* inside and outside of a simulated circular zone of high humidity. Arrows show direction of movement as the isopod 'leaves and returns to dark zone'. Dots shows starting points. For the two individuals shown here, the mean apparent response time leading to a change in the course direction from the dry toward the humid zone was 1.9 s ± 0.3 SE (N = 6) after leaving the humid zone.

lated 20 cm diameter circle, and switching it off (for 10 per cent RH) when the isopod moved out of the circle (Sorensen and Bell 1985). Since the air moved downward over the isopod, no directional information was available to the animal in its plane of movement, thus generating temporal, but not spatial cues. Isopods relocated a simulated zone of high relative humidity by turning back to the 'humid zone' after crossing the border into a 'dry zone' (Fig. 15.2).

Mean absolute TAV values for the six forays shown in Fig. 15.2 are significantly higher in the dry zone (48.4°/s ± 39.1 AD) than in the humid zone (22.0°/s ± 14.7 AD) ($P < 0.05$, Mann-Whitney U-test). Circular variance of absolute TAV is also higher in the dry than in the humid zone ($P < 0.05$, F-test for variance ratios). Mean relative TAV indicates a stronger turn bias in the dry zone (+32.9°/s ± 45.7 AD) than in the humid zone (−5.6°/s ± 24.0 AD), and its circular variance is higher in dry than in humid air ($P < 0.05$, F-test).

Measured with a protractor from Figs. 15.2a and b, the large 'strategic' turns leading the isopod from the dry back to the humid zone are approximately 139°, 126°, 84° and 88° in a, and 100° and 103° in b. High resolution analysis (periods of 0.1 s, 0.1 body lengths) of the first and second forays in Fig. 15.2a (Fig. 15.3a and b) and the first foray in Fig. 15.2b (Fig. 15.3c), reveals low-amplitude oscillations left and right with one or more episodes of strong right turning per foray. There is no point in graph c (Fig. 15.3) at which decisive, course-changing turns can be identified. Several relatively small-dimension turns to the right (at arrow) seem to increment the direction of the isopod from a compass azimuth of approximately −120° to approximately +40°. On the other hand, in graphs a and b the decisive points (arrows) seem to be two 90° turns in a and a

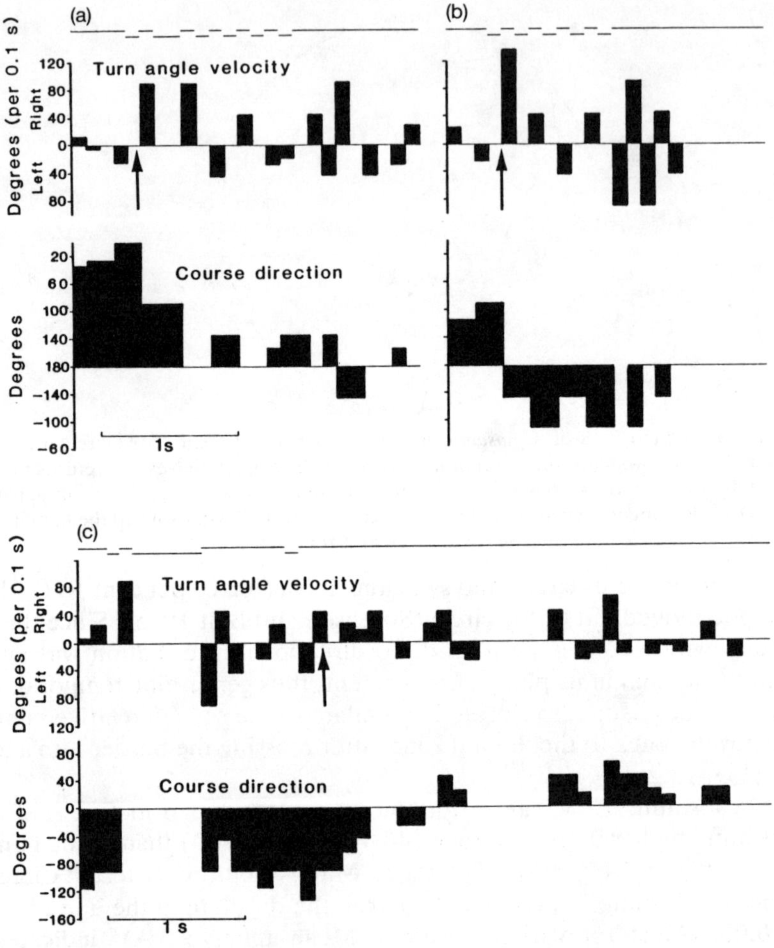

Fig. 15.3. Graphs of stop periods (top trace), turn angle velocity, and course direction at a resolution of 0.1 s for the first two dry periods shown in Fig. 15.2a (graphs a and b), and for the first dry period shown in Fig. 15.2b (graph c).

135° turn in b. The isopod's compass directions are biomodal in dry air, representing movements out of and then back into the humid zone.

15.3.2. Orientation responses of male Trogoderma variabile *in air containing constant sex pheromone and in a simulated sex pheromone plume*

Orientation of *T. variabile* in an air current with uniform and constant sex pheromone (10^{-3} μg *Z*-14-methyl-8-hexadecenal) and in a simulated sex pheromone plume were recorded on a locomotion compensator apparatus (Tobin and Bell 1985).

Figure 15.4a shows the path of a beetle in uniform and constant sex

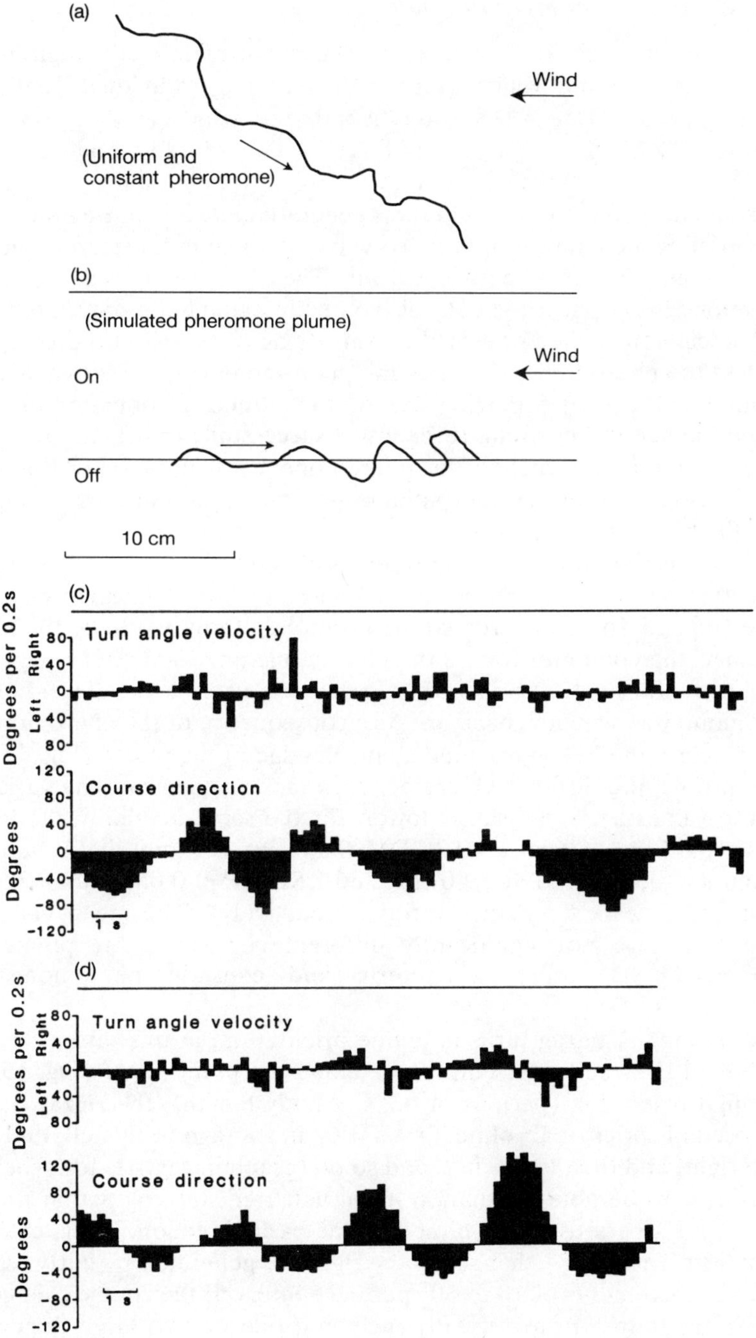

Fig. 4. Paths of male *T. variabile* in (a) uniform sex pheromone in wind, and (b) in a simulated plume of sex pheromone. Parallel lines indicate borders of the simulated 10 cm wide plume. Graphs (c and d) show stop periods (top trace), turn angle velocity and course direction at a resolution of 0.2 s for the path shown in Fig. 15.4a for uniform sex pheromone in wind (graph c), and for the path shown in Fig. 15.4b for a simulated plume of sex pheromone (graph d).

pheromone in a 25 cm/s air current. The mean menotactic angle of the zigzag course to the wind, which differs among individual beetles, is $-23.3° \pm 28.6$ AD ($r = 0.86$) relative to the wind (at 0°). The mean absolute TAV is 26.4°/s ± 26.8 AD. The mean relative TAV (−1.4°/s ± 37.2 AD) reflects a comparatively straight course.

A simulated 10 cm wide plume was generated in a 25 cm/s air stream by programming a computer to switch a valve on to emit sex pheromone into the airstream, or off to emit clean air. The experiment began with the pheromone valve on; if the path of the beetle extended more than 5.0 cm perpendicular to either side of the centre axis of the wind the computer switched the pheromone off. After the pheromone valve was switched off, to simulate the beetle moving out of the plume a computer program tracked the beetle's position. If the insect successfully relocated the plume, the computer then switched the pheromone valve back on. Although a plume was not present in the spatial sense, it was present in the temporal sense (Tobin and Bell 1985).

The beetle responded to a pheromone-off condition by executing a composite turn to the opposite track, heading the beetle back towards the plume (Fig. 15.4b). As it crossed the position determined to be the plume boundary, the computer turned the pheromone on. Soon after re-entering the plume, the beetle again turned, reversing its tack heading after the pheromone was switched back on. As a consequence of the pheromone-on and-off turns, the beetle oriented along an edge of the plume (Fig. 15.4b).

Mean time and distance values between tack changes and the variances of these values are significantly lower, for the same beetle, in the plume (2.1 s ± 0.5 SD, 3.0 cm ± 10.2 SD) than in constant and uniform pheromone (3.0 s ± 2.8 SD, 9.0 cm ± 86.7 SD) ($P < 0.05$, Mann-Whitney U-test and F-test of variance ratios, respectively). The mean values of zigzag turns are not significantly different comparing the plume run (75.6° ± 26.8 SD) and the uniform and constant pheromone run (63.0° ± 24.7 SD).

The 'strategic' zigzag turns in plume orientation, in this case a mean of 75.6°, lead the beetle back and forth along the plume edge (Fig. 15.4b). At a high resolution (periods of 0.2 s, 1 body length), the zigzag path is composed of series of absolute TAVs with an average to the left, and then to the right, and then to the left, and so on (graph in Fig. 15.4d). There is, however, considerable fluctuation in the left/right turning within a given series of 0.2 s periods that is biased in one basic direction. In the example shown here, the course changes to the right are generated by fairly regular 'staircase' increments of 10 to 30° per 0.2 s interval; the course changes to the left are more abrupt, usually including one or two large dimension turns. The switch to the left may be more abrupt owing to the insect's left-turn tendency, or 'preferred menotactic angle', which for this individual is 23° to the left of the wind. Course changes in uniform and constant sex pheromone in the wind are much less regular (Fig. 15.4c), as suggested above by the relatively large angular deviation of TAVs, and are not characterized by incremental turns.

15.4. Discussion

The two categories of spatial orientation that seem most appropriate to this discussion are object orientation, relating to resources or stress sources in the environment, and habitat orientation, relating to physical or biotic factors associated with habitat (Jander 1975). Orientation patterns in both ecological types are controlled by (1) idiothetic information stored by an individual during previous movements, (2) genetic information stored by a given species, (3) learned information, (4) external sensory information, and (5) stochastic information.

15.4.1. Locomotory functions controlling object and habitat orientation patterns

All habitat and object orientation patterns can be generated by regulating the sequence of changes and the number of steps for which: (1) the locomotory rate, (2) the turn angle velocity, and (3) the sign of turn (left or right) are held constant (Bell unpublished).

In *A. nasatum* returning to the preferred humid zone, an orientation pattern is generated by a periodic shift, over a few steps, to a higher TAV value with the sign held constant. The brief shifts to a higher TAV are punctuated by periods of relatively straight walking, generating composite turns between 90° and 180°. The sign of turn (left or right) is maintained throughout each composite turn, even though smaller-scale turns oscillate left and right as in relatively straight walking. Even in the same individual in the same experiment, however, the TAV value and the number of steps for which it was held constant varied considerably, and in only a few cases (as in Fig. 15.2b) did the isopod seem to return to the preferred zone by executing only one (about-face) turn.

The high TAV that the isopod switched to was not maintained for many steps as in looping, even if it failed to locate the favourable condition for some time (Sorensen and Bell 1985). Bouts of turning decreased over time, generally after approximately 10 s, if the preferred zone was not located. The isopod then moved in a relatively straight path in the unfavourable stimulus condition (Sorensen and Bell 1985), which is indicative of escape orientation (Bell *et al*. 1983).

High-resolution data are also available for the response of *T. molitor* to a temporal change from dry to humid air (Havukkala and Kennedy 1984). As with the isopod, a temporal stimulus change from a favourable to an unfavourable condition led to turn sequences with the same sign, consisting of one large turn of approximately 180° or two turns of approximately 90° (resolution = 1 point per second). Turning bouts lasted for approximately 4–5 s, after which the path straightened out. In Havukkala's study, although it was not possible for a beetle to 'relocate the dry zone', measurements of the tracks showed that the beetles' orientation would have led back to the dry zone. Male *Blattella germanica* stimulated by contact sex pheromone also exhibit a brief increase in TAV between periods of relatively straight running (Schal *et al*. 1983): after wing raising and a courtship turn, the cockroach runs approximately 60 cm, turns abruptly after stopping or in several smaller turns while moving, and runs back approximately to the site of stimulation.

The *T. variabile* data show that an orientation pattern, based on responses to temporal stimulus changes, can control a dynamic process such as plume orientation. When *T. variabile* oriented in an air current uniformly permeated with sex pheromone, an internal 'oscillator' increased the TAV for a few steps, with relatively straight walking between these turning episodes. Each time the TAV increased, the tack relative to the wind switched, as though the oscillator has a 'flip-flop' circuit. In this stimulus condition there was substantial variation in the mean time and distance between tack changes. Composite turns were organized with considerable changes in TAVs among 0.2 s periods, and in fluctuation in left/right turning.

In the simulated plume, the temporal change from sex pheromone to clean air and vice versa seems to interrupt the oscillator, switching the tack of the beetle in the wind soon after the stimulus change. There was less variation in time and distance between composite turns than in constant pheromone, but the composite turns were not significantly different from those in uniform and constant sex pheromone. These data support Kramer's (1975, 1978) model of three competing elements to explain zig-zagging upwind. The first is an upwind-turning command whose value increases with deviation from the upwind direction (as seen in Fig. 15.4a when *T. variabile* changed tack after orienting at a relatively large angle to the wind). The second is a rather constant turning command, whose sign changes from time to time stochastically (as in Fig. 15.4a) with constant pheromone in wind), and the third is an increase in the probability of a sign change with decreases in odour concentration (as in Fig. 15.4b during temporal stimulus changes).

15.4.2. Relationships and problems between 'longitudinal klinotaxis' and 'orientation patterns in object and habitat orientation'

Since understanding the behavioural ecology of an animal is the goal of most orientation studies, especially those of habitat and object orientation, it seems that quantifying and defining the orientation patterns relative to its ecological function is more meaningful than classifying the pattern as to longitudinal klinotaxis, klinokinesis or somewhere in the continuum between. The recent definition by Havukkala and Kennedy (1984) that longitudinal klinotaxis is 'a change in the location and/or the number of peak frequencies in the angular distribution of turns' is quite complicated, and is probably too specific since it is based primarily on *T. molitor*. The 'pattern' scheme, with its emphasis on information sources and locomotory control functions, allows for a continuum from klinotaxis to klinokinesis, and therefore includes a subset of ecological terms such as 'local search and ranging', 'foraging', and 'area-restricted search', as well as 'escape', 'attraction', and 'arrestment'.

On the other hand, Kennedy's (1978) rule-of-thumb definition, that in longitudinal klinotaxis an insect turns through a definite angle relative to its immediate previous direction, is general enough to apply to most examples. The only qualification would be that the 'turn' in longitudinal klinotaxis may actually be composed of several turns separated by periods of relatively straight walking, a point made also by Havukkala and Kennedy (1984). Havukkala's (1980) qualification that klinotaxis entails a change in the mean direction(s) of the circular distribution of absolute TAVs holds for most examples, except that circular variance increases in some examples as well as TAV (Bell unpublished). If circular variance does increase, it is due to a mixture of relatively straight walking and turning; when it does not increase, either (1) the animal always walks with a relatively high circular variance of TAV, or (2) the variance around the new TAV is the same as that around the previous TAV. The most significant characteristic of klinotaxis is the turning on of a constant, average sign of turn.

The ecological correlates of object and habitat orientation mechanisms described here are quite straightforward. Increased TAV and a constant sign of turning over a few steps can lead an animal back towards its previous position, in this case suitable habitat conditions in its environment. What has been described as 'self-steering' in longitudinal klinotaxis, the actions that lead an animal towards its previous position in the environment, can be defined as the use of idiothetic (proprioreceptive) information which is compared to some internally stored number of steps, to generate a composite turn between 90 and 180°. The exact number of steps required and the TAV that is specified, although expected to have a certain quantity of variation, represent genetic information. In a pheromone plume, temporal stimulus changes related to the edges of the plume interrupt an animal's internal tack change oscillator, leading to zigzag orientation. When this is coupled with an animal's tendency to move upwind, the resource object—a pheromone-secreting individual—can be efficiently located.

One conclusion that can be drawn from these studies is that the individual composite turns and sign changes that generate zigzagging along the plume border are mechanistically identical to those in the simpler temporal humidity stimulus change responses. In the plume there is a constant directional cue, the wind current, which polarizes the series of composite turns in a specific direction.

Acknowledgements

The author is grateful to T. R. Tobin and K. Sorensen for the data base used in this analysis, and to M. K. Tourtellot for many computer programs and comments. Supported by the National Science Foundation Psychobiology Program (BNS-82-03986).

16 Orientation by insect parasitoids

R. L. Jones

16.1. Introduction

Parasitoids are basically not unlike other insects, for example herbivores, in the manner of their discovery of a suitable host or mate. The fact that primary parasitoids occur at the third trophic level instead of the second (Price 1981), provides an added dimension to the host-finding process but does not necessarily alter the mechanisms involved.

Parasitism behaviour has been viewed by Doutt (1959) in a model which includes: (1) host habitat finding, (2) host finding, (3) host acceptance, and (4) host suitability. These categories of behaviour are similar to those often used in description of herbivore behaviour (Miller and Strickler 1984).

Orientation to hosts by parasitoids can be described as movement that provides contact with the maximum quantity and quality of the physical and chemical signals that are, over time, most reliably associated with the host. If we analysed the environment in which a potential host (a herbivore, for example) is located, we could prepare an extensive list of potential signals associated with the system. In addition to hundreds of chemicals emanating from the plant, the host, and associated organisms, there are electromagnetic, air pressure, and heat signals available.

A parasitoid's response to such an environment will be constantly reinforced with each signal detected that it recognizes as a cue. Those signals recognized as cues will have been determined via natural selection as those that have the highest correlation with the host, and those cues that are most stimulatory will be those that have proved to be the most reliable indicators.

As previous reviews of parasitoid behaviour have not done so, it will be useful to review the literature regarding behavioural mechanisms elicited by chemicals in insect parasitoids. So, though this discussion will centre around the use of chemicals by insect parasitoids to locate hosts, in keeping with the 'orientation' theme of the volume, I will arrange the chapter by the type of behavioural response cued by various chemicals. Additionally, I will include a short section on structure–activity relationships among kairomones, followed by a discussion of behavioural chemicals and 'context'.

With the exception of Waage (1978), most studies reporting observations of parasitoid behaviour were not designed to disclose behavioural mechanisms, but have been oriented around chemical identifications, impacts on parasitism rates or classifications of responses based on types of species interactions—kairomones, synomones, etc. (Brown *et al.* 1976; Nordlund and Lewis, 1976) or Dethier's attractant, stimulant categories (Dethier *et al*. 1960). Nevertheless, quite a bit of information can be assimilated concerning directed or undirected tactic or kinetic responses.

In either mate finding or host finding, a parasitoid is often faced with the problem of locating the target at a point (B) remote from its own position (A). The process of moving from point A to point B is teleologically referred to as searching. The impetus to initiate this search is unknown. Previous authors (Lewis *et al*. 1976) suggest that initial activity is innate and is released by the physiological condition of the insect (e.g., maturing eggs) and certain physical cues (light, temperature, etc.). This initial activity (flight perhaps) has also been called a random search. To date there has been no documentation of a chemical causing arousal or elicitation (Atkins 1980) in a resting parasitoid. Our work with sex pheromones of *Microplitis croceipes* and *Macrocentrus grandii* indicates that these responses occur, since the presence of the pheromone increases the number of males in flight in an olfactometer. However, we have not quantified these responses.

16.2. Orientation

Orientation to hosts and mates by insect parasitoids involves both flight and ambulatory activities.

Chemically cued orientation in flying insects involves inducement of upward anemotaxis and/or chemotaxis. Additionally, chemical cues perceived in flight can elicit casting, flight cessation, and landing responses. There is much documentation and discussion of these phenomena in insects but little, if any, by parasitoids, particularly in host finding. There are a number of reports of 'attraction' and trap catches, but no defined behavioural mechanism studies. For example, *Biosteres longicaudatus*, in a moving airstream, is attracted to ethanol and acetaldehyde released from rotting fruit, where tephritid fruit fly larvae (their host) are likely to be encountered (Greany *et al*. 1977). *Brachymeria intermedia* females are attracted to traps containing gypsy moth pupae (Barbosa *et al*. 1978); *Trichopoda pennipes*, a tachinid parasitoid, is attracted to traps containing its host, male southern green stinkbugs (Mitchell and Mau 1971); and *Aphytis melius* is attracted to traps containing its coccid host (Sternlicht 1973). Smits (1982) reported that *Trichogramma evanescens* landed more often on cabbage leaves containing the odour of *Mamestra brassicae*, a host, and Sato (1979) reported that he observed attraction of *Apanteles glomeratus* to cabbage leaves.

In regard to sex pheromones, males of *M. croceipes* and *M. grandii* exhibit upwind anemotaxis in response to odours of virgin females (R. L. Jones unpublished data).

Thus, although there have not been studies designed to prove the point, there is good evidence that chemical cues elicit either anemotactic or chemotactic, flight cessation, and landing responses in flying insect parasitoids.

After landing, parasitoids use walking responses to locate their host. These responses have been the subject of most studies of parasitoid behaviour. The type of walking behaviour elicited and the type of cues involved are correlated with the nature of the parasitoid host, that is egg, pupa, confined larvae, or mobile larvae.

In the case of egg parasitoids, chemicals from the host serve as cues indicating the presence of host eggs, whereas chemicals from the plant provide cues as to the likelihood that host eggs will be present. There is no indication that, in the walking phase, egg parasitoids use chemicals as directional cues for locating a host.

Host-associated chemicals induce certain behavioural changes in egg parasitoids. Laing (1937) first noted this when she reported the increased time (3×) spent by female *T. evanescens* in an area containing odour of the host, *Sitotroga cerealella*. She also reported that the females waved their antennae more violently and entered the treated area twice as often. She did not comment about rates of movement.

Increased time spent in an area (arrestment) could be due to decreased rates of movement (orthokineses), increased rates of turning (klinokineses), or mechanisms wherein the parasitoid chemotactically monitors a boundary of odour. With *Trichogramma pretiosum*, Lewis *et al*. (1979) have indicated that no attraction is involved in its response to odour of its host. These authors reported increased klino- and orthokinesis in the presence of host chemicals—mechanisms that resulted in increased parasitism rates by this parasitoid. In addition, this behaviour is elicited by hydrocarbon components of moth scale for *T. evanescens* (Jones *et al*. 1973) and for *T. pretiosum* (Jones *et al*. unpublished), organic acids for *T. pretiosum* (Gueldner *et al*. 1984), and pheromone components of *Heliothis zea* for *T. pretiosum* (Lewis *et al*. 1982). Also, *T. evanescens*, in response to the odour of virgin female *M. brassicae* and *Pieris brassicae*, exhibits a positive walking anemotaxis in an olfactometer (Noldus and van Lenteren, 1985a). The parasitoid also spends more time in areas containing odour, but the mechanism was not reported. Nordlund *et al*. (1984) reported a similar anemotactic response by *Telenomus remus* in response to the pheromone components of the host, *Spodoptera frugiperda*.

T. evanscens also is induced to search longer on leaves of cabbage that contain odour of *Pieris rapae*, *P. brassicae* or *M. brassicae* than it does on clean leaves (Noldus and van Lenteren 1985b). This effect is similar to the retention effect reported by Laing (1937) and Gross (1981) for *T. evanescens* in response to various other host odours. Female parasitoids searched leaves containing moth odour for 9–22 minutes (time varied with host species) compared with 4 minutes for control leaves (Noldus and van Lenteren 1985b). The behavioural cause of the retention was not reported.

Close-range orientation by *Trichogramma* spp. is apparently cued visually. This is likely to be true for other egg parasitoids as well. Salt (1935) first reported this in his studies with *T. evanescens*, and further evidence was provided by Boldt *et al*. (1973) who showed that *T. pretiosum* exhibited a preference for several species of lepidopteran eggs that was correlated with the size of the egg (although measurement by contact was not ruled out).

In the case of pupal parasitoids, host location involves many of the same responses seen in egg parasitoids. Edwards (1954) showed that the odour of muscoid pupae caused increased orthokinesis and a resultant arrestment in a pteromalid parasitoid. Within 5 mm of the host, location is cued by vision. *Coccigomimus turrionella*, a gypsy moth pupal parasitoid, responds to the odour of its host with a positive anemotaxis (Sandlan 1980). The female exhibits a sinusoidal pattern in this response and there is no preference for the edge of the plume. A positive orthokinetic effect was also reported. These responses are not present in females in the preoviposition state. *Ventura* (=*Nemeritis*) *canescens* exhibits a positive anemotaxis in the presence of the odour of its host *Ephestia* pupa or oatmeal (Thorpe and Jones 1937).

Thus, similar to egg parasitoids, pupal parasitoids respond upwind and exhibit behavioural responses that result in arrestment and search stimulation in the presence of cues associated with the host.

With regard to larval parasitoids, chemical cues with activity responses similar to trail pheromones play an important role (Price 1970). Once in an area or on a plant containing a host larva, parasitoids sample the substrate repeatedly with the antennae while walking. Chemical cues left on the substrate by the host larvae in salivary secretions and faeces, and, in the case of herbivores, chemicals emitted by damaged leaves, are all active. Upon detecting these same cues, the parasitoid sharply increases its rate of antennation and often alters its turning rate (klinokinesis) and movement rate (orthokinesis). The parasitoid chemotaxically follows the trail of the larvae until contact is made, at which time it uses chemical and visual cues to assess acceptability for parasitism. In the case of hidden larvae, these same cues may act to elicit probing with the ovipositor, in which case decisions about acceptance are made on the basis of cues detected by the ovipositor.

There are a number of examples of detailed descriptions of behaviour of larval parasitoids in host location in the literature (van Lenteren *et al*. 1980; Schmidt 1974; Waage 1978). *M. croceipes* responds to salivary and faecal cues left by its host, larvae of *H. zea*. The active components (a series of mixed monomethyl isomers of hentriacontane—the most active isomer being 13-methyl-hentriacontane) are detected by contact with the antennae and serve as trail kairomones. Although influenced by concentration, the response of the female to these hydrocarbons is one of decreased orthokinesis and intensified antennation. Although intensified antennation is always cued, when coupled with other as yet unidentified chemicals in host frass or saliva, the locomotor response may show increased klinokinesis. When a female *M. croceipes* perceives a host to be present, she immediately ceases locomotory activity and 'freezes' momentarily. After a few seconds, she will lunge and sting the host larva. If in the process of search she stumbles across the larvae otherwise unperceived, she will sting it immediately.

Encarsia formosa is unable to detect its host larva from a distance and the presence of host larvae does not affect locomotion in any respect even within 1 mm (van Lenteren *et al*. 1976). The host is detected only by contact with the antennae.

Sato (1979) reported that *A. glomeratus*, a parasitoid of *P. rapae*, responded to chemicals emitted by leaves of Cruciferae by visiting the leaf and searching. Their closed-system bioassay suggests that the females exhibit a chemotactic response, as 70–90 per cent of the females visit crucifer leaves or filter paper-containing leaf extracts, while 10–40 per cent visit clover leaves. Additionally, an arrestment response was observed to the crucifer leaves, as they sustained a mean search time of 10–12 min by the parasitoid, whereas clover was searched for 3–6 min. The responses were more pronounced in damaged leaves.

The tachinid, *Eucelatoria* sp., exhibits a plant preference via positive anemotaxis in an olfactometer (Nettles 1980). *Leptopilina clavipes* exhibits both klinotactic and anemotactic responses to decaying mushrooms, the habitat of its host, *Drosophila* larvae (Vet, 1983). Chemotaxis was observed also in a static olfactometer.

16.3. Chemical cues

A review of the identified chemicals involved in the parasitoid orientation process can perhaps provide some insight into the physiological aspects of olfaction in this group of insects. There are a number of reports in the literature in which specific chemicals have been demonstrated to cue some types of orientation in parasitoids (Jones 1981). Some of the more interesting cases involve the use of proteins as cues. The mechanisms involved in protein detection by the insect antennae have not been reported.

There are a few studies in which a series of chemicals have been bioassayed, and a look at these can provide some insight into the nature of receptor mechanisms.

Jones *et al*. (1971) evaluated several positional isomers of *X*-methylhentriacontanes as kairomones eliciting antennation by *M. croceipes*. Bioassays produced the following order of activity (scores in parenthesis) to 150 ng of chemical: 13-methyl analogue (2.32), 15-methyl analogue (1.68), 12-methyl analogue (0.48), 11-methyl analogue (0.22), 9-methyl analogue (0.00). The data indicate an acceptor site that is very specific for the methyl position. Tests for mixed effects of these isomers indicated a dilution effect. A score of 1.21 was obtained with 50 ng of the

13-methyl analogue, 0.63 with 50 ng of the 15-methyl analogue, and 0.92 with 50 ng each of the two together. This indicates that these analogues compete with each other at the same acceptor site. Lewis and Tumlinson (personal communication) have found that *M. croceipes* shows no differential response to the optical isomers of these methyl-branched hydrocarbons.

Vinson *et al*. (1975) provide a more detailed study with the response of *Cardiochiles nigriceps* to a series of *X*-methyl hen- do- and tritriacontanes. These data provide information on chain lengths as well as positional isomers and support those of Jones *et al*. (1971) with respect to non-synergism by chemicals with methyl groups in variable positions. They also show a bell-shaped dose–response curve, with optimum responses at 5 μg of chemical. The optimum methyl positions did vary with chain length and were 11-methylhentriacontane, 16-methyldotriacontane, and 13-methyltritriacontane. The most interesting feature of the study was a marked synergism observed when chemicals of varying chain lengths were mixed. Mixtures of several methyl isomers of hen- , do- , and tritriacontanes exhibited synergistic activity, with dose–response curves having optimum activity at the 500 ng level and very good activity at the 50 ng level. These results provide strong evidence that the parasitoid perceives hydrocarbons of varying chain lengths at different acceptor sites and that the different acceptor sites have different optima for methyl positions. There is evidence in the data that there might be a constancy in optimum methyl position from the other end of the molecule, but more precise studies would be needed to clarify this. For example, the 12-methyldotriacontane is very active and the relative position of the methyl group from the other end of the molecule in 11-methylhentriacontane, 12-methyldotriacontane, or 13-methyltritriacontane is the same.

Hendry *et al*. (1973) reported that *Orgilus lepidus* responded optimally to heptanoic acid among a series of alkanoic acids ranging from acetic to decanoic acids.

Mudd and Corbet (1982) assessed the response of *V. canescens* to a series of 2-acylcyclohexane 1,3-diones isolated from the mandibular gland of the host, *Ephestia kuehniella*. Their results showed a response at the 100 ng level with a linear relationship between concentration and activity. Activity varied considerably with the nature of the substituent in the 2 and 4 positions of the ring. The optimum activity was obtained with a *Z,E*-12′,14′-hexadecadienoyl substituent in the 2 position and a hydroxyl group at the 4 position. Removal of the hydroxyl group resulted in a minor loss of activity. When coupled with the hydroxyl at the 4 position, an oleyl substituent at the 2 position was also quite active. No tests of mixtures were reported.

16.4. Discussion

Due to the fact that parasitoids are at the third trophic level, thus having to locate not only an appropriate habitat (a plant species for example) but a suitable host within the habitat, they might be capable of detecting a wider range of chemicals than insects at the second trophic level. As the parasitoid's host may shift habitats phenologically and as a parasitoid may be polyphagous, there is even more reason for the parasitoid to be versatile at chemical detection. These situations also provide an environment conducive to the selection of associative learning, a well-documented phenomena in insect parasitoids (Vinson 1984).

Parasitoids' versatility is supported by their sensory system organization. The data reported by Chapman (1982) indicated that the olfactory sensilla of hymenopteran parasitoids are of a general nature (few sensilla, but each with many neurones). This circumstance makes parasitoids versatile in their response spectrum but rather insensitive, both characteristics that are favourable to their lifestyle. This is evidenced by the fact that parasitoids do respond to a wide range of chemicals, especially the more polyphagous egg parasitoids such as *Trichogramma* sp. It is obvious that, coupled with their associative learning capabilities, this versatility enables insect parasitoids to be very exploitive of a complex and dynamic ecosystem.

This conclusion has several ramifactions. Firstly, it highlights the high efficiency of laboratory bioassays wherein an insect is given a choice of a chemical versus a sterile glass environment containing 'pure' air. Such bioassays actually give an insect a choice between a void world that it has never experienced and a chemical that might serve as a cue to a real world. It will probably be possible to demonstrate responses to thousands of chemicals in these types of bioassays, especially if we are careful to test a wide range of concentrations. We must realize that such bioassays only tell us which chemicals might be important in the real world. The important question is what the relative contribution of a given chemical is in the behaviour of a parasitoid. This can best be answered with bioassays that remove the chemical from the whole rather than adding it to nothing.

Secondly, it reminds us that even an agricultural ecosystem is not nearly as complex as the natural ecosystem in which a parasitoid might have evolved. Consequently, a large variety of chemicals might serve to increase the total cue input that a parasitoid receives and thus increase the likelihood that the parasitoid will remain in a system longer and perhaps be more active in host finding.

Thirdly, it is possible to reason that instead of a behaviour sequence in which a parasitoid responds definitively and serially to cues to habitat, host damage, host, etc., parasitoids respond to all of these cues accumulatively and their response at any given moment is dictated by the total of all cues impinging on its receptors. We can postulate a hierarchy of responses to cues ranging from nothing to: (1) any plant, (2) preferred plant family, (3) preferred plant species, (4) preferred plant part, (5) any animal cue, (6) any insect, (7) preferred order, family, species, etc. Cues from any of these sources act in an additive fashion (via neural convergence and integrative mechanisms) in the host selection process. Thus the detection of any plant odour might lower the response threshold to a host cue, although the optimum lowering might be observed with the odour of the proper plant species.

17 The role of pheromones in the control of behavioural sequences in insects

P. E. Howse, J. C. Lisk and J. W. S. Bradshaw

17.1. Introduction

Insect behaviour may be envisaged as a series of successive subroutines, each triggered by particular environmental stimuli, dependent upon sensory feedback for their execution, and influenced to a greater or lesser extent by the after-effects of preceding actions. When we examine orientation to olfactory stimuli, we often see sequences of separate actions which we recognize as (for example) activation, orientated locomotion, arrestment, and some kind of consummatory response. If the olfactory stimulus is a pheromone, the insect is usually responding to a multicomponent mixture. Sequences may, theoretically, be due to the insect responding successively to different components of the mixture. Alternatively, it may be that the blend of compounds represents a *Gestalt* stimulus so that each component contributes only to the release of a complex fixed-action pattern, which then continues until consummation, guided by various external stimuli. In fact, these are only two extreme possibilities among many, and one may suspect that some kind of dual control system is more likely to be present.

The characteristics of the chemical stimulus itself may assist in the production of a sequence of behaviour in the receiver. For example, over short distances in relatively still air different components of a blend may diffuse out at different rates, or different chemical messages may be released from separate glands at different times. On the other hand, the insect receiving the signal may differ in its sensory or behavioural thresholds to different concentrations of a blend of compounds, or to different concentrations of certain components of that blend. This may result, in the extreme case, in the receiver having discrete behaviour patterns cued by single compounds in a blend, when successive sharp thresholds fall within a narrow range of concentrations. Possibly, thresholds may change in the short term—for example, during upwind flight—as a result of sensory adaptation or 'habituation'. A further possibility is that the reception of one stimulus may be used to modify or 'switch on' a response to a chemical stimulus that the insect has already detected. Modification can take various forms, for example inhibition, summation, or synergism.

Thus, there are various ways in which a behavioural sequence may be controlled as a result of changes in the stimulus with time and/or differences or changes in the sensory or behavioural thresholds of the receiver. Such mechanisms appear to reduce the necessity for central neural programming, with its attendant inflexibility. Examples of different means of stimulus control will be considered below.

17.2. The African weaver ant, *Oecophylla longinoda*

This notoriously aggressive ant builds leaf-nests, made by suturing the edges of living leaves together with silk from the young larvae, which are used as shuttles. The worker caste is dimorphic. The major workers are mainly occupied in foraging and territorial defence outside the nest, while the minor workers normally remain inside the nest with the brood. An alien intruding into the territory of the ants, such as another arthropod, representing potential prey, or an ant of another *Oecophylla* colony or of another ant species, is usually quickly detected. Nest-mates are then rapidly recruited and the intruder attacked. It is normally immobilized by ants biting and pulling, so that it is pinioned on the substrate.

Bradshaw *et al*. (1979a) showed that the signals for alarm and short-range recruitment originate from the paired mandibular glands of the major workers, which open at the base of the mandibles. If the intruder is not easily subdued, poison gland secretion may be sprayed from the tip of the gaster, along with a series of hydrocarbons from Dufour's gland (also in the gaster). The mandibular gland secretion is a mixture of more than 30 different compounds, mostly straight-chain hydrocarbons. The structure of a number of these components from fractions controlling recognizable behavioural responses has been elucidated, and responses to the synthesized compounds compared with responses to the natural secretion. Figure 17.1 illustrates the behaviour elicited by a mixture of four of the principal active components presented at a point source. Hexanal alerts the major workers, which rush around with their jaws open and become very sensitive to moving visual stimuli. They do not, however, orientate towards a source of hexanal in still air. 1-Hexanol attracts the ants, which may circle the source before approaching it more closely. The less volatile components, 3-undecanone and 2-butyl-2-octenal, also attract at short range, but elicit prolonged biting responses. A mixture of the four components releases a sequence of behaviour, proceeding from alerting to attraction and biting. This is exactly similar to the behaviour released by the natural pheromone, and is governed by the active spaces of the various components as they diffuse out in still air from a point source.

Ants that are attacking commonly raise the gaster high and spray the target. Again, there are more than 30 different volatile components in the spray (Bradshaw *et al*. 1979b). The major ones are formic acid, from the poison gland, which is the main toxin, and *n*-undecane, from Dufour's gland. The latter, with other hydrocarbons in the series, facilitates spreading of the secretion over the arthropod cuticle and penetration through it. The spray is rarely used in aggressive encounters, possibly because the ants

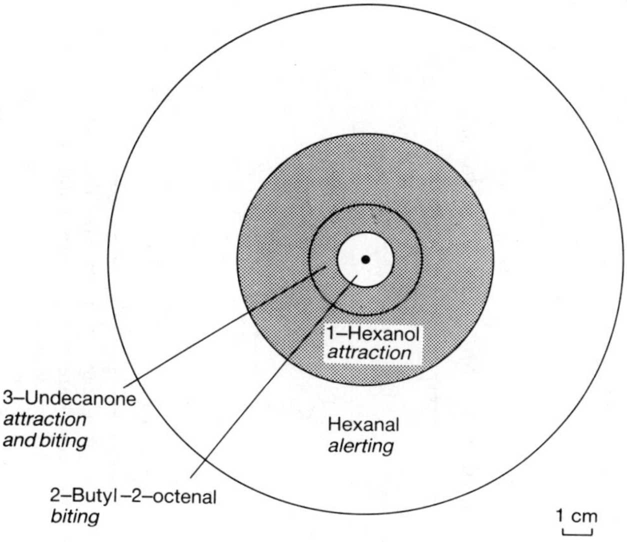

Fig. 17.1. Active spaces of the mandibular gland secretion of major workers of the ant *Oecophylla longinoda*, 20 s after deposition at a central point on a flat surface.

risk contaminating themselves, and because prey that has been sprayed may be less palatable to them, but the general effect is to increase the apparent aggressiveness of the ants, eliciting a mass attack. A source of the venom affects major workers within a 10 cm radius. They open their mandibles wide, approach the source and attack it with repeated bites. Presented separately, undecane elicits opening of the mandibles, raising of the gaster, and approach, while formic acid elicits approach and attack. When the two compounds are presented together, there is a synergistic effect between them in inducing approach.

17.3. The pine beauty moth, *Panolis flammea*

This noctuoid moth has occurred in pest proportions in Scottish pine forests in recent years, stimulating the development of a pheromone-based monitoring trap. The three major components of the female sex pheromone are: (*Z*)-9-tetradecenyl acetate (*Z*-9-TDA), (*Z*)-11-tetradecenyl acetate (*Z*-11-TDA), and (*Z*)-11-hexadecenyl acetate (*Z*-11-HDA), which are present in the ratio 100:5:1 (Baker *et al.* 1982). Trapping experiments carried out in northern Scotland showed that only the *Z*-9-TDA alone, or in combination with one or both of the *Z*-11 compounds, attracted moths into sticky traps (Bradshaw *et al.* 1983). Combinations of *Z*-9-TDA and *Z*-11-TDA became progressively less attractive as the percentage of the latter was increased above 25 per cent. These results led to a series of experiments to ascertain the role of the three components by presenting them singly and in combinations in a wind-tunnel of the type described by Jones *et al.* (1981), in which odour can be

TABLE 17.1

Responses of male P. flammea *in a wind-tunnel to components of the female sex pheromone*

Compounds released	No. of Males tested	No. of upwind tracks (50 cm)	% landing	% attempting copulation
Z-9-TDA	30	0	0	0
Z-9-TDA + *Z*-11-TDA	30	8	88*	14
Z-9-TDA + *Z*-11-HDA	30	14	7	0
Z-9-TDA + *Z*-11-TDA + *Z*-11-HDA	30	16	81*	77†
Z-11-TDA + *Z*-11-HDA in cloud of *Z*-9-TDA	15	>20	0	0
Z-9-TDA + *Z*-11-HDA in cloud of *Z*-11-HDA	15	11	91*	70†

Amounts used to form plumes: *Z*-9-TDA – 500 μg, *Z*-11-TDA – 5 μg, *Z*-11-HDA – 40 μg, but 2.5 mg of *Z*-9-TDA and 25 μg of *Z*-11-TDA were used to form a cloud. Mean values indicated by the same symbol are not significantly different (X^2, $P < 0.01$) (Further information in Bradshaw *et al*. 1983).

presented either as a narrow plume, or as a diffuse cloud permeating the air stream, or as a plume within such a cloud. Upwind tracks of more than 50 cm made by the male moths were recorded on videotape for analysis.

Z-9-TDA presented alone in plume form in the wind-tunnel was ineffective in attracting males upwind, but when all three components were released in a plume there was a high percentage of males flying upwind, landing, and attempting copulation with the odour source (an impregnated rubber septum) (Table 17.1). The percentage of landings decreased markedly in the absence of *Z*-11-TDA. In the absence of the *Z*-11-HDA there was a high percentage of landings, but few subsequent copulation attempts. Thus, the presence of *Z*-11-TDA lowers the threshold for landing, and the presence of *Z*-11-HDA lowers the threshold for copulation.

Further tests were carried out in which the *Z*-9-TDA was presented as a diffuse cloud containing a plume of the other two components, and the *Z*-11-TDA was presented diffusely with a plume of the other two (Table 17.1). The results showed that the *Z*-11 compounds did not themselves elicit zigzag upwind flight, but in their presence this occurred to a plume containing *Z*-9-TDA, when a higher percentage of landings and attempted copulation also occurred. *Z*-9-TDA in diffuse form provoked many rapid flights up and down the tunnel, but no plume following: it therefore appears to activate the males.

When males were exposed to components of the female pheromone in still air, as single sources or in combinations, they were observed to attempt copulation with the rubber septa only when *Z*-9-TDA and *Z*-11-HDA were presented together. The moths then orientated themselves mostly towards the lure containing the *Z*-9 compound, but the latter alone had no effect. Catches in the field with traps baited with *Z*-9-TDA alone were obtained probably because trace levels of the *Z*-11 compounds were in the atmosphere, released from neighbouring traps.

17.4. Discussion

In these two examples we see a variety of phenomena that play a role in the control of behaviour sequences. In the weaver ants, sequences are controlled by changes in the distribution of pheromone components in space and time. It appears that the insects have relatively low thresholds to the compounds concerned, and that their active spaces, when released from a fixed point in still air, depend mainly upon their volatilities. The alarm–defence behaviour can be intensified by release of further pheromone components from abdominal glands, in which there is synergism between the contents of the poison gland and Dufour's gland.

In the pine beauty moth the volatilities of the different components of the female sex pheromone have a negligible effect on their distribution in space, but their concentration within the blend is mainly determined by their release rates from the gland. It appears, then, that downwind, where the dispersion of the pheromone is high because of the effects of turbulence *en route*, male moths are stimulated to fly upwind. The *Z*-9 and *Z*-11 components have different effects on behaviour, the former acting as a marker for the aerial trail, and the latter switching on guidance mechanisms that keep the insect within the plume. Close to the source, the odour plume will be narrower and more coherent (see Murlis, (Chapter 3)) so the insect will begin to experience a higher rate of stimulation of antennal receptors. It appears that the higher concentration of *Z*-11-TDA occurring near to the source, where the turbulence is low, then increases the probability of landing, in which the *Z*-9-TDA still acts as a guide. Closer to the source it continues to guide, but the *Z*-11-HDA in higher concentration releases attempts at copulation.

The interpretation of events above assumes that the male moth has a relatively low threshold to *Z*-9-TDA throughout but is very sensitive to changes in the levels of *Z*-11 compounds that it detects. Responding to them at relatively fixed thresholds, it uses the *Z*-11 components as releasers for different orientation mechanisms to the *Z*-9-TDA, which then appear in sequence in the normal response to a pheromone plume. It follows that the relative and absolute concentrations of the pheromone components on release are critical factors in the control of the behaviour sequence. Addition of the 'secondary' *Z*-11 compounds to the *Z*-9-TDA will not just strengthen the later transitions in the sequence, but will also potentiate the early stages of upwind flight. The sequence that occurs may be viewed equally, therefore, as a succession of actions triggered by separate stimuli, or as a fixed action pattern released by a pheromone blend of precise constitution. Taking the latter view is likely to lead to a less heuristic approach in the future.

18 Manipulation of orientation to pheromones

J. R. McLaughlin

18.1. Introduction

Insect pheromones are classified by Djerassi *et al.* (1974) as one of a group of biorational chemical agents for insect control. This taxonomy distinguishes them from the chemical (hard pesticide) and biological (microbe, predator, parasite) agents. Biorational chemicals are natural products or chemicals that mimic natural products with virtually no toxic properties and relatively little persistence in nature.

There are in general four methods for exploitation of sex pheromones in insect pest management. Two of these, trapping for population assessment (biomonitoring) and annihilative trapping, require a normal behaviour from the target species. A third method is designed to disrupt the communication process by the direct use of pheromonal chemicals. The sex pheromones also act as cues to parasites and predators that attack target species, and these behaviours can be manipulated to our advantage. These procedures may be used in combinations and are compatible with the chemical and biological control agents.

18.2. Population assessment

There are many levels of biomonitoring from the simple survey and detection of a species to the use of trapping information to drive complex predictive models. Pheromones and similar attractants have proved very effective for the former and less useful as a predictive tool. This is because prediction requires knowledge of a number of factors that affect the magnitude of trap captures and because the adult population may not be the best parameter to measure. This is especially true for sex pheromones that generally attract only one sex.

The sex pheromones of moths and many beetles are volatile multichemical messages that attract one or both sexes to the site of evaporation, provided the rate and ratio are biologically correct and the chemicals are released at the proper time and place. These pheromones are highly species specific in the end result, copulation; however, they are often less specific in eliciting behaviours that occur early in the aggregating or premating sequence. The objectives of any pheromone identification project must, therefore, include the identity of the chemicals, a characterization of their rate of release as a total blend, and the ratio of the components in the volatile blend. With such information and the ability to duplicate the blend parameters with formulations of synthesized compounds the behaviourist can determine the role of blend components, design effective traps, and build trap/bait formulation combinations that fit specific needs.

Release rate and ratio studies are often conducted empirically in laboratory wind-tunnels or with field traps. These studies are valid when the pheromone components have very similar vapour pressures. Unfortunately most studies of systems with components having differing vapour pressures are conducted using the ratio of components placed into the formulation to estimate the release ratio, and with the assumption that equal total blend doses will have equal release rates. Heath *et al*. (1986) have developed methodology for determining the relative vapour pressure of most of the common chemicals found in moth pheromones from their retention indices on liquid-crystal columns. Using the gas laws, one can then calculate the release ratio of any blend and the amount of the blend required to achieve a given release rate. They have also determined that the release ratios of mixtures of pheromone components formulated in the polymer matrices commonly used as trap baits also follow the ratios calculated for a pool of free liquid.

We have applied this finding to wind-tunnel and field studies with the sex pheromone of the velvetbean caterpillar moth (VBC), *Anticarsia gemmatalis* Hübner, which contains a *C*20 and a *C*21 hydrocarbon (Heath *et al*. 1983). The *C*20 material is about four times more volatile than the *C*21. Thus, a mixture of 70:30 (*C*20:*C*21) will give a release ratio of about 80:20 (*C*20:*C*21), and 1.2 mg of this mixture will evaporate at the rate equivalent to 1 mg of the *C*20 material alone. The female insect produces the materials with a ratio near 5:3, which would evaporate at a ratio of about 75:25. This knowledge enabled us to obtain data on dose–response for the VBC male that was similar to that reported by Bellas and Bartell (1983) for a moth in which the two pheromone components have similar vapour pressures. We had not obtained good dose–response relationships for the VBC using load ratios to estimate release rates and ratios. This method has become even more useful and important in studies of insects with three or more pheromone components.

The most effective ratio and release rate of the VBC pheromone also attracts males of the genus *Mocis* into field traps, where they are difficult to distinguish from VBC. We, therefore, selected a rate and ratio that captured VBC males in lesser numbers than the ideal formulation, but rarely captured *Mocis* males. This allows more rapid and accurate assessment of trap captures by less-experienced personnel. The formulation chosen detects the arrival of the migrant species in local crops well and tracks the major population fluctuations. It would not be the formulation of choice for extremely critical situations, for annihilative trapping, or for mating disruption. A more thorough analysis of the VBC pheromone might reveal a component that inhibits the response of *Mocis* males, in which case it could be added to the more ideal ratio for capturing VBC males.

The placement of pheromone traps for biomonitoring is not well defined for most cases. Most species are best captured with traps in or near hosts, but an absolute requirement for the presence of a host (presumably host odour) before a male moth will respond to the sex pheromone of its conspecific female seems rare. Host odours are an integral part of the mass attack behaviour of bark beetles, and the beetle-produced odours (aggregation pheromones) operate together with these odours and appear to regulate the attack response. Their role in mating behaviour is not clear.

The possible aggregation odours from hosts are not well researched for moths, and deserve more attention. In our work with the corn earworm, *Heliothis zea* (Boddie), we find that the capture of males rises and falls in adjacent blocks of sweet corn, planted two weeks apart, as the corn in each block tassels and then matures. It would seem that if we could simulate the host conditions that influence the capture of males during this period, it would provide a more sensitive and stable biomonitoring or mass-trapping tool.

In most cases moths are best captured in traps placed at the height of the vegetation. Forest and orchard pests, however, frequently stratify their response behaviour. An example is two moth pests of peach trees: the larvae of one bores into the trunk and large roots at the ground line, and the other bores into the upper trunk and limbs. The adult females emerge and almost instantaneously attract males. Males of each species are captured in greatest numbers at traps placed at the respective heights at which their conspecific females will emerge (McLaughlin 1979).

The pheromone release and response behaviours of most insects are very rhythmic. No serious attempt has been made to regulate the time of release of a pheromone formulation to coincide with the periodic activity of a target species, although this would conserve the bait and might serve to exclude non-target species. Such a design is probably not feasible since in most species the rhythms are regulated on a daily or seasonal basis by ambient conditions, particularly temperature and photoperiod. Certainly the design of formulations and traps, and the evaluation of trap captures, must involve consideration of the meteorological conditions that prevail during the usual period of activity of the target species.

The influence of visual cues on the response of males to pheromone traps is not documented for most species. It is common practice to use white or galvanized metal traps. We have suggested (McLaughlin *et al.* 1975) that traps for nocturnal species should have low reflectance in the two regions of the spectrum in which airglow radiation (the principal source of night sky energy on a moonless night) exhibits peak energy levels. These regions, 320–360 nm and 540–570 nm, correspond to the regions of maximum spectral sensitivity of nocturnal moths. Traps with low total reflectance in these spectral regions captured more males of two nocturnal looper species (Noctuidae) than did traps with high reflectance. The influences of other physical factors of trap design on the physical nature of the pheromone plume are reviewed elsewhere (Cardé 1979).

18.3. Annihilation

The development of optimal bait–annihilant combinations has been the goal of insect pest suppression programmes for many years prior to the development of sex pheromones. In recent years attractant–pesticide combinations utilizing the powerful and behaviourally enigmatic male lures for tephritid fruit flies have shown much promise. Also, the aggregation pheromones of certain scolytid and curculionid beetles have been used successfully to protect timber and crops. These programmes either exploit mass attack behaviour or the mass broadcast of a lethal attractant formulation.

The attempts to utilize moth sex pheromones have involved mass trapping with attendant complications of trap design, trap spacing, attractant formulation, and complex logistics. Recently, commerical research and development projects have begun to explore formulations of attractant–pesticide combinations that can be mass-broadcast. The synthetic pheromonal mixture is designed to elicit a complete response so that a male will attempt to copulate with the dispenser and receive a lethal or debilitating dose of a pesticide incorporated into the adhesive that holds the dispenser to the plant. Also, efficacy has been claimed for the synergistic action of pheromone and pesticide applications made independently and concurrently to the same field. This is possibly due to the reduced rate at which males locate females (see mating disruption), which would increase search time and also the exposure of both males and females to pesticide-coated surfaces. Sublethal poisoning might also reduce the ability to perceive, respond to, or release pheromone, since it is claimed that reduced amounts of the pesticide are required to achieve efficacy.

18.4. Disruption of sex pheromone communication

The history and possible mechanisms involved in the disruption of sex pheromone communication through permeation of the air with the pheromonal chemicals of the target species has been well documented (Bartell 1982; McLaughlin and Mitchell 1981). When the research team in Professor Shorey's laboratory met in 1971 to determine the terms to be applied to this new technique, they were required to adopt nomenclature that did not imply a mechanism. Unfortunately this remains the case today.

Through empirical research, primarily with moths, a good deal has been learned of the parameters that must be considered in order to manipulate the orientation of a male to a pheromone-releasing (calling) conspecific female. The primary factor is the chemistry of the sex pheromone of the target species; however, it is not yet clear if the entire pheromone blend is essential. In the case of *Heliothis zea* we have yet to find a consistent indication that the four-component blend of the identified pheromone is more effective in reducing mating than is the major constituent alone. The so-called minor constituents of this pheromone are poor disruptants. Their behavioural role appears to be as subtle modifiers of close-range approach, and it may be more difficult to disrupt this action effectively.

Air-permeation-mediated mating disruption is a dose-related phenomenon, and the transition from a high degree of effect to a low degree occurs very rapidly at some critical dose. Hagstrum and Davis (1982) concluded that air permeation reduces the distance from which a male can respond to a female but does not affect other parameters of male flight. This would account for the relative ineffectiveness of permeation with materials that effect only close-range (a few centimetres) behaviour. It also reveals why disruption is less effective with increasing moth density and why the effect would be quickly lost at dosages that would allow males performing their usual search patterns to enter routinely the active space of females.

There is a limit to the dosage of pheromone that can be achieved. It cannot exceed the saturation vapour pressure and thus, even though the rate of release at a given point may be continually increased, the overall molecular density of the pheromone must, at some dose, cease to increase. The flux of a pheromonal treatment has rarely been measured, and our knowledge of the actual distribution of pheromone in a treated area is largely speculative.

It seems particularly noteworthy that small-scale disruption test plots within larger untreated fields frequently achieve levels of mating reduction far greater than those achieved when the test is scaled up. Small tests usually involve hand-placed evaporators, each with a relatively high rate of release compared with a female and spaced at regular intervals. Large tests have usually involved the mass distribution of evaporators with low release rates (from about 0.5 to 10 times that of a female). The dosage of the large tests is usually based on the estimated rate of release per unit area of pheromone from a small test plot. This dosage is then increased to allow for the inefficiency of mass application, a parameter that is often poorly estimated. The small and large-scale tests may present the moth with quite different sensory stimuli. In the small plot there should be greater sensory contrast to males that fly from unaffected areas into the plot or the plume emanating from the plot, and to males within the plot who enter and leave plumes emanating from the high-rate evaporators. A male in the large treated field is in an area that should be more uniform with respect to the odour stimulus, and contrast might often be provided by a calling female. It appears that air permeation does not inhibit the duration or pattern of male flight in the permeated atmosphere of a room (Hagstrum and Davis 1982). However, the pattern does seem to be affected in field experiments, and males are frequently observed orienting to evaporators of pheromones (attractive single chemicals or blends) in plots where mating is reduced. Non-attractive or poorly attractive chemicals have also been shown to reduce mating, but rarely as well as the pheromonal constituents. It appears that mating disruption is achieved through a complex of effects which may affect species differentially depending upon their ecology and behaviour. Both neurophysiological and behavioural mechanisms should be assumed, and disruption tests should attempt to reveal the degree to which they contribute and to maximize both. The difficulty arises in our limited ability to understand and measure the neurophysiological events.

18.5. Integration

Mass annihilation and mating disruption are not to be relied upon as the sole means to control an insect pest in most instances. They are inherently inefficient because they are usually directed at males and they are most effective when used as a prophylactic in anticipation of a damaging population that may or may not develop. They can be effectively integrated with other population suppression techniques. A recent and exciting finding is that *Trichogromma* and other egg parasites respond to the sex pheromone (kairomone) of the host females (Lewis *et al*. 1982). The pheromone apparently causes the parasite to search in areas where it is present (presumably as a residue on the foliage), and other odour cues are utilized to locate the eggs. Synthetic pheromone can be placed in an area to induce the parasite to search at host densities below the normal search threshold. We have conducted preliminary tests that indicate that levels of pheromone that will disrupt mating communication in the host, *Heliothis zea*, will also increase parasitism by *Trichogramma*.

Mating disruption does not completely stop mating. It decreases the rate at which mating occurs. Hopefully it is reduced to the point where females die before mating can occur. Thus, any parameter that increases mortality or that further interferes with the communication system will complement this control approach. Even seemingly opposed methods for control that rely on the normal male pheromone response (e.g. mass trapping or sterile insect technique) can be integrated with communication disruption.

We should seek to use the pheromones of insects in new and innovative ways as safe biorational insect control agents. This can be done only through a complete understanding of the mechanisms by which they are perceived by the organism and the manner in which the sensory information is translated into a behavioural event.

19 Correlations between electrophysiological and behavioural responses elicited by pheromone

R. J. Bartell† and E. R. Rumbo

Two components of the female sex pheromone of the light-brown apple moth, *Epiphyas postvittana* (Walk.) (Lepidoptera: Tortricidae), have been identified as (*E*)-11-tetradecen-1-yl acetate (compound I) and (*E,E*)-9,11-tetradecadien-1-yl acetate (compound II) (Bellas *et al*. 1983). These two components together elicit precopulatory behaviour in males over a wide range of ratios, optimally converging at 17:1 (I:II) (Bellas and Bartell 1983), but neither compound alone elicits such responses.

Studies on behaviour elicited by the two-component blend of compounds I and II showed that responsiveness of males could be reduced by their previous experience of the same blend during a single diel cycle (Bartell and Lawrence 1973, 1976, 1977b). Moreover, the degree of reduction in responsiveness was dose related and could persist over five hours or more after initial stimulation. It was also found that after a pre-exposure to discrete pulses the subsequent reduction in responsiveness was greater if the intervals between pulses were at least 2 min, above which the effect decreased (Bartell and Lawrence 1977b).

From these results it was concluded that at least two processes were involved in reducing responsiveness: a short-term effect probably related to the adaptation/disadaptation time-course of the pheromone receptors, and a longer term habituation of the central nervous system, with dishabituation extending over some hours after the initial stimulus.

In order to discriminate between these possibilities it was thought appropriate to examine the electrophysiological properties of the receptor cells responding to individual pheromone components.

Rumbo (1981) developed a technique for accumulating and averaging the spike output generated by cells of the antennal sensilla trichodia, which yielded more information content than had hitherto been possible. A typical profile obtained from a cell responding to 10^{-8} g of compound I, in which the spike counts have been accumulated over 26 successive stimulations into class intervals of 50 ms, is shown under the first stimulus (0 to 1 s) in Fig. 19.1. The profile defines the temporal development of the spike train elicited by the pheromone stimulus. It consists of a sharp phasic peak approximately 200 ms wide, followed by a slowly decaying tonic phase which is terminated by a sharp fall in spike frequency at the end of stimulation.

†We are deeply grateful to Roger Bartell for participating in our meeting and for preparing his chapter, even though he knew he did not have long to live. He died on 22nd May 1985.

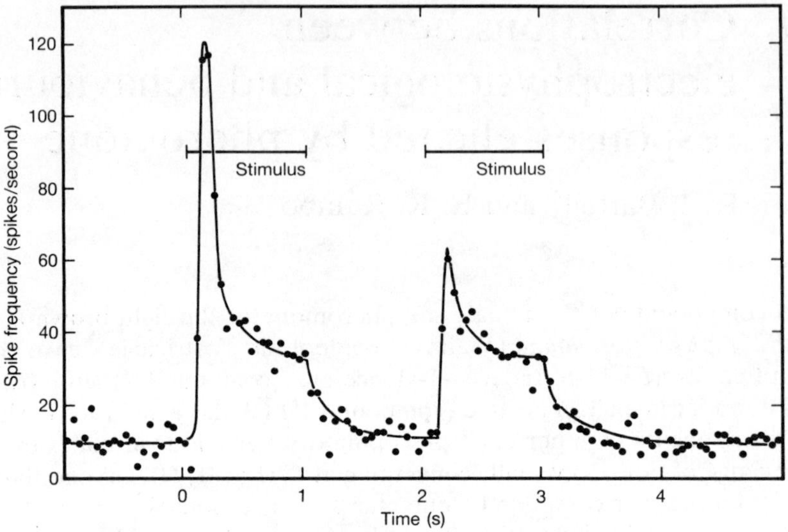

Fig. 19.1. The spike output profile from a single sensillum averaged over 26 recordings. The responses are to pairs of 1 s exposures to 10^{-8} g of compound I, separated by 1 s and repeated at 1 min intervals (from Rumbo 1983).

Since the insect antenna bears a large number of pheromone receptors operating in parallel, we assume that the resultant averaged profiles are representative of the overall pattern of spike transmission to the central nervous system (CNS) arising from the simultaneous stimulation of the whole receptor array.

In his preliminary studies Rumbo used exclusively the major sex pheromone component (I) with repetitive 1 s duration stimulations at 60 s intervals, and the preparations produced reproducible results. When the same programme of stimulation was applied using compound II, it was discovered that only two out of 60 preparations yielded reproducibility and were thus suitable for averaging purposes. The response profiles for these two preparations were indistinguishable from those resulting from compound I stimulation. In the remaining set the response to compound II diminished rapidly after two or three stimuli, and the data had been discarded in the belief that there was some technical impediment. It was decided, however, that this phenomenon was perhaps an inherent property of the compound-II-responding receptors.

Rumbo (1983) examined the time-courses of successive stimulation by compound II, progressively increasing the interstimulus phase from 1 to 5, 10 and 15 min, and found that these receptors were becoming cumulatively adapted. At the shortest interval the responses became progressively reduced, with the phasic response being affected first and then the tonic response with further stimulations. Increasing the interval between stimulations to 5 min allowed first the tonic and then the phasic responses to recover, while widening the intervals to 10 and 15 min effected little or no further recovery. In this more extensive study, Rumbo found no compound-II-responding cells which failed to adapt.

Further experiments on the I-responding cells revealed that the ratio of phasic to tonic responses fell from between 4:1 and 5:1 at primary stimulation to 2:1 for a second stimulus applied after 1 s (Fig. 19.1). A 2 s interval between stimuli allowed the ratio for second stimulation to recover to between 3:1 and 3.5:1, and with 4 s interval the ratio became 4:1, with the phasic peak about 90 per cent of that elicited by the first stimulus.

The foregoing results suggested the possibility that some information processing might take place at the peripheral (sensory) level, and, if so, the characteristics of sensory cell response to different components of the sex pheromone might be reflected in subsequent patterns of behaviour. The results also showed that the time-courses for disadaptation of the receptors for the two pheromone components were substantially different. Knowledge of this difference provided the basis for a series of experiments to investigate the carry-over of pheromone-mediated sexual responses into the post-stimulus phase, and the effect of subsequent stimulation by one or both pheromone components within an appropriate period following initial stimulation.

Since neither component alone elicited sexual behaviour, an *adequate* stimulus for naive males was defined as a 20:1 blend of compounds I and II at a dose rate of 10^{-9} g + 5 × 10^{-11} g at source (approximately 10^{-1} female equivalents) (Bellas *et al*. 1983), applied for 5 s under standardized bioassay conditions (Bartell and Lawrence 1976; Bartell and Shorey, 1969).

Under such stimulation the number of males exhibiting precopulatory behaviour typically rose to a peak at approximately 6 to 7 s after presentation of the stimulus source, with a median response rise time (= time taken for half the ultimate number of males to respond) of 3 to 4 s, and a latency to respond of about 2 s. The latter is in part due to the time taken for pheromone leaving the source to reach the chamber containing the males, and therefore a correction value is required for the former parameters. After the peak response the number of males which continued to respond declined to around zero level at 100 to 120 s, although individual short bursts of the designated behaviour continued for at least another 240 s (median decay time—the time taken for half the formerly responding males to cease responding = 30 s). This decline in activity was less rapid under conditions of continuous stimulation, where the baseline was not reached within 360 s of the start of stimulation (median decay time = 64 s) (Bartell unpublished).

When either compound I or II was removed after 5 s, leaving the alternate compound still present, the resultant decay patterns were markedly different. Removal of I resulted in a response profile closely similar to that when both compounds were removed at 5 s, whereas removal of II matched the profile under continuous stimulation by I and II (median decay time = 90 s). We speculate that the males *remember* the original presence of II once they have experienced an initial adequate stimulus, but there is apparently no similar memory for I. It is noteworthy that the decay time to baseline under constant stimulation, and that after the removal of II, closely matched the 300 s time-course for disadaptation of II-responding cells in Rumbo's study (1983).

The question then arose as to whether the memory for compound II could be integrated with current sensory input from I-responding cells under stimulation. It was already known that male responsiveness to adequate stimulus would recover substantially, provided that the interval between brief stimulations was long enough (Bartell and Lawrence 1977b). The response profiles for three such treatments are shown in Fig. 19.2a. Each set of data shows the initial response peak followed by the post-stimulus decay, then at 120 s, 180 s or 300 s after the first stimulus the groups of males were again exposed for 5 s to the same adequate stimulus. Second response levels were considerably reduced for the two shorter intervals, falling to 47 per cent and 54 per cent of first response (adjusted for carry-over baseline) at 120 s and 180 s respectively, but recovering to 83 per cent at 300 s. Fig. 19.2b shows the results of second 5 s stimulations by compound I alone at the same three intervals. In this instance, males which would not normally respond to compound I did so. The second response peaks were 58 per cent, 47 per cent and 19 per cent at 120 s, 180 s and 300 s, showing a reverse trend in the numbers of males responding to that for the repetitive adequate stimuli. A comparative set of treatments using compound II (Fig. 19.2c) shows no responses to any of the second stimulations.

In the first two sets of data, the time-courses for recovery of responsiveness to I + II (a) and decline of effect of I (b) are the same, and coincide with the time course for II-responding cell disadaptation.

Because of the close correspondence of the time-courses in the two lines of study, it is tempting to infer that there is a memory effect complementary to changes in the degree of adaptation of the II-responding cells at the peripheral level. Due to the shortness of the disadaptation time for the I-responding cells (<4 s) it is not possible, within the constraints of the bioassay technique, to discover during that phase whether a similar phenomenon might pertain when compound II stimulation is provided.

The fact that males respond to compound I alone after receiving an adequate stimulus shortly before, contrasts with an earlier finding (Bartell and Lawrence 1977b) where short-interval alternation of stimulations by both compounds elicited no precopulatory responses. In the latter case, a regime of such alternate stimuli resulted ultimately in depression of responsiveness to an adequate stimulus equivalent to that produced from a sequence of stimulations to both compounds together. It can be inferred from this, either that the pathways to the CNS are separate for each com-

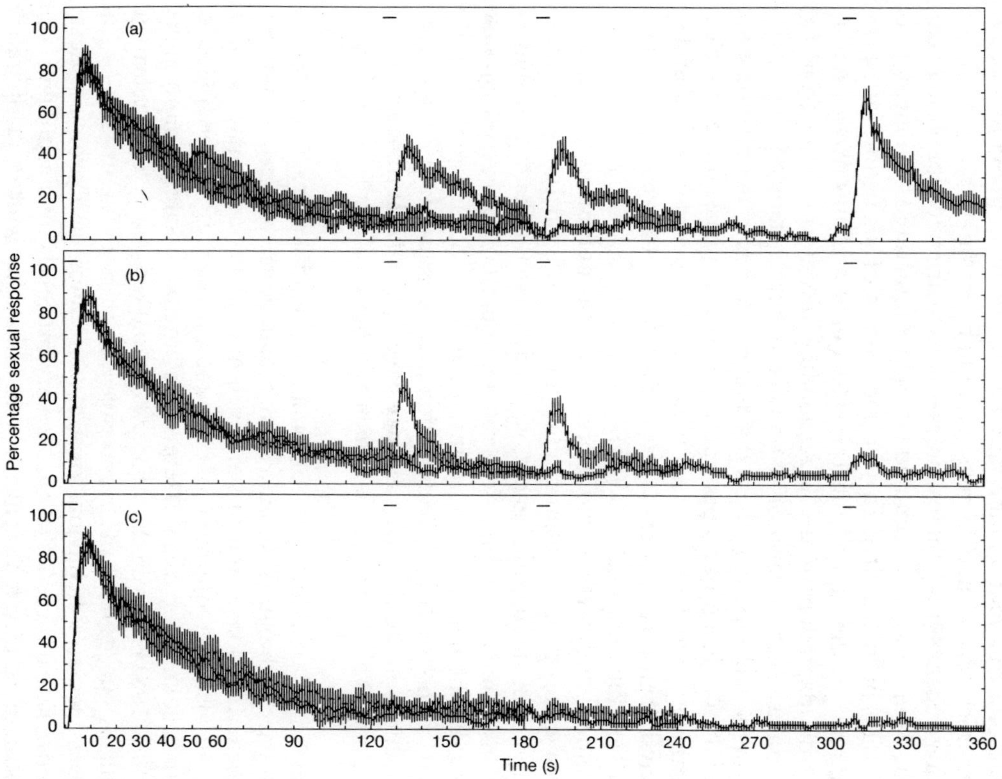

Fig. 19.2. Behavioural response profiles following stimulation of male moths by a blend of 10^{-9} g of compound I + 5×10^{-11} g of compound II. Each graph window contains three sets of data in which second stimulations are applied at 120 s, 180 s or 300 s after the initial stimulus (shown by the short horizontal bars above each graph), as follows: (a) both compounds applied simultaneously; (b) compound I applied alone; and (c) compound II applied alone (from Bartell 1984).

pound, or that the output from the receptor cells is coded in some way so as to allow the relevant information to be separated in the CNS. The sensory input could also be integrated in at least two ways: (a) when both compounds are presented simultaneously in the correct ratio to elicit an appropriate behavioural response, and (b) if only one compound is presented at a time in an alternating sequence of both compounds, the sensory input from one compound alone may be integrated with memory of the former alternate compound to give rise to habituation.

Our most recent set of studies indicates the further possibility that sensory input from only one compound (I) can be integrated with memory of the other (II) provided that: (a) the memory is of a previous adequate stimulus, and (b) the input from the subsequent stimulus arrives before complete disadaptation of the II-responding receptors arising from the initial stimulus. Such integration is only likely within the CNS.

There are some impediments to our understanding of the phenomenon described above. From our work and the work of others, there is a clear relationship between dose and responses, whether these be behavioural or electrophysiological (Baker *et al*. 1981; Bartell and Lawrence 1976; Den Otter 1977; Linn and Roelofs 1983).

In Rumbo's study (1983) the source dose was optimized for the total number of spikes generated during the stimulus phase at 10^{-8} g for both I and II, and each stimulation lasted 1 s. In the related behavioural studies, stimulations lasted 5 s, and the dose rate was different for each compound as dictated by the natural ratio for the species and within a range defined for optimal response. If the disadaptation times are also dose related, then the correlations in the time-courses may have been adventitious in the two lines of study. This is an avenue which will have to be researched more fully.

The inclusion of additional components (identified from the female gland) as part of the initial stimulus may modulate or change the character of the elicited response, and this also will have to be investigated.

There is also the question of the appropriateness of the piece of behaviour selected for assay. Here there might be safer ground, since Baker *et al*. (1981) showed that there was a marked degree of similarity in the changes in levels of responsiveness for different phases of pheromone-mediated behaviour over a wide range of pheromone doses and blends. This indicates that the sensory processes involved might well serve to mediate different sequences of behaviour while still operating within a constantly defined set of operating parameters.

It is clear that before we are able to define a mechanism, a wider range of studies must be embarked upon. These should involve examination of the characteristics of other behavioural parameters (such as orientations), the roles of other pheromone components, and neurobiological studies on the sensory pathways to, and integration centres within, the CNS.

20 Epilogue: behavioural mechanisms

R. T. Cardé

20.1. Introduction

Among animals the quintessential paragon of chemical communication is the silkworm moth, the male of which often remains motionless from emergence until death, unless enveloped by the female's pheromone. Some 200 molecules of her pheromone have been estimated as sufficient to evoke his response: wing fluttering and an upwind sojourn. So legendary is the male's perceptual power that some accounts now repute that the male silkworm can detect a calling female more than 11 km downwind and navigate a flight to her side (Halliday 1980), despite the fact that as a domesticated species *Bombyx mori* has long been flightless (Darwin 1868).

The use of airborne chemical signals embraces more than sexual recruitment. Among many insect groups such cues are used to recognize hosts, to signal alarm, indeed to mediate a wide range of behaviours. In social insects the repertory of reactions elicited is a catalogue of the varied behaviours that define social interactions. Pheromones thus mediate diverse behaviours in a variety of habitats, and in many comparisons these traits presumably evolved independently, as in sexual recruitment in moths and beetles. Given this heterogeneity, does a comparative approach offer insight into the mechanisms and evolutionary causes of a behaviour, or does it camouflage the uniqueness that may characterize each system? Dethier (Chapter 1) has cautioned that current studies of insect responses to airborne and other chemical signals often seem to be undertaken without referencc to studies of chemically-mediated behaviour in other organisms. Surely the comparative approach will yield principles and insights more readily than a parochial perspective.

In the previous presentations the focus has been on the mechanisms used by insects to discover the source of an airborne chemical cue, with an emphasis on the manoeuvres of flying insects. Not only has the current view of how insects perform these behaviours been summarized, but so have the holes in our knowledge. J. S. Kennedy (Chapter 2) has set forth an elegant synopsis of the manoeuvres and the presumed control mechanisms that seem verified, as well as avenues of investigation that are apt to resolve the outstanding questions. It is now appropriate to develop an evolutionary and ecological approach to understanding these manoeuvres.

20.2. Evolution of attraction to airborne semiochemicals

Besides unravelling the proximate (or physiological) mechanisms that enable an organism to perform orientation manoeuvres, research may be directed to asking why a particular behavioural pattern (through a genetic template) persists, and why another solution may be less appropriate. There is, for example, a considerable body of information on the specificity of sex pheromones: closely allied species can partition the chemical communication channel, either by employing differing ratios of the same chemical components, or by slight alterations of the pheromone structure (Cardé and Baker 1984; Borden *et al*. (Chapter 10); Löfqvist (Chapter 13); Van Der Pers and Löfstedt (Chapter 27). Yet how such differences evolved cannot be observed, even though it is obvious that these differences in the chemical communication channel now serve as formidable reproductive isolating mechanisms.

One ultimate factor that may be considered is the premium placed upon rapid location of the associated resource. In moths, typically the male is attracted to the female-emitted pheromone (but see also Schneider (Chapter 21), an arrangement that presumably evolved because of the disparity in parental investment (Trivers 1972). Generally a female's success is dependent upon the number of eggs produced, whereas a male's success can be gauged by the number of females inseminated. Thus rapid location of a pheromone-emitting female should be a favoured tactic among males. Unfortunately there are few studies bearing on this contention, perhaps because most moths are nocturnal, which severely restricts the possibility of field observations. In the day-mating gypsy moth, *Lymantria dispar*, a female typically mates with the first male contacting her, and the ratio of males to unmated females is very high (Cardé and Hagaman 1984). In such systems, male success should be enhanced by orientation mechanisms that promote rapidity in mate finding.

Among most scolytid bark beetles, mate location is associated with colonization of a host tree. First-arriving beetles, the so-called pioneers, are not successful in colonizing a host tree unless they attract a sufficient number of conspecifics to overwhelm the tree's defensive tactic of copious extrusion of sap into the beetles' galleries. In some scolytid species the particular tree chosen for attack may be in a vulnerable physiological state (Payne (Chapter 11)). Once an optimum beetle density is exceeded, the chemical signals are altered, terminating the arrival of new beetles. The decline in attractivity of a tree, thought to be due to the switch in the beetles' production of pheromone and emission of 'antipheromones', must be reassessed in light of the auto-oxidation of the host tree volatile α-pinene to beetle attractants and anti-attractants (Borden *et al*. (Chapter 10)). Nonetheless, given the relatively wide time-span during which a beetle (of either sex) can locate the combined resource of a host and a mate, the premium placed upon speedy location of the pheromone source in most scolytids may be reduced in comparison with moths. Rather, beetles should be expected to invest effort in identifying a vulnerable host tree, or one with an optimum level of infestation.

The influence of lipid reserves upon the tendency of beetles to disperse before responding to an olfactory cue offers an example of a physiological basis for variation in response. The possibility that only pioneers release sufficient pheromone to initiate a successful attack, and only colonizers invade trees beset by sufficient beetles to presage the tree's demise, may be contrasted with a strategy of beetles attempting to bore without assessing a host's vulnerability (Borden *et al*. (Chapter 10)). Possible genetic differences among individuals in foraging strategy and pheromone production, and the existence of an evolutionarily stable strategy involving pioneers and colonizers, are at present speculative.

The location of a host, be it a plant, dung, or an insect to parasitize, can generally (unless the parasite utilizes the pheromone of its host) be characterized as finding a resource that is decidedly less ephemeral than a pheromone-emitting potential mate. In most systems there are hosts whose suitability extends over a moderate time-span lasting perhaps minutes in the case of dung, to perhaps the life-span of an adult insect in the case of plant hosts. A premium is thus to be placed upon sensing the suitability of a particular resource—is, for example, it the 'correct' plant species or is the host already parasitized?—rather than upon its speedy location.

The remarkable ability of male gypsy moths to locate within several minutes a pheromone source perhaps 100 or so metres upwind under relatively non-turbulent wind conditions (Elkinton *et al*. 1985) can be contrasted with the cabbage fly, *Delia brassica*, which approaches its plant host by relatively short flights ('hops'), with evident detection of olfactory cues at distances in the range of only several metres (Hawkes and Coaker 1979). There are additional factors that are thought to mould chemically mediated host and mate finding (Miller and Strickler 1984; Cardé and Baker 1984). Phylogenetic constraints, as one example, would assume an efficient, long-distance chemical communication system as a generic solution to mate finding in most moths, whereas, in general, upwind flight to olfactory cues is less prevalent among Diptera. Many Coleoptera utilize pheromones in mate location and aggregation on the host, but there is little evidence on the manoeuvres that these beetles use to locate the olfactory stimulus.

Current concepts of the orientation manoeuvres of flying and walking insects are supported mostly by studies of selected species of moths and flies. Many other major insect groups thought to fly to attractants have not yet been tamed in the laboratory wind-tunnel. Given some diversity in the importance attached to rapid location of the chemical cue and a differing phylogenetic history, a number of orientation strategies remain for us to decipher in these groups.

20.3. Habitat and the distribution of the chemical signal

In considering orientation in terms of the advantages of rapid movement to the source and the limits imposed by the organism's lineage, displacement manoeuvres have remained hazily defined. What is known of response to the spatial and temporal structure of the chemical signal, as well as the other inputs of visual and wind cues, has been gleaned from few studies of even fewer species. Gaussian plume models, widely used by pheromone biologists, depict concentration of odorants over minimum time-intervals of several minutes, whereas an insect's response occurs over a time-span of several seconds to perhaps a fraction of a second (Elkinton and Cardé 1984). This discrepancy was pointed out by Wright (1958), although he did not explicitly state that concentrations generated from the Sutton model were applicable to a minimum time-span of three minutes. Not unexpectedly, field experiments simultaneously monitoring the wing-fanning reaction of gypsy moths and the wind-field, and emitting pheromone at known rates, have demonstrated that the concentrations generated by the Sutton and more recent Gaussian dispersion models are several orders of magnitude lower than those known to evoke wing fanning (Elkinton *et al.* 1984).

Description of the 'instantaneous' flux of odorant impinging upon walking and flying insects is crucial to verifying the nature of the orientation system, because many taxes and kineses are defined in part by the pattern of odour detection in space and time (Kennedy (Chapter 2)). Further, wind velocity must be above a minimum level for the insect to be capable of detecting the wind's direction and thus to navigate a course to the emitter by walking or flying upwind. The structure of habitats differs widely, causing potentially very different patterns of odour dispersion (Perry and Wall (Chapter 9)). Also, the size of an odour source and its position within a habitat determine plume structure. The primordial odour plumes generated from a plant and from an insect's scent gland, for example, will be quite different in size and structure. Approaches to estimating the instantaneous structure of a plume from a point source have been provided by Murlis ((Chapter 3); see also Aylor *et al.* 1976; Miksad and Kittredge 1979; Murlis and Jones 1981), and to describing the meander of the plume by David *et al.* (1982; 1983) and Elkinton *et al.* (1985). Several facets of odour dispersion, as they bear upon flying navigation to a pheromone source, deserve amplification.

The 'fine-scale' distribution of pheromone offers potential information to a responding insect. Turbulent diffusion of pheromone from a point source generates filaments of pheromone-laden air interspersed among parcels of air with below-threshold concentrations of pheromone. The peak-to-trough ratio of pheromone diminishes with increased distance from the source, or, put more simply, the pheromone plume becomes more diffuse and homogeneous. Almost paradoxically, the frequency of these bursts is not greatly dependent upon the distance to the source, at least in the range of tens of metres (Murlis (Chapter 3)). Further, the rate at which a flying insect encounters such packets of pheromone is dependent upon the insect's air speed. In the case of the gypsy moth this is often >50 cm/s (e.g. Cardé and Hagaman 1979), particularly in dilute plumes, as would occur many metres from a female. Such flight speeds render the perception of the fine-scale structure of a pheromone plume more difficult than for a stationary sampler.

Thus there is but a scanty idea of how discontinuous the plume is under natural conditions, and even less of a notion of what a responding insect actually perceives. Relatively small changes in pheromone concentration may be sufficient to alter orientation manoeuvres. The interneurones in the male silkworm moth's ventral nerve cord that exercise part of the neural control of upwind walking alter their output ('flip-flop') given changes in pheromone concentration of only 10 per cent (Olberg 1983b). Further, there is striking evidence in *Adoxophyes orana* (Kennedy *et al*. 1981) and *Grapholitha molesta* (Willis and Baker 1984) that stimulation with a homogeneous cloud of pheromone causes behavioural habituation/adaptation that terminates the characteristic upwind flight pattern. The importance of a flickering signal in the maintenance of orientation manoeuvres in these two tortricid moths implies that definition of these temporal relationships will be required for a full understanding of insect movement toward a chemical signal. Further, the threshold for plume following in these two moths must be reformulated to embrace the fluctuating nature of the signal required to maintain continued movement toward the source.

In the silkworm moth assayed on a servosphere, pheromone signals of 150 ms duration presented 0.3 or 0.6 s apart elevate the speed of upwind walking twofold and result in a more direct upwind course, as compared with orientation in a continuous stream of pheromone (Kramer (Chapter 6)). How closely these intermittent patterns mimic a natural fluctuating signal created by turbulent diffusion remains to be established, but the evident improvements in orientation seen in a pulsed signal indicate that definition of the instantaneous structure of the pheromone plume is mandatory to understanding the orientation process.

Temporal modulation of the signal also can be achieved by rhythmic pulsation of the pheromone gland, as has been documented in several families of moths, most notably the arctiids (Cardé and Roelofs 1973; Conner *et al*. 1980). Several functions for pulsing have been proposed: these may be broadly categorized as those invoking a communicative role for a temporally modulated signal (e.g. distance to the source, species specificity, lowering of the threshold, preventing adaptation/habituation) and those suggesting that pulsing influences the rate of pheromone emission (conserving pheromone or elevating release) (Cardé *et al*. 1984; Conner *et al*. 1980). In the arctiid *Holomelina lamae*, rhythmic pulsing of the gland elevates the rate of release of the major pheromone component, 2-methylheptadecane. Freely pulsing females release 24 ng of this hydrocarbon in the first 10 min of calling, whereas females whose glands are forcibly extruded emit but 6 ng (Schal and Cardé 1985). It would appear that the principal function of gland pulsing in *Holomelina* is to elevate the rate of emission, but the potential role of pheromone puffs in providing information requires further investigation.

Also in need of resolution is the degree to which an odour plume's centreline is aligned with wind direction. Previously the pheromone plume had been envisioned as a meandering cloud whose long axis was coincident with the wind direction. David *et al*. (1982, 1983) demonstrated that 'parcels' of pheromone-containing air that emanate from a point source in an open field travel in straight paths over 20 m. However, in the forest habitat of the gypsy moth, the trajectory of 'puffs' of pheromone is non-linear, with a mean total change of over 100° within 20 m (Elkinton *et al*. 1985). The generally sinuous path of a pheromone plume in a forest implies that males in such habitats flying directly upwind will often be aimed away from the source, even though the mean direction of the wind tends toward the source.

One explanation for successful gypsy moth flight to pheromone over many metres is that progress occurs when the long axis of the pheromone plume and the wind are closely parallel. For a moth encountering a plume aligned across the instantaneous wind direction, flight upwind would carry the moth out of the plume. However, the wind direction is often consistent for intervals of several or more seconds (David *et al*. 1983; Elkinton *et al*. 1985), and it is during these episodes when plume segments are aligned with the instantaneous wind direction that substantial progress toward the source can occur. In this view it is the consistency of wind direction and not the linearity of puff trajectory that determines the success of moth navigation (Elkinton *et al*. 1985).

As wind speed increases, the plume becomes straighter and the proportion of gypsy moths successfully locating an upwind source is elevated, at least at moderate wind velocities. Thus a major influence upon the effective distance of communication is the decrease in the variability in wind direction that accompanies increases in wind speed, thereby generating longer tracks of puffs travelling in the same direction. This limitation on odour communication suggests that a gypsy moth could not substantially increase its distance of attraction by elevating pheromone emission or by lowering threshold, because, although both strategies would increase the distance of detection, few moths could locate the emitter over such expanses (Elkinton *et al*. 1985). At higher wind velocities, communication may be limited by consequent declines in pheromone concentrations (Bossert and Wilson 1963), although pheromone flux is not affected in the same way (Elkinton *et al*. 1984).

The structure of the pheromone plume within several metres of its source could provide crucial cues to the process of localizing the pheromone's point of origin. Female gypsy moths typically call while perched on tree trunks, and the mechanical turbulence caused by air flowing around the trunk generates eddies several centimetres across that tear apart the comparatively narrow plume issuing from the female's pheromone gland. The degree to which such turbulent diffusion broadens the plume depends in part upon where the moth is situated relative to the wind direction. Although an occasional male will follow a plume directly to a female, most males will engage in vertical flight near the tree bole, often contacting the trunk, before landing at a site often as much as a metre or more away from the female.

A landed gypsy moth can walk directly upwind to a calling female, but it is infrequent that a male is so ideally situated downwind of a female after landing on a tree. Instead a male typically must engage in a 'pre-programmed' zigzag walk until he either enters a pheromone plume or touches a female. Recognition of the female and attempts to couple appear mediated by tactile cues perceived tarsally. Visual detection of the female evidently plays no part in the male's selection of a specific landing site on the tree, walking toward the female, or recognition of the female at close range (Charlton and Cardé 1985a, b).

These observations establish that the alignment of the pheromone plume with both the instantaneous and average wind directions is crucial to mate finding in the gypsy moth. Simply flying upwind once an above-threshold concentration of odorant envelopes a moth will not be a routinely effective strategy for locating the odour source.

In yet another habitat, namely a wheat field, Perry and Wall (Chapter 9) have suggested that the 'discrete' plume (derived from a point source) is maintained for only a short distance; further downwind a continuous field of pheromone stimulates pea moth, *Cydia nigricana*, males to fly, on average, up the mean wind direction. Pheromone adsorption and re-release from the plants seem to play an important role in this system by providing a diffuse emission of pheromone from a wide source (Wall *et al*. 1981; Wall and Perry 1983). Besides downwind displacement, in a tropical rain forest Schal (1982) has documented that during the night convective currents carry the pheromone plume upward.

A great deal more information on the instantaneous structure of plumes in a variety of habitats and the nature of the visual cues (Prokopy (Chapter 8)) available for the optomotor response (David (Chapter 5)) will be necessary before we can properly duplicate field conditions in the laboratory and thereby systematically dissect the importance of various features of plume structure. A related problem, not considered here, is the ability of moths to fly along a pheromone plume in still air (Kuenen and Baker 1983).

20.4. Orientation mechanisms

Kennedy (Chapter 2) has provided, in the framework of the Fraenkel and Gunn (1940) classification scheme, a historical retrospective on the evolution of the terminology describing orientation manoeuvres and, more important for future sutdies of orientation to chemical cues, a reformulation of the types of orientation manoeuvres that result in displacement toward an odour source. Bell and Tobin (1982) have pointed out limitations in the Fraenkel–Gunn classification system: (1) terms are not comparable because some are based upon the type of sensory input, and others the form of the locomotory path; (2) simultaneous use of more than one orientation manoeuvre is neglected; (3) internally stored information is not considered, so that displacement relative to the last position can be termed random; (4) rarely is sufficient information available to verify the use of these mechanisms; and (5) input from a single sensory modality is usually emphasized.

Bell and Tobin (1982) suggest that an enhanced understanding of the mechanisms involved would be promoted by specifying: (1) the information available; (2) the type of information processing; (3) the locomotory output; and (4) the nature of the guidance mechanism. To date, however, there are relatively few cases that would allow such sophisticated analysis.

The view that taxes and kineses and their derivative terms are no longer useful is belied by their continued use (and reformulation) in the literature. Some of these concepts, such as the orthokinetic reaction, in which the speed of movement is modulated by the intensity of the stimulus, remain essentially unaltered from the Fraenkel and Gunn (1940) formulation. For example, several moth species lower their airspeed as pheromone concentration increases, seemingly a simple orthokinetic response. In a given concentration of pheromone (using a time-average calculation), moths adjust their airspeed to maintain a constant flow of the visual field and hence groundspeed. But as Sanders (Chapter 12) points out, in different pheromone concentrations the spruce budworm, *Choristoneura fumiferana*, alters its airspeed, seemingly to maintain a constant flux of pheromone. The concept of orthokineses camouflages the complexity of these interactions.

Continuing alterations in definitions of taxes and kineses reflect advances in our understanding of what inputs the organism encounters and presumably perceives and, as well, the nature of the orientation manoeuvre. Thus such changes in terminology are to be welcomed, but repeated revisions also generate a confusing literature. The zigzag path of a moth advancing upwind along a pheromone plume, as an example, was termed 'reversing anemomenotaxis' by Kennedy (1977) and was thought to be controlled by course reversals when the flyer reached the margin of the plume. Subsequent innovative experiments by Kennedy *et al*. (1981) using a wind-tunnel containing a homogeneous cloud of pheromone showed that zigzagging persisted for several seconds, even if the moth did not contact clean air. Now this manoeuvre is viewed as being regulated by an internal pattern generator (a 'preprogrammed' reaction).

The traditional distinctions between klinotaxis and klinokinesis have become blurred. Havukkala and Kennedy (1984) have argued that the recent use of longitudinal klinotaxis to denote 'a turn though a definite angle relative to its immediately previous direction of steering' as the organism detects odour differences along its path (Kennedy 1978) is incorrect. Rather, longitudinal klinotaxis is not a taxis at all, because it is not steered by a stimulus gradient. The coining of the term schemakinesis (Kennedy (Chapter 2)) to replace longitudinal klinotaxis circumvents this contradiction. Yet the conclusion that the initiating stimulus does not provide 'feedback control of the subsequent path', and thus it cannot be a directed reaction or taxis, is evidently confuted by the alteration in the zigzag path found when *Trogoderma* beetles reach the edge of a plume (Bell (Chapter 15) and discussed below).

The manoeuvres that insects use to 'home in' on an olfactory cue thus remain in some question, though not for want of either theoretical models or empirical tests. Insects in continuous flight cannot detect the wind direction by mechanoreceptors; instead they assess progress by appraising the movement of their visual field. The nature of the visual field impinging upon an insect flying upwind, downwind, or zigzagging, and the probable optomotor feedback mechanisms allowing these flight trajectories are expounded by David (Chapter 5). Evidence with the gypsy moth seems to support an internal turn-generating mechanism for zigzagging, as hypothesized for other moth species.

Priess and Kramer (Chapter 7), using a tether and a projection system to simulate the visual sensation of movement, have in gypsy moths documented flight courses in pheromone aimed 'straight upwind', without a zigzag path or menotactic orientation. Cross-wind excursions within a range of 45° were attributed to random 'internal noise functions'. This view of pheromone-stimulated flight contrasts with the mainly horizontal zigzag path documented in the same species in free flight in the wind-tunnel (e.g. Cardé and Hagaman 1979).

The walking manoeuvres of *Trogoderma* beetles in wind to pheromone in plumes and clouds (simulated on a servosphere) is a zigzag course upwind. In a 10 cm wide plume, beetles zigzag along the boundary between pheromone-laden and clean air. In a uniform cloud, beetles also zigzag upwind, but the legs of the zigzag are longer and the turns less frequent than when beetles walk along a plume's edge. The temporal change from pheromone to clean air tends to switch the tack of the beetle, although the next turn occurs within the plume (Bell (Chapter 15)). This system thus is interpreted to involve both an endogenous turn generator and an increase in the probability of turning when a sharp decrease in pheromone concentration is encountered.

The contention that counterturning is an error-correcting device, supposed to adjust the moth's aim toward the upwind direction (Priess and Kramer (Chapter 7)), contrasts with counterturning both as a reaction in part to loss of the scent and as an output of an internal turn generator. Some of these differences could be due to the insect species studied, and whether the manoeuvres are executed by walking, tethered flying, or freely flying organisms. Resolution of these controversies awaits further experimentation. Three explanations for zigzagging need to be assessed: (1) an internally generated programme (Baker (Chapter 4); Kennedy (Chapter 2); (2) a schemakinetic manoeuvre coupled with an internal turn generator (Bell (Chapter 15)); and (3) a course correction manoeuvre without either an internal turn generator or a menotactic component (Priess and Kramer (Chapter 7)).

Further, the generic importance of zigzagging in maintaining an intermittent signal and thus preventing behavioural habituation (Baker (Chapter 4)), and, in flying insects the role of zigzagging in increasing the visual information available to assess wind direction (Cardé 1984), both remain to be documented.

The olfactory stimulus has been treated, so far, as a unitary chemical signal, although most pheromones and host volatiles inducing attraction consist of more than a single chemical. The possibility that either a portion of a pheromone blend (or an imprecise ratio of components) has as low a threshold for upwind movement as the complete pheromone, with subsequent orientation behaviours governed by the addition of components (or the natural ratio), has been developed by Howse (Chapter 17) and critiqued by Baker (Chapter 4). One theoretical advantage of the major pheromone constituents or a broad range of component ratios evoking the initial upwind movement is that it could extend the distance of communication by reducing the information required and thereby lowering the threshold for upwind movement (Cardé and Charlton 1985).

The difficulty involved in documenting that the initial upwind movement need not require detection of the entire pheromone blend at its natural ratio is the unambiguous demonstration of which components and ratios elicit attraction at the lowest threshold. To verify the existence of 'nested' active spaces it will be necessary to define the behaviours precisely and avoid consideration of spurious behaviours not involved in successful orientation (Baker (Chapter 4); Cardé and Charlton 1985). An added complexity is the possibility that the threshold of quiescent insects, as would be assayed in the wind-tunnel, may be higher than that of flying insects (Cardé and Charlton 1985).

20.5. Terminology and jargon

Language that can be deciphered only by devotees of a particular speciality can be classified as jargon. Often it is difficult to translate jargon into concrete meaning; indeed it may denote muddiness of thought. The continued, uncritical use of terms such as chemotaxis in the literature substantiates this view. A related sin in animal behaviour is teleology: as Kennedy (Chapter 2) categorizes it, 'the explanation of a behaviour by its end-result'. As an example Kennedy cites 'appetitive' as it has been employed by Cardé (1981) to describe the flight activity of a male moth before it has encountered a pheromone plume. In this instance the term was used to describe behaviour in relation to what may follow, considering in part both the locomotory aspects and the presumed lowered threshold to cues (largely olfactory) evoking plume following. Appetitive as a term dates to Craig (1918) who used it both to specify a sequence of behavioural events and to signify what was thought to be the goal-seeking aspects. It thus has unabashedly teleological origins. Kennedy proposes to substitute the term 'foraging'. But foraging is linked with optimization theory, and it is by no means clear that such behavioural patterns are optimal solutions to resource location (Elkinton and Cardé 1983). Also, foraging too is defined in part by the behaviours that follow and the goal sought, and thus it has teleogical roots, although it may be a less controversial term.

Does the use of such terminology relegate us to the anthropomorphic level of Romaes (1882) who, viewing 'the mental states of an insect', concluded 'the nearest conception that we can form of their true nature is that which we can form by assimilating them to the pattern of the only mental states with which we are actually aquainted'? As Kennedy intimates, it is the preciseness of the terms as they are defined in their present usage and the accuracy of the descriptions of the behaviour's supposed sensory inputs and control mechanisms that will be useful in analysis. That some terms are in part defined by their place in a sequence of behaviours does not constitute teleology. As Kennedy (1972) puts it, 'the trouble comes when the function of a piece of behaviour is confused with its mechanism'.

20.6. Outstanding questions in insect movement to chemical signals

The preceding presentations have delineated the current understanding of the mechanisms that selected insect species appear to use in flying and walking toward a chemical cue. Understanding the processes and cues involved has burgeoned since Kennedy (1972) admonished against labelling chemicals that induce insects to accumulate at a source as 'attractants' when little was known of what directional cues, if any, the chemical offered. It is now timely to ask what are the unresolved questions and what approaches may prove useful to their denouement?

The orientation mechanisms used for location of a chemical signal has been inferred from studies of a few species in even fewer phylogenetic groups. The challenge of defining the potential diversity of solutions to finding the source of an odour remains an exciting prospect, although it is clear that the number of logical mechanisms is limited. The redundancies and divergences between manoeuvres of insects that both fly and walk to an odour source may prove especially illuminating in the understanding of optimal designs of navigational systems.

Although optomotor-modulated anemotaxis coupled with a programme of self-steered turns is the favoured explanation for upwind flight to pheromone in moths, the extent to which the pheromone plume's boundary with 'clean' air provides information that flying or walking insects use in navigating a path along the plume's axis remains in doubt. The classification of the manoeuvres presumed to be used in spatial orientation to a chemical signal continues to be refined, largely because of an increased appreciation of the role of internally-stored (idiothetic) information. As Kennedy (Chapter 2) notes, at least some of the current conflict over the use of terms, and the dissatisfaction with the Fraenkel and Gunn system of taxes and kineses, is dictated by the level of analysis sought. Those in quest of a cybernetic explanation will be less contented than the behavioural ecologists.

The temporal and spatial distribution of the instantaneous odour plume remains largely uncharted, yet the nature of the signal as it impinges upon the insect appears to be a crucial factor in maintenance of the insect's continued movement toward the source. Two tortricid moth species must experience an intermittent pheromone signal to continue their flight upwind, and the silkworm moth walks more rapidly and directly upwind in a pulsed pheromone signal. The importance of signal flickering in the orientation manoeuvres of other insects remains to be established. And the frequency with which the instantaneous wind direction is aligned with the plume's long axis in various habitats will bear on the effective distance of communication. The importance of understanding pheromone dispersion in the natural habitat of the insect cannot be overemphasized. The relevance of turbulent diffusion of odours to orientation mechanisms demands a collaboration between micrometeorology and ethology.

The adaptive significance of an insect moving toward a chemical signal is thus usually well understood, although in a few species attraction to a chemical remains enigmatic (Prokopy and Roitberg 1983). However, the intensity of selection for the ability to locate a chemical source rapidly and the types of orientation manoeuvres that might be appropriate, given an organism's phylogeny and the nature of odour dispersion in its habitat, are very often unknown. The actual spatial manoeuvres used by several species in movement to a chemical source are now under some scrutiny, as is the terminology appropriate to describe these manoeuvres. But the very terms used to categorize orientation manoeuvres can dictate the questions asked and the experiments performed. It would thus seem more illuminating to attempt to see what occurs during orientation under controlled conditions that reflect the natural distributions of chemical, wind, and visual information, followed by experiments that systematically dissect out the role of each input. This level of cybernetic analysis offers a substantial challenge, and as well the prospect that the principles underlying these complex manoeuvres will emerge.

Part II
Physiological mechanisms

21 Physiology of insect olfaction—reflections on the last 30 years

D. Schneider

More than 30 years ago and more than five years before the first identification of a sex-attracting pheromone (bombykol: Butenandt, *et al.* 1959), I was encouraged by a chemist to develop an electrophysiological bioassay, or—what it turned out to be eventually—to study the function of insect odour receptors. The first outcome of all this was the detection of the electroantennogram, the EAG (Schneider 1957). Before this time, comparative physiologists had verified earlier anecdotal reports on odour reactions of insects (see examples in the textbooks of W. von Buddenbrock 1952; V. B. Wigglesworth 1972). Until 100 years ago the question of the anatomical locus of the odour receptors of insects had stirred controversial opinions, and it was first of all Gustav Hauser (1880) with his behaviour studies of antennectomized insects (which failed to respond to odours) which cleared the question. Then followed Karl von Frisch (1921) who reconfirmed in the bee that the insect's antenna is its 'nose' and furthermore proposed that the pore plates are the odour-sensitive sensilla.

But the conclusive proof for this prediction of the pore plate function needed neurophysiological recordings (Lacher 1964; Lacher and Schneider 1963) like those from antennal hair sensilla in Lepidoptera and beetles (Boeckh 1962; Schneider and Boeckh 1962; Schneider *et al.* 1964).

The history of our understanding of insect odour receptor function and structure—as seen from my laboratory—has recently been outlined in some detail (Schneider 1984). Several factors supported this development, most importantly modern electrophysiological and histological methods. Furthermore, we biologists were, after the first pheromone identification (1959), now also able to take crucial advantage of the synthetic pheromones as prepared by the chemists. Thus, the identification of the odour receptor cells and the study of their qualitative and quantitative properties were first made possible. Compared with other branches of sensory physiology, olfactory research is technically difficult because of the complicated olfactometry. Nevertheless, following the bombykol identification, the field developed rapidly, and a wealth of pheromones became known, many of which were also used in receptor studies.

Detailed investigations of the odour receptors in species other than moths remain few. Species studied include beetles (see for example Behrend 1971; Boeckh 1962; Dickens and Payne 1977; Light and Birch

1979; Mustaparta 1975, 1977), Hymenoptera (Dumpert 1972; Kaissling and Renner 1968; Lacher 1964; Vareschi 1971), cockroaches (Sass 1976; Selzer 1981), locusts (Boeckh 1967a, b; Kafka 1970), flies (Boeckh 1967a; Kaib 1974), and mosquitoes (Davis 1976, 1977; Lacher 1967, 1971).

Much of our functional understanding is based on experiments with receptor cells of male moths, which respond specifically to a particular component of the sex attractant of their females (Kaissling 1974; Kaissling and Priesner 1970; Priesner 1980a; see also the reviews by Kaissling 1971, and Priesner 1973). Central processing of odour messages is a relatively unexplored area of research (see Boeckh 1974; Boeckh and Boeckh 1979; Boeckh *et al*. 1984; Matsumoto and Hildebrand 1981; Olberg 1983b). Another little-studied field is the reception of odours of food, hosts, competitors, rivals, predators, and of the environment at large (see for example Boeckh and Ernst 1983; Hansen 1983; Sass 1976)

As in other fields of animal behaviour, an understanding of the reception process is only the beginning of our insight into the whole processing chain from the detection of signal to the consummatory action (Craig 1918). In the male moth, which flies after his mate, this final act is the mating with the calling female. Neurophysiological attempts to unravel this chain in *Bombyx* are done by following the message from the antennal receptors via the relay stations in the olfactory brain to the motor ganglia (see Olberg 1983a; M. Koontz unpublished data; Light (Chapter 33)), but I confess that progress is slow. On the other hand, cyberneticists study details of the odour-induced orientation process, which is a function of several modalities. Here, the odour is mainly the search-command, while vision (in effect: detection of wind direction) gives the vectorial information (Cardé and Hagaman 1979; David 1979b; Kennedy and Marsh 1974; Kennedy *et al*. 1981; Kramer 1975; Preiss and Kramer 1983, 1984).

Even in the Lepidoptera, the range or scope of olfactory functions goes beyond responding only to female moth attractants. As long as one hundred years ago, naturalists had already reported and speculated on both male and female Lepidoptera behaviour (Fabre 1878; Hauser 1880; Müller 1877). With respect to the male odours in the Lepidoptera, interests were revived by a monograph of R. Barth (1960) and a short paper by the Oxford Professor G. C. Varley (1962). One of Varley's students in the nineteen sixties was our host for the seminar out of which this book arose, our colleague Martin Birch, who took the challenge to study male moth pheromone systems and opened this field of research. Among the Lepidoptera, male pheromones control a variety of biological functions (Baker and Cardé 1979a; Birch 1970, 1974a; Boppré 1984; Brower *et al.* 1965; Conner *et al.* 1981; Guerrero 1981; Schneider 1975, 1980, 1983, 1984).

Comparison of female and male pheromone systems in the Lepidoptera allows some very basic distinctions. On the structural side it needs to be stated that all female glands (producing either attractants or so-called inhibitors) are specialized epithelial cells with glandular properties. In contrast, the male pheromone-producing organs are of great variety and are all composed of glandular cells which also form a scent-dissipating scale. Physiologically, most female moths do not have odour receptors for their products. Male scents—as far as have been studied—have their corresponding receptors cells on the antennae of both sexes.

While my senior associates in Seewiesen continue their morphological and physiological studies on male moth odour responses I have, with students and postdoctoral fellows, for a number of years studied male pheromone systems in butterflies and moths. In particular we looked into such systems in the Danainae, following the advice of Thomas Eisner (Cornell University). With the danaines, plant-produced alkaloidal precursors of pheromones were of particular interest. The same final products in the male sex were then again found among arctiids (Boppré 1984; Conner *et al*. 1981; Culvenor and Edgar 1972; Schneider 1983, 1984; Schneider *et al*. 1982).

It was a long way from the first bombykol electroantennogram to our present understanding of insect olfaction. The more we know, the more we see the complexity of this system, but this seems to be the normal situation in experimental biology. The great biologist Paul Weiss is said to have once stated that the 'nature of biology is not simplicity'. One cannot help but think of this statement when one studies odour signal production and reception, and odour-controlled behaviour of insects.

22 Temporal characteristics of pheromone receptor cell responses in relation to orientation behaviour of moths

K.-E. Kaissling

22.1. Introduction

Pheromone stimuli may not only elicit the approach flight of a male moth, but may also control its orientation behaviour until the odour source is found. Within a natural odour plume a moth is exposed to frequent and extensive changes of odour concentration (Murlis and Jones 1981; Murlis, (Chapter 3)). Kramer (Chapter 6) found that repetitive odour pulses are more effective stimuli for the orientation behaviour of *Bombyx mori* moths than are constant odour concentrations when the interval between stimuli is below 0.7 s. These and earlier studies (Bartell and Lawrence, 1977a; Cardé *et al*. 1984; Kuenen and Baker 1981) raised the question of how single receptor cells respond to pulsed odour stimuli (see also Boeckh, 1962).

In this study, nerve impulses of single receptor cells in response to relatively low stimulus intensities, with stimulus durations between 0.5 s and 10 min, were recorded and the response characteristics studied. Repetitive odour pulses of 20 ms duration were applied at higher odour concentrations, such as would be expected near the natural odour source. This paper presents preliminary results obtained from male moths of *B. mori* and *Antheraea polyphemus*.

22.2. Methods

The first of two methods of pheromone stimulation used in this study allowed control of stimulus intensity by using radioactively labelled compounds. The odour molecules were released from a 1 cm^2 filter paper which had been impregnated with the stimulus compound and placed in a glass cartridge with an inner diameter of 7 mm. An airstream of 100 ml/s was blown through the cartridge towards the antenna or the isolated antennal branch (Kanaujia and Kaissling 1984; Schneider *et al*. 1967). This method produced an odour stimulus contained within a turbulent airstream and was, therefore, not appropriate for stimuli with a duration less than 0.5 s.

The second stimulation method permitted reproducible stimuli of much

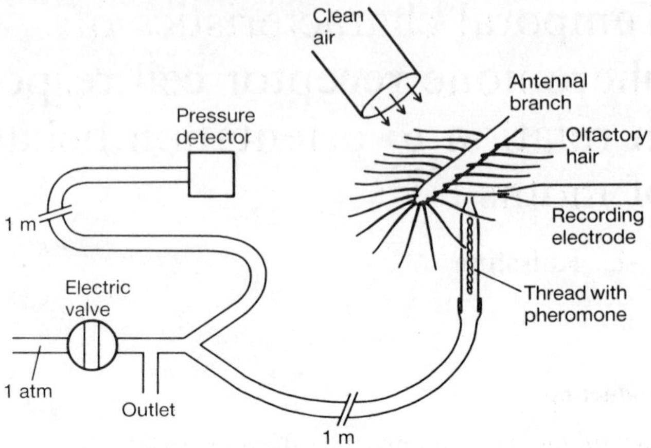

Fig. 22.1. Diagram of the set-up for local stimulation with short odour pulses.

shorter duration (20 ms) and higher repetition rate (above 10 per s) to be given. A 1 cm nylon thread was impregnated with the odour compound and placed in a glass capillary (outer diameter 1.5 mm) as used for micro-electrodes. The tip of the glass capillary was broken back squarely to a diameter between 20–30 μ. The outlet was placed near to a single olfactory hair. For stimulation, the inside of the capillary was connected to a compressed air source at a presence of 1 atm that was operated via an electric valve (Fig. 22.1). The pressure increase in the capillary was monitored by a fast pressure detector. The strength of the stimulus was estimated by comparing the magnitude of the receptor potential or the impulse number with the responses elicited by the first method of stimulation.

Electrophysiological recording of extracellular receptor potentials and nerve impulses from single hairs (after cutting the hair tip, Kaissling 1974) was done using standard methods (Boeckh 1962). The nerve impulses were analysed using a computer (Nicolet Med 80). The experimental animals were bought as pupae from breeders in Taiwan, Italy, Germany (*B. mori*) and the United States of America (*A. polyphemus*). The stimulus compounds were supplied by H. J. Bestmann, Erlangen.

22.3. The stimulus intensity dependence of the nerve impulse response

The response pattern of nerve impulses found in pheromone receptor cells of both species investigated depends on the stimulus intensity in a way similar to that described for vertebrate olfactory cells (Gesteland 1976). At low stimulus intensities the impulse frequency rises to a constant level within about one second after stimulus onset and declines within a few seconds after the end of stimulation. With increasing stimulus intensities the response becomes more phasic–tonic, showing an initial burst of impulses and a subsequent lower frequency level (10^{-4} and 10^{-3} μg)

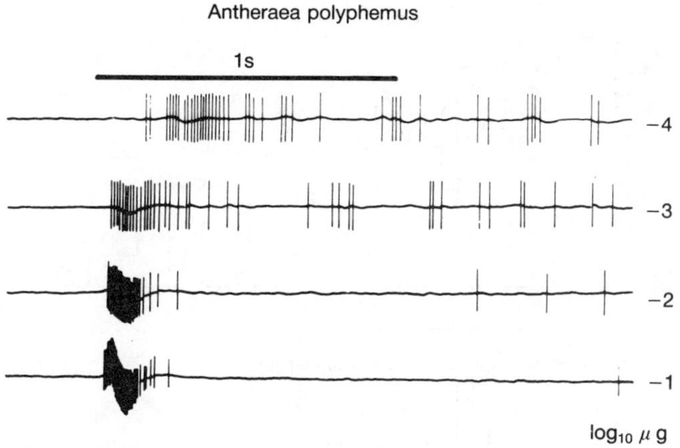

Fig. 22.2. Nerve impulses elicited by 1 s stimuli (bar) with increasing amounts of *E*-6, *Z*-11 hexadecadienyl acetate on the odour source (AC amplification). The response type is tonic at low stimulus intensities (below 10^{-4} μg, not shown), phasic–tonic at medium intensities (10^{-4}, 10^{-3} μg) and purely phasic at higher intensities (10^{-2}, 10^{-1} μg).

(Fig. 22.2). At very high stimulus intensities (10^{-2} and 10^{-1} μg), the response is phasic with a rebound of impulses after stimulation. A brief initial burst of impulses is followed by a period where no impulses are visible (Fig. 22.2). After the end of the stimulus the impulses reappear, and their frequency rises above the level of spontaneous activity and lasts over several seconds or even minutes.

22.4. Single-cell and behavioural responses at lower stimulus intensities

For the bombykol receptor cell, the latencies of the very first nerve impulses of responses to low stimulus intensities varied considerably and could reach up to two seconds (Kaissling and Priesner 1970). Since each nerve impulse elicited at weak stimulus intensities can be considered as a response to a single pheromone molecule, the distribution of response latencies can be determined by taking into account all nerve impulses following a stimulus. Ideally, the stimulus duration should be much smaller than the expected latencies between the arrival of the molecule at the antenna and the elicited nerve impulse. Figure 22.3a shows the distribution of nerve impulses following a 1 s stimulus, with latencies ranging from values less than 100 ms to values above 2 s. Long and scattered latencies may be due to the transport of the molecules from the site of adsorption towards the receptor molecules, as discussed by Getchell *et al*. (1980) for vertebrate olfactory cells (Kaissling and Thorson unpublished).

The reaction time of the behavioural response can be less than the average latency of the nerve impulses. More than 80 per cent of *B. mori* males had already started wing vibration 100–500 ms after onset of stimu-

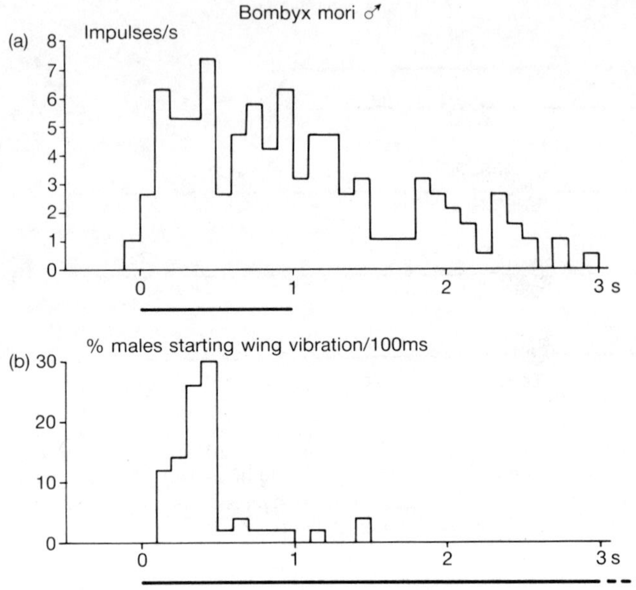

Fig. 22.3. (a) Averaged temporal pattern of nerve impulses elicited by 14 bombykol stimuli in six receptor cells (tonic response type; stimulus duration 1 s; stimulus load 10^{-2} μg of bombykol). (b) Reaction times of the fluttering response of 50 *B. mori* males stimulated with 10^{-2} μg of bombykol (from Kaissling and Priesner 1970) (stimulus duration 10 s).

lation with 10^{-2} μg of bombykol on the odour source (Fig. 22.3b; Kaissling and Priesner 1970). This means that the behavioural response was triggered by the nerve impulses fired within the initial response phase. It remains unknown whether the subsequent impulses have a behavioural significance, for example on the duration of the wing vibration response in *B. mori*.

22.5. The phasic–tonic response

At medium stimulus intensities the impulse response shows the typical phasic–tonic behaviour, with an almost constant (tonic) frequency level maintained until the end of the stimulation (Fig. 22.4). Stimuli lasting longer than a few seconds are usually followed by one relatively large impulse interval of one or a few seconds (Fig. 22.4). This transient suppression in impulse frequency is equivalent to a phasic off-response. The impulse response then returns and falls to the level of spontaneous firing within several seconds, or even minutes, depending on stimulus strength and duration. The off-response is much less pronounced than the on-response and could be detected in the central nervous system (CNS) only if simultaneous off-responses of a large number of cells are evaluated.

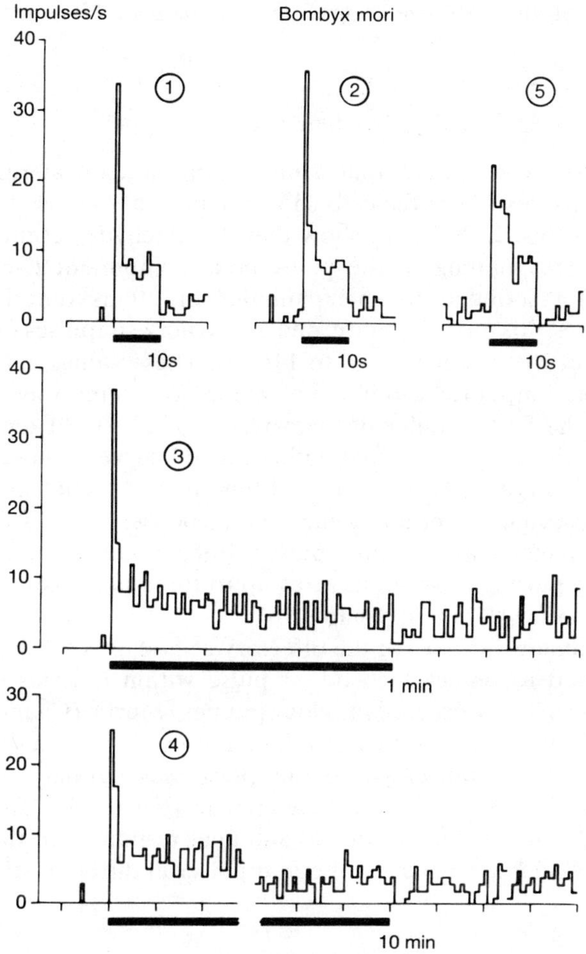

Fig. 22.4. Nerve impulse response of a single bombykol receptor cell by five subsequent stimuli of duration 10 s, 1 min, and 10 min. The intervals between stimuli 1 and 2, and 2 and 3 were 10 min, those between 3 and 4, and 4 and 5 were 30 min. This cell exhibited the phasic–tonic response type at 10^{-2} μg of bombykol and the odour source. Note the transient suppression of impulse frequency following the end of each stimulus.

The full on-response can be obtained again only after the impulse frequency caused by a preceding stimulus has fallen to the level of spontaneous frequency. Return to the resting level occurred about 10 min after the 10 s stimuli and about 30 min after the 10 min stimulus in the experiment shown in Fig. 22.4. In other words, long exposure to a constant stimulus blocks the following on-response for a long time. This effect may explain why constant stimuli caused less habituation than pulsed stimuli in males of the light-brown apple moth, *Epiphyas postvittana* (Bartell and Lawrence

1977a) and of the cabbage looper *Trichoplusia ni* (Kuenen and Baker 1981).

22.6. Response to pulsed pheromone stimuli

Responses to pulsed pheromone stimuli were studied at relatively high stimulus intensities where the cells show the phasic response type. Preliminary results (Figs. 22.5, 22.6.) show that the aldehyde receptor cell of *A. polyphemus* (responding to the pheromone component *E*-6, *Z*-11 hexadecadienal) responded to each stimulus of 20 ms duration up to 3 stimuli per s. At this repetition rate, one or two nerve impulses followed each stimulus onset with a delay of 40 to 110 ms. These values indicate considerable adaptation, since a minimum latency of 25 ms was measured in response to the first stimulus of the series (Fig. 22.5). At repetition rates above 4 stimuli per s not every stimulus elicited nerve impulses. However, the observed nerve impulses remained time-locked to the stimulus onset. Therefore, averaging over many such cell responses would still enable the animal to detect each odour pulse. Behaviourally, *Bombyx* males responded to pulsing at stimulus rates up to three per second, the highest rates tested so far (Kramer (Chapter 6)).

The described properties of the olfactory cells show that these receptors are well suited to detect each odour pulse within a series of repetitive odour pulses such as observed in odour plumes (Murlis (Chapter 3)). Even at the second pulse of a series of odour pulses given at a high rate, the number of nerve impulses per odour pulse was strongly reduced as a consequence of the phasic–tonic behaviour (Fig. 22.5). This reduction could simplify the evaluation of the antennal responses in the CNS. The maximum detectable frequency of odour pulses at different stimulus inten-

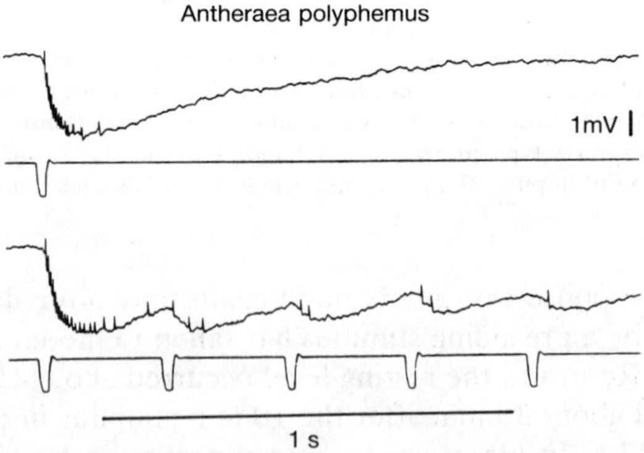

Fig. 22.5. Extracellularly recorded receptor potentials and nerve impulses of a receptor cell of male *A. polyphemus* responding to *E*-6, *Z*-11 hexadecadienal. The second and fourth trace from the top show the stimulus pulses of 20 ms duration as monitored by a presssure detector (upper half: single stimulus; lower half: five stimuli; stimulus frequency: 3.3/s).

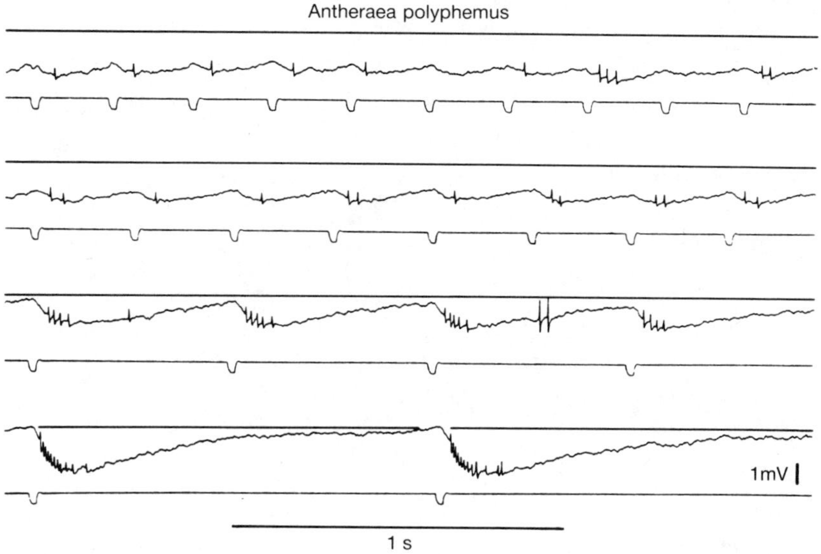

Fig. 22.6. Responses of the same receptor cell as in Fig. 22.5 to repetitive odour pulses of 20 ms duration, each with varying repetition rates (from top to bottom, 4.2/s, 3.4/s, 1.7/s and 0.8/s). Each series of stimuli started 3 s before the section shown. Each section consists of three traces. The top trace represents the zero level of the extracellularly recorded receptor potential. The middle trace is the DC-coupled receptor potential with superimposed nerve impulses. The bottom trace shows the stimulus pulses.

sities remains to be determined. It is likely that there are differences in the temporal resolution of stimulus pulses between the different types of receptor cells within the same moth species (see also Baker (Chapter 4)). From the faster time-course of extracellular receptor potentials, as well as of impulse responses, we expect that the bombykal receptor in *B. mori* males and the aldehyde receptor cells in males of *A. polyphemus* and *A. pernyi* are able to detect higher stimulus pulse frequencies than the bombykol and acetate receptors of these species (Kaissling 1979). Behavioural experiments using high rates of odour pulses may, therefore, clarify the behavioural significance of compounds like bombykal which so far is not satisfactorily explained (Kaissling *et al*. 1978).

Acknowledgements

I thank Gloria Adamek and Ulla Klein for valuable suggestions and help in correcting the English.

23 Pheromone reception: a kinetic equilibrium

R. G. Vogt and L. M. Riddiford

23.1. Behaviour and pheromone reception

Male moths show at least two discrete behaviours during pheromone-elicited flight orientation: zigzag upwind flight while within a pheromone plume, and lateral casting flight on losing the pheromone plume (David *et al.* 1983; Kennedy and Marsh 1974). Behavioural (Preiss and Kramer 1983) and physiological (Olberg 1983a) evidence suggests that a major role of pheromone during flight orientation is to regulate the optomotor response, allowing the appropriate visually controlled flight mode, depending on the relative presence or absence of pheromone. The switch between these two behavioural modes is very rapid, occurring within 1 second (Krammer 1975; Marsh *et al.* 1981).

When an animal flies out of a pheromone plume, the external drop in pheromone concentration must be mirrored by a rapid internal drop. There must exist a fast inactivation mechanism within the sensory hair to prevent the build-up of confusing background pheromone levels. Furthermore, the mechanism allowing transport of pheromone to the dendritic membrane must coexist in space with the inactivation mechanism. Somehow these two processes are kept separate so that receptor binding tends to occur before inactivation.

23.2. Pheromone-binding protein and sensillar esterase

Our work has focused on the biochemical events involved in pheromone reception in the pheromone-sensitive sensory hairs of the male silk moth *Antheraea polyphemus*. We have demonstrated the existence of two proteins uniquely associated with the receptor lymph in these hairs: a pheromone-binding protein and a sensillar esterase (Vogt and Riddiford 1981b). Both proteins are extracellular and soluble.

When ^{3}H-pheromone was incubated with antennal homogenates, high levels of radioactivity were found associated with a 15 000 dalton protein that was the major protein constituent of the receptor lymph (Vogt and Riddiford 1981b). The concentration of this pheromone-binding protein was found to be about 20 mM (Vogt *et al.* 1985), equivalent to a 30 per cent protein solution, suggesting that the receptor lymph can not be considered a simple aqueous solution. Such proteins appear to be common features of pheromone-sensitive lepidopterous sensilla (Kaissling and Thorsen 1980; Vogt and Riddiford 1981a).

The relative concentration of the esterase (75 000 daltons, U. Klein personal communication) is at least four orders of magnitude below that of the binding protein, or less than 2×10^{-6} M in the receptor lymph (Vogt *et al*. 1985). Kinetic studies of the sensillar esterase (Vogt *et al*. 1985) showed the K_m to be 2×10^{-6} M and the V_{max} per antennal unit of the enzyme to be 5×10^{-12} mol/l/s (pH 7.2, 30 °C). The pheromone half-life in the presence of a physiological concentration of esterase was estimated to be 15 ms over pheromone concentrations ranging from one molecule per hair, the suggested physiological threshold, to 10^6 molecules per hair (Vogt *et al*. 1985). This is for an idealized situation where there are no other proteins present to interfere with the reaction or availability of substrate.

Our estimated half-life is enormously shorter than the four-minute half-life Kasang and Kaissling (1972) estimated from whole-antenna degradation studies with *Bombyx mori*. The reason for the apparent discrepancy between the *A. polyphemus* and *B. mori* pheromone half-lives presumably lies in experimental design differences. We studied a purified enzyme that was restricted to the receptor lymph, whereas Kasang and Kaissling (1972; Kaissling, personal communication) applied tritium-labelled pheromone to intact antennae, then subsequently solvent-extracted these antennae. They observed degradative events averaged over the entire antenna, rather than events restricted to the sensory hair interior. The fact that their observed slow degradation failed to match the physiological and behavioural time courses may suggest that a large proportion of pheromone never enters the hair lumen, where it should be rapidly degraded, but may be lost elsewhere on or in the antennal cuticle.

23.3. Molecular models of pheromone reception

In 1974, Kaissling proposed a linear or one-step-at-a-time model summarizing what was known of the pheromone reception process. In this model, a pheromone molecule arrived at a presumed receptor site on the membrane of the sensory dendrite by diffusing down a pore-tubule tunnel. These pore-tubules were perceived to contact the dendritic membrane (Steinbrecht and Müller 1971). The pheromone molecule interacted with a receptor to initiate the transductional events, and was subsequently inactivated by some non-enzymatic process which Kaissling termed 'early inactivation' (Kaissling 1974). Enzymatic degradation of pheromone was considered too slow to be a part of the reception process (Kaissling 1974; Kasang 1971; Kasang and Kaissling 1972).

Our early studies of the site-specific pheromone-binding protein and pheromone-degrading esterase led us to suggest that these proteins could be the mechanisms of Kaissling's inactivation process (Vogt and Riddiford 1981b). In this scheme the binding protein acted as the 'early inactivator', sponging up pheromone from the receptor, while the esterase acted as the apparently slow-acting enzyme (Fig. 23.1a).

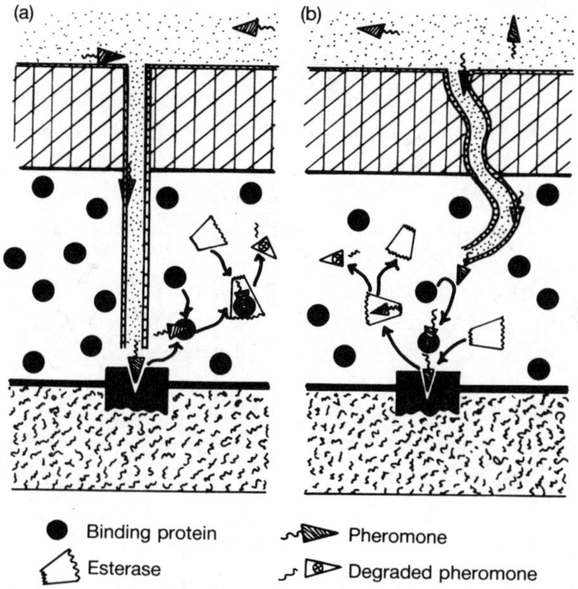

Fig. 23.1. (a) A molecular model in which pheromone arrives at a receptor via a pore-tubule, then is inactivated by first binding to binding protein and subsequently becoming degraded. (b) A molecular model in which pheromone arrives at a receptor via a binding protein acting as a carrier, and a sensillar esterase functions as principal inactivator. (Figure provided by M. Meyer.)

In light of the kinetic properties of the purified enzyme, we have been forced to reconsider the roles of these two proteins. The aggressive properties of the sensillar esterase suggest that it must be the principal inactivator of pheromone. What then is the function of the binding protein? We have proposed that the binding protein might function in part as a carrier (Vogt *et al*. 1985). In such a view (Fig. 23.1b), pore-tubules would deliver pheromone only to the hair interior, not all the way to the dendritic membrane. Binding protein would carry pheromone across the receptor lymph space from pore-tubule to dendritic receptor, and sensillar esterase would rapidly degrade pheromone post-transduction.

This 'carrier model' differs from Kaissling's (1974) model only in that the binding protein takes on some of the function of the pore-tubules, and the enzyme assumes the role of major inactivator. Such a 'carrier model' still possesses the traits of a one-step-at-a-time model, the tunnel aspect remaining intact. But there are some serious questions that such a model raises. If the binding protein can pick pheromone from the tubule and deliver it to the receptor, what prevents the protein from picking pheromone from the receptor and recycling it to another receptor? This problem is compounded if we assume that, as a carrier, the binding protein should protect pheromone during transport. What insures that this protection occurs only enroute to the receptor, and not post-transduction?

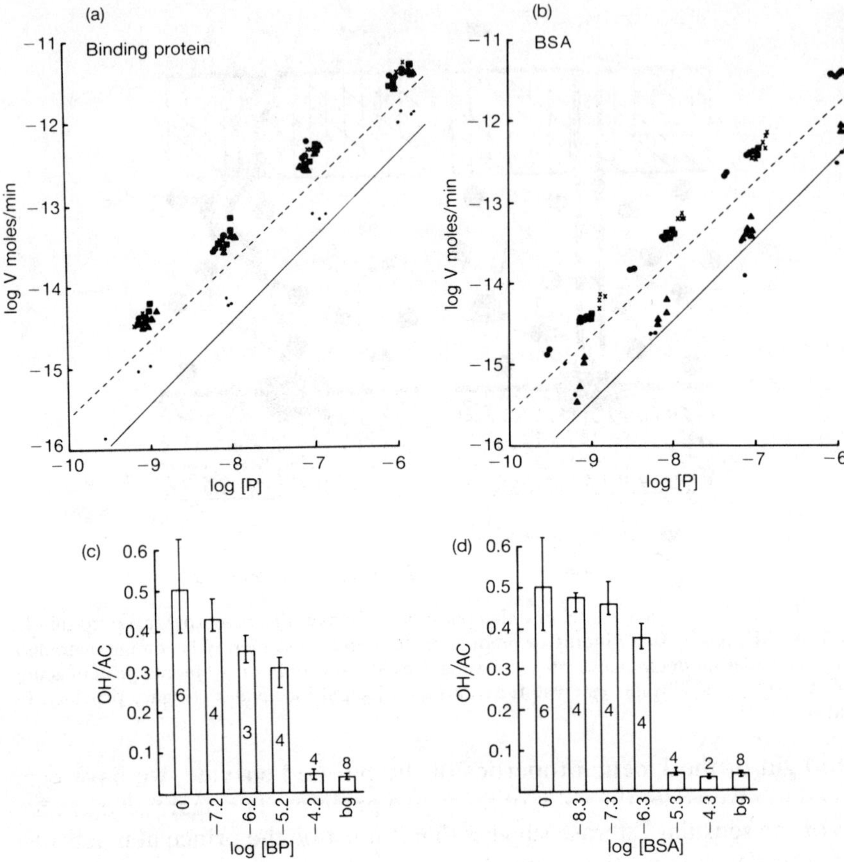

Fig. 23.2 The effect of different concentrations of binding protein or BSA on the rate of pheromone degradation by sensillar esterase. Binding protein and esterase were electrophoretically purified (Vogt *et al*. 1985). Enyzme + protein were incubated with ^{3}H-pheromone ([11,12-$^3H_2$]-(*E,Z*)-6,11-hexadecadienyl acetate; 58 Ci/mM; Prestwich *et al*. 1984) for 10 minutes (50 μl each, pH 6.5, 20 °C). The reaction was stopped by addition of 150 μl of ethyl acetate, and 3 μl was assayed for product formation by thin-layer chromatography (Vogt *et al*. 1985). The ratio of the highest binding protein concentration to sensillar esterase in these experiments was about 15 times higher than exists in the sensory hair.

Figures 2a (binding protein) and 2b (BSA) represent velocity versus substrate relationships for the different protein concentrations. (X = buffer in place of added protein; ● = 6.6×10^{-1} M binding protein or 5×10^{-5} M BSA; decade dilutions of these are represented by ▲, ■, • respectively; dashed line represents transitional boundary for second-party protein effect; solid line represents background level of radioactivity at product R_f on thin-layer plates.)

Figures 2c (binding protein) and 2d (BSA) represent product/substrate versus protein concentration relationships for a fixed pheromone concentration (10^{-8} M). ('O' represents the case where buffer replaces added protein; 'bg' represents background level of radioactivity at product R_f on thin-layer plates; error bars represent spread of actual data; numbers under error bars represent '*n*' value.)

23.4. Second-party protein effects on esterase activity

To study the carrier nature of the binding protein, we have now examined the ability of the pheromone-binding protein and a known carrier protein, bovine serum albumin (BSA), to reduce the rate of pheromone degradation by sensillar esterase. As the amount of protein was increased, the rate of degradation by the esterase remained constant up to a certain transition concentration. Above this transition, degradation dropped to near background levels (Fig. 23.2). When binding protein was increased from 6.6×10^{-6} M to 6.6×10^{-5} M, and when BSA was increased from 5×10^{-7} M to 5×10^{-6} M, the rate of degradation decreased about fivefold for binding protein and about tenfold for BSA. Considering the difference in molecular weights of these two proteins (15 000 binding protein; 67 000 BSA), these transitions occurred at approximately the same weight/volume concentration for both proteins.

This phenomenon can be seen in greater resolution in Figs. 23.2c (binding protein) and 23.2d (BSA), which show changes in product/substrate ratios. These ratios, which reflect the pheromone half-life, drop abruptly as the transitional protein concentration is crossed. In addition to the transitional inhibition effect, these figures show that there is a slight protein effect on degradation below the transitional concentrations, degradation slowing as protein concentration increases.

We suggest that the transitional concentration reflects the binding constant between pheromone and these second party proteins. Thus, the dissociation constant (K_d) for pheromone-binding protein would be in the range of 10^{-5} M, and the K_d for pheromone-BSA would be in the range of 10^{-6} M. Protein concentrations above this transitional point would ensure maximal interaction with pheromone independent of pheromone concentration. Considering the high *in situ* concentration of binding protein (20 mM), a pheromone molecule within a sensory hair would spend almost all of its time associated with binding protein molecules.

The observation that the effect of BSA is similar to the binding protein effect suggests that protein-pheromone interactions do not take place through architecturally specific binding sites, but rather involve properties inherent to soluble proteins. The fact that the transition effect relates more closely to the total amount of protein than to the molar amount present supports this suggestion. The observation that these associations are weak suggests that there is a very rapid on/off reaction, and that pheromone may be extremely mobile within the binding protein matrix.

23.5. Kinetic equilibrium model

When a pheromone molecule enters a sensory hair, it confronts a multitude of competing binding sites (Fig. 23.3): pheromone-binding proteins, ravenous enzymes, receptors and lipophilic environments such as dendritic

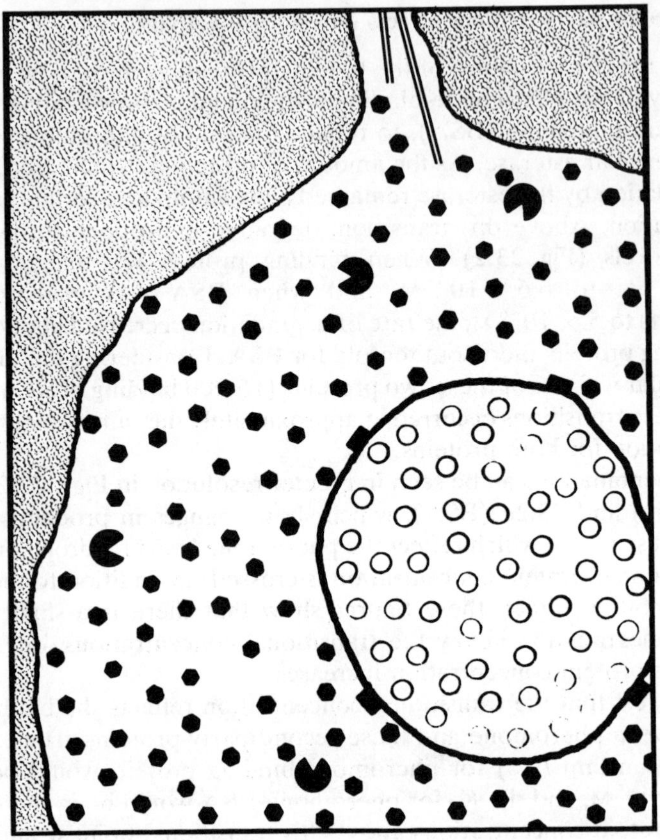

Fig. 23.3. Diagram of the biochemical environment encountered by a pheromone molecule on entry into sensory hair via pore-tubule (●: binding protein (20 mM); ◖; sensillar esterase (<2 μM). Other potential binding sites include the receptor proteins in the dendritic membrane (■), and the lipophilic dendritic membranes and pore-tubules.

membranes and pore-tubules. In light of this picture of the molecular environment of the receptor lymph space, and our recent findings, we have reconsidered our model of pheromone reception in terms of a dynamic kinetic equilibrium (Fig. 23.4).

In this kinetic equilibrium model, all potential interactions are possible at any time and in any order. Pheromone enters a hair via pore-tubules, and to reach a receptor the pheromone must pass through a screen of binding protein and enzyme. Pheromone is effectively removed from the picture upon encountering an enzyme. Three major routes exist for this passage: pore-tubule—binding protein—receptor—enzyme, pore tubule—binding protein—enzyme, and pore-tubule—enzyme. Relative to other interactions, only pheromone encounters with enzyme are considered to be irreversible.

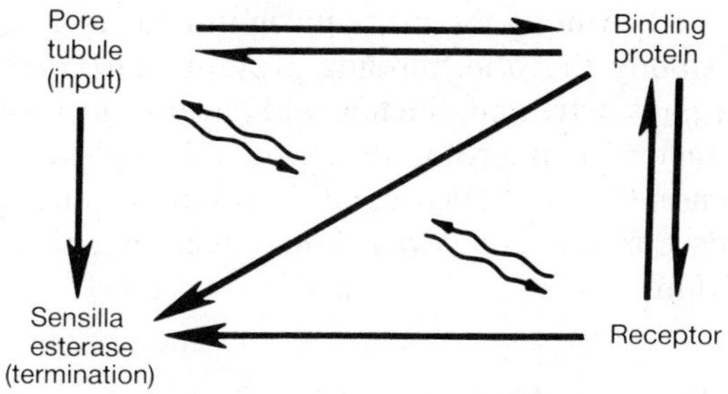

Fig. 23.4 Kinetic equilibrium model of the biochemical interactions involved in pheromone reception (see text for details).

The rate constants in this scheme, and hence the net direction of pheromone movement, would depend on the concentrations of the various species involved, and on the molecular scale diffusion around three- and two-dimensional interfaces (Berg and Purcell 1977; Futrelle 1984). In this model, we suggest that the high concentration of binding protein acts like the solid-phase matrix of a chromatography column, with the water present acting as the solvent. The binding protein acts to phase partition pheromone into the hair lumen, and the pheromone molecule is then chromatographed along this protein matrix. The high binding-protein concentration relative to enzyme concentration insures that, probabilistically, there will be more interactions with binding protein than with esterase. Binding protein protects the pheromone from the esterase by mass action blockage, that is, by getting in the way of the enzyme. In time the pheromone will be degraded, but the high binding-protein concentration insures that this time is sufficient for the bulk of pheromone to reach receptors and thus initiate transduction.

Support for phase separation by the binding protein comes from a re-evaluation of our early electrophoretic studies (Vogt and Riddiford 1981a). In these studies we incubated *A. polyphemus* ^{3}H-pheromone with various tissues from different moth species, and were able to demonstrate association of pheromone with male antennal-specific proteins. But we also observed its association with many other proteins. We now suggest that, as the proteins concentrated in the sample wells of these electrophoretic gels, the pheromone molecules became phase partitioned into the region of high protein concentration. As these proteins migrated through the gel, the pheromone migrated with them, showing a phase preference for high protein concentrations over low protein concentration.

23.6. Predictions of the kinetic equilibrium model

The kinetic equilibrium model suggests that binding protein can interact with pheromone either before or after the receptor, and that individual pheromone molecules might interact with receptors more than once before becoming degraded. Among the more intriguing ideas that the model suggests is the possibility that the 'binding protein—enzyme' pathway could become differentially activated. Such would occur if the capacity of binding protein to interact with pheromone approached saturation due to high levels of pheromone. Under such conditions excess pheromone might be shunted towards enzyme and away from receptors, thereby preventing receptor saturation.

23.7. Pore-tubules versus binding protein

Pore-tubules appear in this model only as the principal mode of entry into the hair lumen. For pore-tubules to play a larger role in pheromone transport, there must be direct contact between tubule and dendritic membrane. Such contacts have been observed in some electron micrographs (Keil 1982; Steinbrecht and Muller 1971), but they are observed only when the dendritic membrane is very close to the inner cuticular wall of the hair (Keil 1982), and their observation is highly dependent on the mode of tissue fixation (Steinbrecht 1980; Steinbrecht and Müller 1971). Certainly, pore-tubules do occupy a considerable amount of space within the hair lumen (Keil 1982), and a certain number of tubule–dendrite contacts can be expected whatever their nature or function. The kinetic equilibrium model argues that when a pheromone molecule enters a sensory hair, it interacts with the molecular components of the hair in a manner consistent with molecular interactions. Our evidence suggests that pheromone will interact very strongly with the exceptionally high concentration of binding protein within the hair.

23.8. The uniqueness of the pheromone-binding protein

At first glance, the pheromone-binding protein seems to possess no special or specific binding properties in its interactions with the pheromone. Why then do these animals produce this uniquely situated protein? Perhaps the answer has to do with the size and physical properties of the protein. Relatively few low-molecular-weight proteins appeared on any of our gels. A protein of about 15,000 daltons is small enough to diffuse rapidly, and large enough to show tertiary structure conducive to certain low-affinity protein–ligand interactions. We suggest that only a small protein of low specificity and at high concentration could possess the multifunctional properties ascribed to it by this kinetic model of pheromone reception: solubilizer, carrier and protector.

24 Neuropharmacological effects on olfactory perception

C. E. Linn and W. L. Roelofs

24.1. Introduction

One of the many important problems in pheromone research concerns the nature of the thresholds that determine the response specificity of male insects to multicomponent sex pheromones (Roelofs 1978). Male moths appear to be tuned to a narrow range of ratios and release rates of pheromone, with the complete complement of components being a critical factor for optimal behaviour (Linn and Roelofs 1983; Linn *et al*. 1984).

In part the parameters that define the limits of male detection and discrimination of the signal are fixed by genetic and developmental constraints. It is also well known, however, that male sensitivity to the chemical signal is not constant throughout the photoperiodic cycle, but rather occurs optimally during a particular time of the photoperiod (Baker and Cardé 1979c; Castrovillo and Cardé 1979). This periodicity appears to be governed by a circadian rhythm that also regulates general flight activity, and which is significantly influenced by temperature (Cardé *et al*. 1975). This pattern, of fluctuating behavioural thresholds to stimuli eliciting flight activity and response to the chemical signal, suggests that the activity of elements in the central nervous system that are involved in signal recognition and control of flight motor programs is modulated by endogenous factors that are influenced by the photoperiod and temperature regime.

A variety of compounds that act as neuromodulators have been identified (Feldman and Quenzer 1983). These compounds act either to change the quality of transmitted information or to change the spontaneous activity of the target cells. One class of compounds, biogenic amines, are widespread in the insect nervous system (Evans 1980; Hildebrand and Maxwell 1980) and have been found to have roles as neurotransmitters and neuromodulators in a number of invertebrate systems (Evans 1980; Kandel and Schwartz 1982). In one of the best studied, the locust (Evans 1982), octopamine (OA) is released from the dorsal unpaired median nerve and acts to alter the response of the extensor tibia muscle to inputs from excitatory and inhibitory motorneurones. The action of OA changes the extensor muscle in such a way that instead of favouring maintenance of posture it favours rapid movements. Octopamine, serotonin (5-hydroxytryptamine, 5HT), and dopamine (DA) have also been implicated in modulating thresholds in a number of sensory modalities, such as olfaction in honey bees (Mercer and Menzel 1982), circadian-regulated flight activity in nocturnal moths (Hinks 1967), and feeding in blowflies (Long and Murdock 1983).

We report here on experiments to determine the effects of neuroactive compounds on thresholds for male-pheromone-stimulated flight. We first observed the effects of a variety of compounds (Linn and Roelofs 1984). These included insecticides with different modes of action, as well as known neuromodulators. The success of this screening study led to experiments designed specifically to determine the effects of biogenic amines on the circadian rhythmicity of flight activity and thresholds for response to the pheromone signal.

24.2. Screening of neuroactive compounds with the oriental fruit moth

Male oriental fruit moths (OFM), *Grapholitha molesta,* were tested in our flight-tunnel after topical treatment with: the insecticides chlordimeform, permethrin, carbaryl, dieldrin; the biogenic amines OA and 5HT; and two OA antagonists, yohimbine and cyproheptadine (Linn and Roelofs 1984). In general, sublethal concentrations of insecticides did not alter the sensitivity of males to the chemical signal or the time of response in the photoperiod, but rather disrupted specific phases of the flight response, each compound inducing a unique effect. Males treated with OA, however, displayed a marked increase in sensitivity to the concentration of the signal, taking flight almost immediately upon contact with the odour plume and exhibiting lengthy periods of stationary flight near the release cage prior to flying vertically out of the odour plume. This response was very similar to that observed in untreated males tested with a high concentration of the pheromone (Linn and Roelofs 1983), and suggested that the threshold for upwind flight had been significantly lowered by treatment with OA. This was supported by the observation that yohimbine and cyproheptadine alone decreased the number of males that became activated to the signal and that both compounds reversed the effects of OA, restoring the normal response pattern to the optimal blend and concentration. In contrast to the effects observed with OA, male response was not affected by 5HT. We hypothesized that this was due to poor penetration of the molecule into the cuticle as a result of using a topical application.

24.3. Effects of biogenic amines on male cabbage looper moths

The screening study with male OFM demonstrated the sensitivity of the male response to a variety of neuroactive compounds. To pursue the idea that candidate endogenous neuromodulators might affect response thresholds, we initiated experiments to examine the effects of the biogenic amines on the cabbage looper moth, *Trichoplusia ni*. We have recently re-examined the sex pheromone of this species, in part as a result of biosynthetic considerations, and showed that females produce and release a six-component blend to which males are very responsive (Bjostad and Roelofs 1983; Linn *et al.* 1984). Male *T. ni* are also larger than OFM, allowing injection of the compounds. Results will be presented here from preliminary studies with OA and 5HT, with a detailed report to be presented elsewhere.

24.3.1. The effect of octopamine and serotonin on general flight activity

Octopamine and 5HT were injected into four-day-old male *T. ni* 3 h prior to the initiation of an eight-hour scotophase, in amounts of 0.1, 1, 10, and 100 μg in 1 μl insect saline. In the absence of pheromone, untreated males exhibited a peak of flight activity from the fourth to sixth hours of the scotophase, with less than 5 per cent of the males active in the hour prior to or succeeding the scotophase (Fig. 24.1). Octopamine did not alter the circadian activity, except with the 100 μg amount, which eliminated all activity in moths throughout the observation period. These moths recovered from the influence of the drug after 24 h, which suggested that the lowered activity was not due to a lethal effect. Increasing amounts (0.001–10 μg) of 5HT, however, increased the number of males that were spontaneously active during the scotophase. With 10 μg of 5HT, 85–100 per cent of the males were active throughout the entire eight-hour scotophase. Similar to OA, the 100 μg concentration of 5HT eliminated all flight activity.

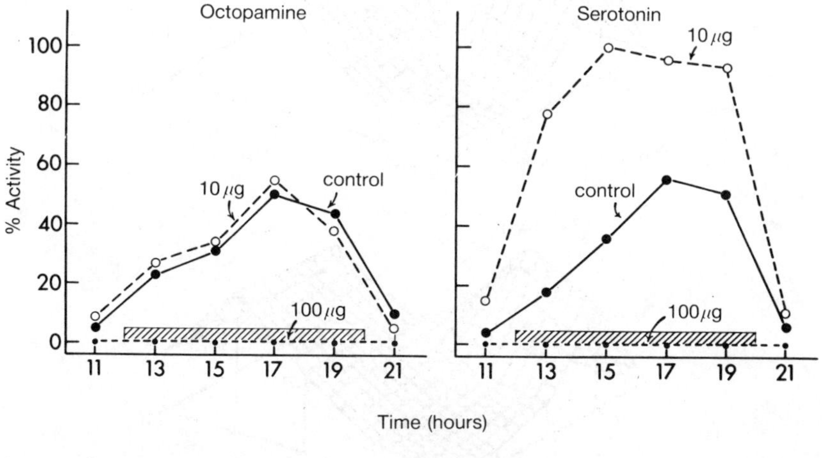

Fig. 24.1. Percentage of male *T. ni* exhibiting spontaneous flight activity during the scotophase period (12 to 20 h) in the absence of pheromone. Solid line is for untreated moths, while the dashed lines show the response to males treated with 10 and 100 μg OA or 5HT (n = 25 for each time and concentration of drug).

24.3.2. *The effect of octopamine on pheromone response at the peak time of activity*

Male *T. ni* were treated as in Section 24.3.1 and tested at the midpoint of the scotophase with a series of pheromone concentrations (Fig. 24.2). Increasing amounts of OA significantly increased the number of males taking flight to the lowest concentrations of pheromone tested (0.01 and 0.001 μg on filter paper). With 10 μg OA the peak concentration of pheromone was shifted from the normal 1 μg to 0.01 and 0.001 μg, indicat-

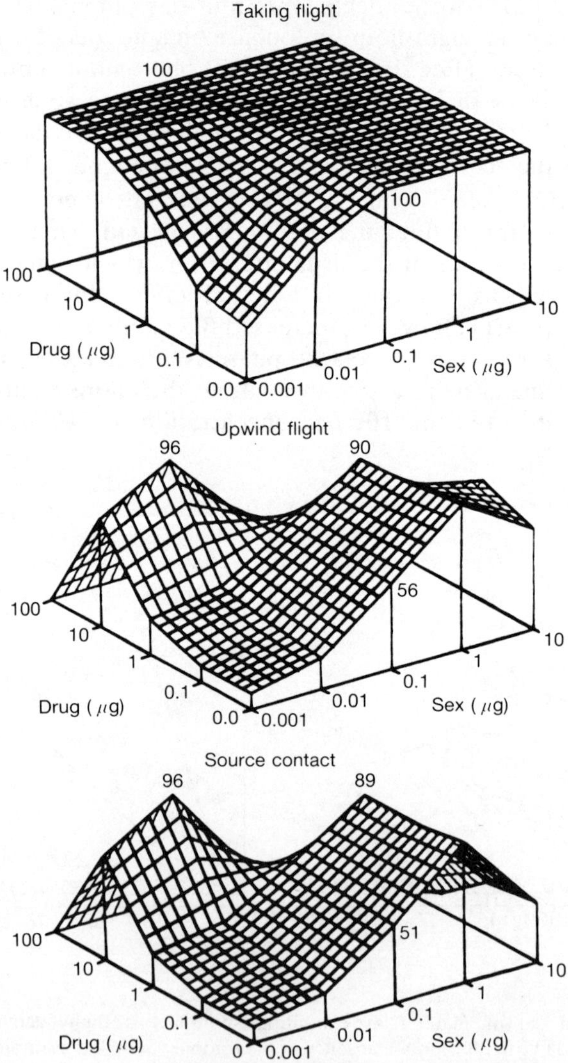

Fig. 24.2. Response of male *T. ni* to the sex pheromone (μg applied to filter paper, Linn *et al.* 1984) after injection of OA (μg in 1 μl saline). Observations were made during the fifth hour of scotophase. Values indicate the percentage response for each behaviour ($n = 30$ for each pheromone–drug combination).

ing that, as with male OFM, a shift in threshold had occurred. With 100 μg OA males were activated by the pheromone (in contrast to the total lack of spontaneous activity), but when attempting to take flight they went to the floor of the tunnel and exhibited a rigid posture with wings extended and held nearly vertical over the body. These results suggest that increasing amounts of OA initially affected neural elements involved in signal recognition and then coordination of flight motor units.

24.3.3. Effect of octopamine on male response over the scotophase period

Male *T. ni* were injected with 10 μg OA and then tested at two-hour intervals over the scotophase period (Fig. 24.3). The control response showed

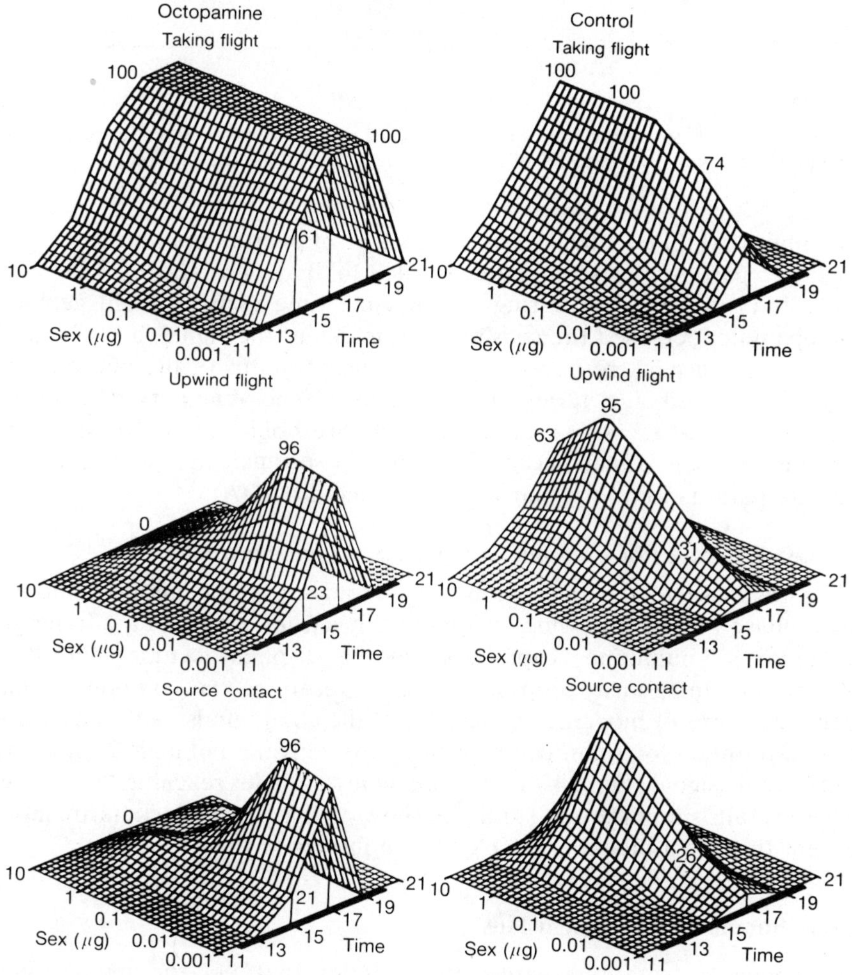

Fig. 24.3. Response of male *T. ni* to the sex pheromone (μg applied to filter paper) at six times in the photoperiod following injection of 10 μg OA three hours prior to 11.00. The dark bar from 12 to 20 hours indicates the duration of scotophase. Values indicate the percentage response for each behaviour (n = 35 for each pheromone–time combination).

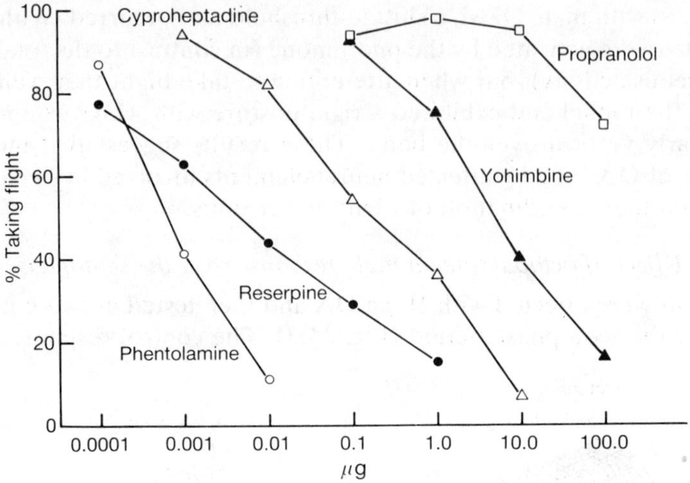

Fig. 24.4 Percentage of male *T. ni* taking flight to a 1 μg concentration of the sex pheromone after treatment with OA antagonists, and reserpine (n = 20 for each data point).

that males exhibited a peak response to 1 μg of pheromone during the mid to latter part of the scotophase, similar to that observed by Linn and Roelofs (1984), and occurring at the same time as peak flight activity. Octopamine increased the sensitivity of males to the signal, as evidenced by the shift in peak response to lower concentrations of the pheromone, but this occurred only during the normal peak time of activity. The results indicate that whereas OA affected male thresholds for response to the pheromone, this effect occurred during the normal circadian-regulated activity period, a pattern that was not affected by OA.

24.3.4. Effects of octopamine antagonists on male activation

Male response to the pheromone was observed at the midpoint of the scotophase period following treatment 5 h earlier with alpha adrenergic antagonists yohimbine, cyproheptadine, and phentolamine; the beta adrenergic antagonist propranolol; and reserpine, a compound that depletes stores of biogenic amines. All of the compounds, with the exception of propranolol, significantly reduced the activation of male *T. ni* to the pheromone signal (Fig. 24.4), resulting in fewer males reaching the source. These results support other studies in showing that alpha blockers are more potent than beta blockers at OA sites in insects (Evans 1982).

24.4. Summary and conclusions

We conclude from our exploratory studies that (1) the pheromone-mediated flight response of male moths is very sensitive to a range of neuroactive compounds and is a new and useful probe for pharmacological studies, and (2) that biogenic amines might be endogenous modulators of pathways involved in the flight response. The different effects observed with OA and 5HT, and selected OA antagonists, support this conclusion. Caution is added, however, to the second point since we recognize that behavioural assays cannot provide definitive proof for the role of a proposed neuromodular (Evans 1980, 1982). Cytochemical, physiological, and biochemical studies would be needed to provide conclusive evidence. We feel, however, that a sensitive behavioural test can be an important element in pharmacological studies. We propose to conduct extensive studies with additional antagonists and agonists of OA, as well as for 5HT, DA, and other compounds known or suspected of neuromodulatory activity (Hildebrand and Maxwell 1980). These studies will hopefully provide us with a database for detailed analysis of the effects of changing photoperiod and temperature on the circadian rhythmicity for flight activity in these insects, particularly as it relates to the thresholds for pheromone response.

25 Electrophysiological responses to pheromone blends in single olfactory receptor neurones

R. J. O'Connell

25.1. Introduction

In many insect species, the detection and localization of potential reproductive partners is accomplished with a coupled system of specific chemical signals and high-sensitivity olfactory receptor neurones. An original supposition held that this form of communication involved only one species-specific chemical signal and only one type of olfactory receptor neurone (Fabre 1913; Schneider 1957).

In the last few years we have come to realize that this form of chemical communication is, in reality, much more complicated (Bjostad and Roelofs 1983; Priesner 1983). For example, a behavioural sequence like male orientation to receptive females may involve various blends of several female-produced odours, each of which may individually (Bradshaw *et al.* 1983), or in concert (Linn *et al.* 1984), modulate one or another of the steps in the sequence of male behaviour. Subtle changes in the absolute or relative composition of the pheromone blend may have profound effects on the ultimate sequence and staging of the behavioural response (Klun and Robinson 1972; Priesner *et al.* 1980).

In large measure our knowledge of the sensory limb of this communication system derives from work in an insect which had been thought to have a single-component sex attractant pheromone. Accordingly, it has long been traditional to explore the response properties of single olfactory receptor neurones with single-component stimuli. To the extent that blend perception has been addressed electrophysiologically, it has usually been found to involve several different types of olfactory receptor neurones, each specialized for one of the individual components of the relevant blend (Benn *et al.* 1982; Hansen *et al.* 1983; Priesner 1980b). The ultimate orchestration of the behavioural sequence elicited by activity in these specialized receptor neurones was thus thought to reside primarily in the combinatorial properties of more central components of the olfactory system. The extent of this central processing is thought to depend upon the level of behavioural complexity which is to be supported.

It should be obvious that as the number of behaviourally relevant components in a pheromone blend increases, so does the requirement for additional private channels of communication. Theoretically, generation of less specific but more broadly tuned olfactory receptor neurones with the capacity for across-fibre patterning is a more efficient mechanism for encoding increasing amounts of blend information than is the generation of additional classes of specialized receptor neurones (O'Connell 1981, 1985; Pfaff 1975). Several alternative mechanisms for increasing channel capacity lie between these two extremes, including one which could involve direct modification of primary olfactory receptor neurone activity by blend components. For example, compounds in a blend may not have their own specific receptor neurones and thus are ineffective stimuli when presented alone. However, these compounds may alter the ability of a specialized receptor neurone to interact with its parent compounds if the relevant odorants are present simultaneously. Preliminary electrophysiological experiments demonstrated the existence of such a mechanism in the red-banded leafroller moth (O'Connell 1972). A series of behavioural and electrophysiological studies have provided sufficient information to postulate a similar mechanism in the cabbage looper moth (Linn *et al*. 1984; Mayer 1973; O'Connell *et al*. 1983).

The pheromonal communication system of *Trichoplusia ni*, the cabbage looper moth (CL), involves at least seven behaviourally relevant odour components. Six of these have been identified in the female pheromone gland (Bjostad and Roelofs 1983; Linn *et al.* 1984). The major component of the gland, (*Z*)7-dodecenyl acetate (*Z*7-12:AC) is essential for flight initiation, whereas its alcohol, (*Z*)7-dodecenol (*Z*7-12:OH), significantly arrests upwind flight. The saturated compound, dodecyl acetate (12:AC), enhances several of the responses shown by males close to a pheromone source, including landing and search behaviours (Bjostad *et al*. 1980). The remaining identified components are required in order to produce a synthetic pheromone blend which is competitive with calling CL females. However, except for *Z*7-12:AC, which is essential for flight initiation, all of the remaining female-produced compounds seem to act in unison to elicit the full gamut of male behaviour because one or two of them may be removed, very nearly at random, without noticeably reducing the communicative efficiency (Linn *et al*. 1984). Although there are specialized olfactory receptor neurons for *Z*7-12:AC and *Z*7-12:OH in certain of the sensilla on male CL antennae, none have been described for 12:AC (O'Connell *et al*. 1983).

Here, I would like to outline our preliminary attempts to unravel blend perception in the CL moth. We asked if the response properties of single olfactory receptor neurones were differentially affected by single and multiple-component chemical stimuli each presented at a range of different intensities. A special focus was placed on the mechanism whereby 12:AC effects neural outflow sufficiently to account for its behavioural potency.

25.2. Methods

All of the animals used in this study were derived from the laboratory-reared colony maintained by the Insect Attractants, Behavior, and Basic

Biology Research Laboratory, USDA, Gainesville, FL. Newly pupated animals were sexed and experimental males isolated from further contact with females. Virgin adult males had access to 5 per cent sucrose solution and were typically used as subjects when two to five days old. Animal preparation, tungsten microelectrode construction, extracellular recording conditions and odour cartridge placement and control were standard for this laboratory (O'Connell, 1975; O'Connell *et al.* 1983). Data acquisition, stimulus application and extracellular spike sorting algorithms were controlled and timed by a digital computer system described elsewhere (O'Connell *et al.* 1973). For each stimulus application the computer produced: an event–time histogram, a reconstructed spike record, and the total number of action potentials elicited in the prestimulus, stimulus and post-stimulus intervals for both the A receptor neurone and the B receptor neurone found in each sensillum trichodea. All of the data presented here were obtained from sensilla whose morphological and physiological properties typed them as high-sensitivity (HS) sensilla (O'Connell *et al.* 1983).

The two monounsaturated compounds used as stimuli were obtained from Farchan Chemical Co., Columbus, OH. The saturated compound was obtained from Dr W. Roelofs, New York Stage Agricultural Experimental Station, Geneva, NY. The purity of these materials was evaluated prior to use by Albany International, Controlled Release Division, Needham Heights, MA. Based upon gas chromatographic retention times on polar (SP-2340) and non-polar (SPB-1) columns (60M,WCOT), each calibrated against authentic standards, the level of cross-contamination among the three compounds was below the detection limit for the system (<0.1 per cent). Absolute purity was 87 per cent for *Z*7-12:OH, 93 per cent for *Z*7-12:AC, and 98 per cent for 12:AC. A range of stimulus intensities for each compound was produced by serial dilution with light mineral oil (Aldrich Chemical Co., Milwaukee, WI; 16,140–3). The dilution series varied in decade steps from 0.1 μg/μl to 0.0001 μg/μl. The various one-, two- and three-part stimuli were produced as required by spotting, in turn, 0.5 μl of the appropriate dilution series onto a filter paper (160 mm^2) contained in each glass odour cartridge. Make-up mineral oil, where required to equalize the evaporative surface area, was added to each filter paper so that total liquid volume for each cartridge was 1.5 μl. The amount of 12:AC in each of the binary stimuli was held at a constant 10 per cent of the concentration of the individual unsaturated compounds. This mirrored the natural ratio found analytically in extracted pheromone glands (Bjostad *et al.* 1980). The ternary mixture contained 5 per cent 12:AC and a 1 to 1 mixture of *Z*7-12:AC and *Z*7-12:OH. Completed odour cartridges were sealed with Teflon plugs and caps and stored in individual Teflon-capped glass vials, which were in turn stored in an opaque container continuously purged with dry nitrogen (1.6 l/min). Between experiments the purged container was stored at 4 °C.

Individual stimuli were two seconds in duration and applied by passing purified oxygen-free nitrogen (60 ml/min) over the filter paper contained in each glass odour cartridge. The outlet of the odour cartridge was positioned 8 mm from the antennal surface. In the interstimulus interval the antenna was bathed in humidified pure air (Ultra-zero grade; Matheson; Inc., E. Rutherford, NJ) at a flow rate of 120 ml/min.

Each of the individual single- or multicomponent stimuli was evaluated at several intensities. These had total pheromone contents which ranged in three decade steps from a single component at 0.00005 μg to a ternary mixture containing 0.105 μg/cartridge. In general, the order of presentation for the different components and mixtures was randomized within each of the intensity ranges. The different intensity ranges were themselves presented in increasing order.

Control odour cartridges, which contained only 1.5 μl of mineral oil on filter paper, were interspersed with odour stimuli on a random basis, except that they were always given at least once at each odour intensity level. The occasional response to control stimuli detected inadvertent contamination of the odour line connections as they were always eliminated by changing the relevant Teflon fittings. The response measure reported here was derived by subtracting the number of action potentials produced by a particular neurone during the 10 s immediately preceding the two-second stimulus from the number of action potentials which occurred in the succeeding 10 s.

25.3. Results

A total of 52 sensilla were sampled from 34 individual males. Twenty sensilla produced recordings which were sufficiently stable to sort reliably the electrical activity elicited in their A and B olfactory receptor neurones through the 2.5 to 3.5 h required to evaluate the entire (n = 30) stimulus battery. The remaining 32 sensilla were exposed to only a portion of the stimulus battery. Eleven of these were stimulated with each of the cartridges comprising both the low and midrange stimulus brackets.

Each HS sensillum trichodea contains two olfactory receptor neurones. In this sample of sensilla the A receptor neurones had an average (mean ± SEM) spontaneous activity of 3.0 ± 0.6 impulses/10 s, whereas the B receptor neurones had an average of 3.2 ± 0.6 impulses/10 s. Although there were no systematic differences in the general levels of spontaneous activity between the two neurons within a single sensillum, there was a small but significant positive correlation ($r_s = 0.497$; $P < 0.005$; Siegel, 1956) between the spontaneous levels in the two neurones across the whole sample of sensilla.

25.3.1. Responses to single-component stimuli

Although there was considerable quantitative variation in the absolute levels of discharge obtained in response to the various single-component stimuli from one sensillum to another, there was, nonetheless, a marked tendency for uniformity in the relative pattern of discharge from one compound to another. For example, 19 of the 20 sensilla studied in detail contained A receptor neurones which were more strongly responsive to *Z*7-12:AC than were their companion B receptor neurones. An inverted relationship occurred when *Z*7-12:OH was used as the stimulus in this same set of sensilla. Here, 19 of 20 B receptor neurones were more strongly responsive to *Z*7-12:OH than were their companion A receptor neurones. In each case, the one exceptional sensillum produced responses which were among the smallest obtained in the whole sample. The responses elicited by 12:AC were uniformly small. None of the sensilla sampled contained olfactory receptors which were more responsive to this compound than they were to either of the unsaturated compounds. These response trends are readily apparent in Fig.25.1, which illustrates the average response functions obtained with the indicated single-component stimuli. Over the intensity range examined, the A receptor neurone has sensitivity for *Z*7-12:AC which is, on average, comparable to the B receptor neurones' sensitivity to *Z*7-12:OH at equivalent dilutions. Although

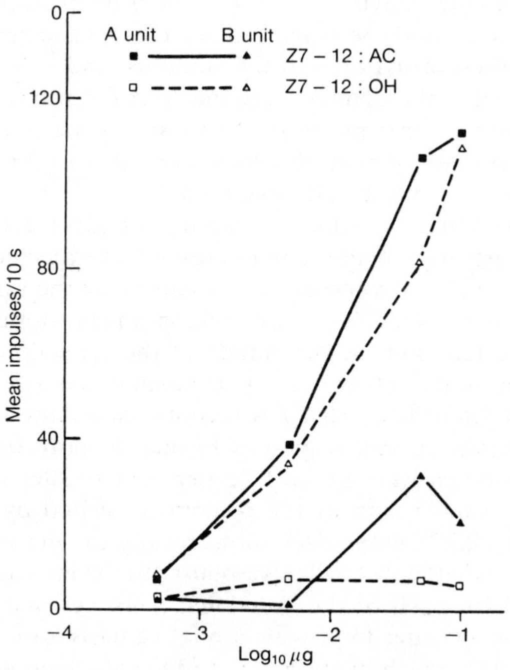

Fig. 25.1. The average response functions of the A and B olfactory receptor neurones when stimulated with (*Z*)7-dodecenyl acetate (*Z*7-12:AC) or (*Z*)7-dodecenol (*Z*7-12:OH). These neurones were contained within 20 HS sensilla trichodea. For purposes of clarity the average response functions for stimulation with comparable amounts of dodecyl acetate (12:AC) are not shown. The largest average A unit response to this compound was 8 impulses/10 s at 0.005 μg and the largest average B unit response was 1 impulse/10 s at 0.1 μg.

the average B receptor neurone responds preferentially to *Z*7-12:OH, it also exhibits a weak sensitivity to the higher intensities of *Z*7-12:AC. This sensitivity seems too large to be explained by assuming that the response is actually caused by contamination of the stimulus with *Z*7-12:OH because the measured level of cross-contamination among compounds is too small to produce the observed response. This result suggests that the average B receptor neurone is more broadly tuned than is the average A receptor neurone. Neither of the receptor neurones in the average HS sensillum is preferentially responsive to 12:AC. In fact, the mean response functions for this compound lie below that plotted for the A receptor neurones' response to *Z*7-12:OH and, thus, for the purposes of clarity are not plotted in Fig. 25.1.

25.3.2. Responses to mixtures

Addition of 10 per cent 12:AC to *Z*7-12:AC and to *Z*7-12:OH results in binary stimuli which tend to elicit enhanced responses in A (for *Z*7-12:AC + 12:AC) and B (for *Z*7-12:OH + 12:AC) receptor neurones when compared with the sum of the responses elicited in the same neurones by the individual components of the mixture. The number of neurones affected and the magnitude of the enhancement was dependent on the amount of material in the stimulus, and was maximal at the middle range of intensities. Nineteen of the 20 A receptor neurones and 13 of the B receptor neurones produced responses to one or more intensities of the appropriate binary mixture which were at least 25 per cent larger than those expected from the additive model.

In contrast, stimulation with the ternary mixtures elicited responses which were consistently smaller than the responses expected from a simple additive model, and in many cases were smaller than the responses elicited by either of the most effective one- or two-component stimuli. The number of neurones affected, and the magnitude of the reduction observed, was again dependent on the intensity range evaluated and was maximal at the middle range of intensities. All 20 A receptor neurones and 18 of the B receptor neurones produced responses to one or more intensities of the ternary mixture which were at least 25 per cent smaller than those predicted from the simple sum of the responses elicited by the individual components. Fig. 25.2 summarizes these results for the middle range of intensities in 31 individual sensilla. Responses to the indicated stimuli were first normalized for each of the individual neurones and then averaged across the sample in order to provide a relative basis for comparison. The addition of 12:AC, which alone is an ineffective stimulus, to *Z*7-12:AC significantly enhances the response of the A receptor neurones. In a like manner, addition of 12:AC to *Z*7-12:OH enhances the response of the B receptor neurones. In both cases little reciprocal enhancement was seen in the companion receptor neurones of a sensillum. The responses to the ternary mixture are significantly smaller than those produced by the

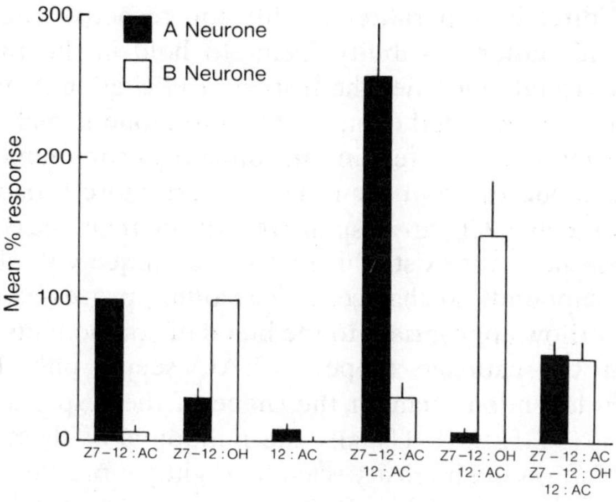

Fig. 25.2. Normalized (mean ± SEM) responses to the indicated stimuli, each presented at the middle range of intensities. These responses are from 31 HS sensilla including the 20 averaged in Fig. 25.1. For each A neurone the response to 0.005 μg of *Z*7-12:AC was set equal to 100 per cent and the responses of that neurone to the remaining stimuli were expressed as a percentage of that response. In a like manner, the response of each B neurone to 0.005 μg of *Z*7-12:OH was set equal to 100 per cent and the responses to the remaining stimuli expressed as a percentage of that response. The key to stimulus compounds is given in Fig. 25.1.

appropriate single- or binary-component stimuli and are equivalent in both receptor neurones. These reductions were not caused by the neurones' inability to generate sufficient numbers of action potentials, but must be ascribed to some event proximal to action potential production.

25.4. Discussion

It is axiomatic that odorants with demonstrable impact on an insects' behaviour must interact uniquely with sensory receptor neurones whose outputs terminate in an appropriate central neural structure. This must be so even if the compound of interest originates in the retort of a synthetic chemist rather than in the insects' own pheromone gland. As the number of compounds with behavioural significance increases, so too does the need for additional unique neural inputs capable of signalling the appropriate messages. This can be accomplished by creating a new primary receptor neurone specifically sensitive to each additional compound of interest, or by modifying the response properties of an existing receptor neurone to accommodate the additional pheromonal traffic. In the latter case there are at least two ways in which the response properties of a neurone may be expanded. In one, the ability to interact with new compounds is coupled directly to the action-potential-generating mechanisms so that an increased number of pheromone components become effective single stimuli. In the second, additional behaviourally significant compounds do not elicit spike activity directly, but rather modify the response elicited by other compounds. This latter possibility seems to hold in the cabbage looper where dodecyl acetate modifies the responses elicited in A and B receptor neurones by the unsaturated components, but alone is ineffective.

When information transfer in an olfactory communication system involves pheromone receptor neurones that are more broadly tuned, and thus, like those seen in CL, are responsive to more than one component of a signal, mechanisms must exist which preserve uniquely the identity of the stimulating compounds so that central decoding processes may elicit the behavioural outflow appropriate to the blend of components present in the periphery. In the cabbage looper, 12:AC seems only to alter in a concentration-dependent manner the shape of the response functions for *Z*7-12:AC and *Z*7-12:OH. Usually the midpoints of these functions are shifted to the left on the intensity scale by slightly more than a factor of 10 so that a given amount of neural discharge is generated in the receptor neurones by one-tenth the amount of compound required when the components are presented alone. This apparent lack of a unique alteration in neural outflow suggests that the presence of 12:AC in the CL pheromone blend does not signal a unique event in the behavioural repertoire of the male, but rather enhances or prolongs the behavioural sequence generated by other compounds. Although this suggestion is supported by behavioural experiments (Linn *et al*. 1984) additional components of the natural blend will have to be evaluated singly and in concert before the coding mechanisms operating in this species are fully revealed.

Acknowledgements

These studies have been supported by grants NS 14453 from NINCDS and BNS 8016395 from NSF. I thank M. Eaton and L. Hager for their assistance.

26 Correlating sensory and behavioural responses in multichemical pheromone systems of Lepidoptera

E. Priesner

26.1. Introduction

Pheromone communication systems composed of multiple chemical components are almost ubiquitous in Lepidoptera. O'Connell (1972, 1975) and Den Otter (1974, 1977) were the first to demonstrate that these multichemical signals are perceived via sensory systems composed of different types of receptor cells, each apparently tuned to one particular blend component. This principle has since received broad support from electrophysiological results obtained from different groups of Lepidoptera. One line of current research is concerned with establishing correlations between pheromonal behavioural actions and patterns of nerve impulse activity generated across these antennal multireceptor systems of male moths (see also Bartell and Rumbo (Chapter 19); Boeckh (Chapter 34); Kaissling (Chapter 22); O'Connell (Chapter 25); Van Der Pers and Löfstedt (Chapter 29)).

These complex neural patterns cannot be assessed from only a few sensillum recordings. Any experimental approach to the process of neural coding must, rather, be carried out in a stepwise fashion. Generally the analysis begins with a search for the different cell types present in the male antennal hair sensilla of the test species, followed by the determination of dose–response relationships to establish the relative effect of each behaviourally relevant test chemical on each cell type. The procedure is to be able to calculate, in a first approximation, the pattern of neural activity generated across the entire receptor system by each particular test stimulus. Unfortunately, as yet there is no consensus among researchers concerning a number of features basic to this stepwise analysis (such as the numbers of different cell types involved in a given test species, the response specificities shown by individual cells towards test chemicals, and the parameters of nerve impulse discharge relevant to the coding process). Evidently progress in this field is somewhat hindered as a consequence of the different experimental procedures used for the collection, analysis, and synthesis of electrophysiological data (see contributions by above authors).

Here I wish to outline briefly the experimental approach used in this laboratory for studying pheromone perception mechanisms in different groups of Lepidoptera. The major analytical steps will be illustrated and the procedure will be applied to the problem of neural coding of minor versus major components in multichemical lepidopterous pheromone systems.

26.2. Cell type combinations

Pheromone-sensitive receptor cells in Lepidoptera are located (on male antennae) within the olfactory hair sensilla known as the sensilla trichodea. These male receptor cells show 'odour specialist' response characteristics (considered further below) and may be categorized according to maximally effective ('key stimulus') compounds. A comparative study by this author of pheromone receptor systems in different groups of Lepidoptera has revealed that any given species uses between three and six different cell types. Furthermore, in all species examined more closely, the 'key compounds' for cell types combined within the same receptor system were generally closely related to each other in chemical structure (the usual modifications including changes with respect to the olefinic double-bond system, the length of the carbon chain, and the kind of functional group attached to a common carbon skeleton).

The prevailing pattern of cell type combinations found in Noctuidae and Tortricidae (the two families studied most extensively) consists of several types of acetate-specific receptor cells combined with one or two further cell types sensitive to a particular alcohol or aldehyde compound. The male summerfruit tortrix, *Adoxophyes orana*, uses a total of five different cell types, specific to the primary pheromone components *Z*9-14:Ac* and *Z*11-14:Ac*, the dienic analogues *Z*9,*E*11-14:Ac and *Z*9,*E*12-14:Ac, and the alcohol *Z*9-14:OH (Den Otter 1974, 1977; Priesner 1983). The latter type of alcohol receptor cell also occurs** in the male red-banded leafroller, *Argyrotaenia velutinana*, in combination with several types of acetate-sensitive cells including receptors for the two primary pheromone components (O'Connell 1972, 1975). In the Noctuidae, the cotton leafworms *Spodoptera littoralis* and *S. litura* both use a combination of specialist cells for *Z*9,*E*11-14:Ac, *Z*9,*E*11-14:Ac, 9-14:Ac, and *Z*9-14:OH (the same as found in *A. orana*), whereas male dotted clay moths *Rhyacia baja* have specialist cells for *Z*5-12:Ac, *Z*7-12:Ac, *Z*9-14:Ac, *Z*11-16:Ac, and *Z*7-12:OH (Priesner 1979a, 1984a). Male gamma moths and alfalfa loopers, *Autographa gamma* and *A. californica*, share the same five types of receptor cells specific to *Z*5-10:Ac, *Z*5-12:Ac, *Z*7-12:Ac, *Z*9-14:Ac and *Z*7-12:OH (Priesner 1980a); and male silver-V moths and tomato loopers, *Chrysodeixis eriosoma* and *C. chalcytes*, the combination of receptors for *Z*5-12:Ac, *Z*7-12:Ac, *Z*9-12:Ac, *Z*9-14:Ac, and *Z*7-12:OH (Benn *et al.*

*(*Z*)-9- and (*Z*)-11-tetradecenyl acetates. Compounds are generally listed with firstly the configuration and position of the olefinic double bond(s) and secondly the number of carbon atoms in the molecule chain (Ac, acetate; OH, alcohol; Ald, aldehyde).

**Throughout this report, data not specified by reference are from unpublished work of this author.

1982). In the pheromone receptor system of the male armyworm *Mythimna unipuncta*, specialist cells for *Z*7-12:Ac, *Z*9-14:Ac, and *Z*11-16:Ac occur in combination with cell types specific to the alcohol *Z*11-16:OH and the aldehyde *Z*11-16:Ald (Priesner 1979b). The combination of these five cell types appears to be common in Noctuidae, recurring in various test species from the subfamilies Agrotinae, Hadeninae and Amphiphyrinae.

Pheromone receptor systems composed only of acetate-specific receptor cells do occur in these two families, although less frequently. The male European fir budworm *Choristoneura murinana* uses a combination of specialist cells for *Z*9-12:Ac, *E*9-12:Ac, 11-12:Ac, *Z*9-14:Ac, and *Z*11-14:Ac, (Priesner and Bogenschütz 1982; Priesner *et al*. 1980), and male moths of *Eulia ministrana* have receptor cells for *Z*7-12:Ac, *Z*9-14:Ac, *Z*11-14:Ac, *E*11-14:Ac, and *Z*11-16:Ac (Priesner 1984b), with no evidence for the presence of alcohol- or aldehyde-sensitive cells obtained during receptor recordings from these two tortricid species. This observation also holds for some noctuid species, considered further below.

The complete set of receptor cell types present in a test species is not generally found within an individual hair sensillum. The electrophysiological data rather indicate a distribution of the various cell types over the male antennal sensilla trichodea as 'incomplete sets', mostly dominated by one or two cell types (the A and B cells) recorded with the largest nerve impulse amplitude. In some species these distributions indicate distinct 'physiological sensillum types', which could be correlated with morphological sensillum subtypes in some cases but not in others (see O'Connell (Chapter 25); Van Der Pers and Löfstedt (Chapter 27)).

26.3. Cell specificity data

The pheromone-sensitive receptor cells in male Lepidoptera are the paradigms of 'odour specialist' cells as defined by Schneider (1969). This concept states that individual cells, maximally responsive to the same ('key stimulus') compound within the same or taxonomically related species, also show a stereotyped pattern of graded responses to altered test chemicals, not differing within experimental limits. The relationship between stimulant dose and nerve impulse activity may be summarized into a 'response spectrum' of equipotent stimulus amounts for each given cell type and set of test chemicals.

Response spectra covering broader sets of systematically altered test chemicals have been reported in particular for receptor cells tuned to olefinic acetate, alcohol, and aldehyde pheromone components in Noctuidae and Tortricidae (Liljefors *et al*. 1984; Priesner 1979a,b, 1980a, 1983, 1984a). These spectra show a gradual loss in stimulatory potency of test molecules upon 'stepwise' alternating molecular features such as chain length, the olefinic double-bond system, or the terminal functional group.

TABLE 26.1

Stimulatory effectiveness of 15 (Z)-alkenyl acetates on four types of receptor cells located in antennal hair sensilla of male Polia pisi *(Noctuidae)*

Test compound	Effect on cell type A	B	C	D
(*Z*)-5-dodecenyl acetate	>1000	>1000	>1000	100
(*Z*)-6-dodecenyl acetate	>1000	1000	>1000	30
(*Z*)-7-dodecenyl acetate	>1000	100	1000	**1**
(*Z*)-8-dodecenyl acetate	100	300	1000	30
(*Z*)-9-dodecenyl acetate	30	100	1000	300
(*Z*)-7-tetradecenyl acetate	>1000	100	1000	100
(*Z*)-8-tetradecenyl acetate	1000	100	1000	
(*Z*)-9-tetradecenyl acetate	300	**1**	100	100
(*Z*)-10-tetradecenyl acetate	30	30	300	
(*Z*)-11-tetradecenyl acetate	**1**	100	100	>1000
(*Z*)-9-hexadecenyl acetate	>1000	300	300	1000
(*Z*)-10-hexadecenyl acetate	1000	300		>1000
(*Z*)-11-hexadecenyl acetate	100	100	**1**	1000
(*Z*)-12-hexadecenyl acetate	300	1000	30	>1000
(*Z*)-13-hexadecenyl acetate	100	1000	300	>1000

The values indicate equipotent stimulus amounts referred to a half-decade scale (see Priesner 1983, 1984a).

For example, with receptor cells tuned to different alkenyl acetates (the most common type of pheromone structure known from these two families), the displacement of the olefinic double bond from the optimum position by two carbon units, or the alteration of the optimum chain length by two carbon units, generally caused a loss in stimulatory effectiveness of the test molecules by 10^2–10^3 fold. Greater losses were observed for more extended alterations in these molecular features.

This may be illustrated by presenting (Table 26.1) response spectra to a set of 15 (*Z*)-alkenyl acetates by the four types of receptor cells located in the antennal hair sensilla of the male broom moth *Polia pisi* (Priesner 1980b). Two cells respond maximally to *Z*11-14:Ac and *Z*9-14:Ac, the primary components identified from the *P. pisi* female sex pheromone (Renou *et al*. 1981), and the two others to the non-pheromonal compounds *Z*11-16:Ac and *Z*7-12:Ac. For each cell type most of the analogues listed were effective within the measurable range as given by a thousand-fold increase of equipotent stimulus amount. The data also show that the 'key compounds' do elicit some responses even from other cell types of the same receptor system.

26.4. Calculating across-fibre sensory excitation

From the dose–response data, determined separately for each cell type and set of test chemicals, attempts can be made to calculate the pattern of relative nerve impulse activity generated across the various cell types of the system by a single chemical or a particular compound mixture. In this attempt the assumption is made that the different test compounds contribute to the nerve impulse outcome of a cell in a simple additive manner (the addition of 100 parts of a hundredfold less stimulatory analogue having the same effect on the cell response as has one part of the 'key' compound). The sum of these calculations, conducted independently for each cell type, is assumed to represent, to a first approximation, the 'across-fibre activation pattern' for this particular test stimulus.

The procedure is illustrated in Table 26.2 for the pheromone receptor system of *P. pisi* with respect to the four 'key compounds' (mentioned above) and some of their binary and ternary mixtures. The single compounds (stimuli no. 1–4), each maximally effective on its particular cell type, also have a measurable effect on the other cell types (as already indicated from the data given in Table 1). Accordingly, the 100/100 pheromonal mixture of *Z*11-14:Ac, *Z*9-14:Ac (stimulus no. 8) at higher concentrations also partially activates cell types C and D. It is as yet unknown whether the weak stimulation of these two 'inhibitory cells' by the attractive pheromone mixture has relevance for signal recognition.

Addition to this mixture of increasing amounts of either *Z*11-16:Ac or *Z*7-12:Ac selectively raises the nerve impulse activity of cell types C and D, respectively (stimuli no. 13–15 and 17–20 in Table 2), without affecting the pattern of activity of the two other cell types beyond that produced by the binary pheromonal mixture alone.

For the 20 stimuli listed, the pattern of relative cell type activation is related to the attractancy of these stimuli to *P. pisi* males as reflected by field trap captures (Priesner 1980b). This illustration will be used below for considering sensory perception of primary and potential secondary components in lepidopterous pheromone systems.

26.5. Sensory perception of major (primary) pheromone components

Usually only one or two chemical components from a multichemical sex pheromone blend released by a calling female moth are required for eliciting long-distance upwind orientation in the responding males. Electrophysiological evidence from different groups of Lepidoptera shows that each of these 'primary sex pheromone components' (Roelofs and Cardé 1977) appears to have a corresponding type of male specialist receptor cell.

There are species (not considered here) in which long-range male attraction appears to rely on the selective activation of only a single cell type

TABLE 26.2

Relationship between receptor and behavioural responses elicited by 20 synthetic test stimuli in male Polia pisi

Test No.	Stimulus amount of: Z11–14 : Ac	Z9–14 : Ac	Z11–16 : Ac	Z7–12 : Ac	Response of cell type A	B	C	D	Attractancy in field
1	100	–	–	–	◯	o	o	∘	–
2	–	100	–	–	o	◯	o	o	–
3	–	–	100	–	o	o	◯	∘	–
4	–	–	–	100	∘	o	∘	◯	–
5	100	3	–	–	◯	o	o	∘	–
6	100	10	–	–	◯	O	o	∘	–
7	100	30	–	–	◯	◯	o	∘	+
8	100	100	–	–	◯	◯	o	o	+ + +
9	30	100	–	–	◯	◯	o	o	+
10	10	100	–	–	O	◯	o	o	–
11	3	100	–	–	o	◯	o	o	–
12	100	100	1	–	◯	◯	o	o	+ + +
13	100	100	10	–	◯	◯	O	o	+ + +
14	100	100	30	–	◯	◯	O	o	+ +
15	100	100	100	–	◯	◯	◯	o	(+)
16	100	100	–	1	◯	◯	o	o	+ + +
17	100	100	–	3	◯	◯	o	O	+ +
18	100	100	–	10	◯	◯	o	O	+
19	100	100	–	30	◯	◯	o	◯	(+)
20	100	100	–	100	◯	◯	o	◯	–

Each of the four chemicals in the 'key compound' for a particular type of sensory cell. Size of circles symbolizes nerve impulse discharge as derived from dose–response data (see also Table 26.1). Field attractancy data taken from Priesner (1980b). Stimulus composition given in arbitrary units relative to major component(s).

from the receptor system. The simultaneous activation of further cell types, by release of additional blend components, terminates the response. In other cases, the electrophysiological and behavioural data indicate that proper activation of both the A and B type cells of the receptor system is required for keeping the male moths in upwind flight over some distance. Thus, male red-backed cutworm moths, *Euxoa ochrogaster*, showed upwind flights in a wind-tunnel only in response to binary combinations of *Z*5-12:Ac and *Z*7-12:Ac (Palanismawy *et al*. 1983), as did male black cutworm moths, *Agrotis ypsilon* in response to combinations of *Z*7-12:Ac and *Z*9-14:Ac (Hill *et al*. 1979). These compounds are the identified primary sex pheromone components of these two species, each activating a particular type of male sensory cell. Similarly in *Argyrotaenia velutinana* the combination of *Z*11-14:Ac and *E*11-14:Ac, in the proper ratio, was required for eliciting male long-range upwind orientation (Baker *et al*. 1976; Roelofs and Cardé 1977; Roelofs *et al*. 1975); the specialist receptor cells for these two compounds present on *A. velutinana* male antennae had already been studied by O'Connell (1972, 1975).

The sex pheromone system of *P. pisi* includes two primary components, *Z*11-14:Ac and *Z*9-14:Ac, isolated from the female secretion in a ratio of close to 1:1 (Renou *et al*. 1981). Detailed behavioural studies on the effects of the two compounds on male orientation responses are lacking, but the total failure of the single chemicals to capture male moths in the field (Priesner 1980b and Table 26.2) suggests that both components are primary.

26.6. Sensory perception of minor (secondary) pheromone components

Many pheromone blends of Lepidoptera are known to include constituents which in behavioural tests were not essential for eliciting upwind anemotaxis in conspecific males but rather induced close-range responses such as landing, wing fanning, and copulatory attempts (Roelofs and Cardé 1977). Although most of these 'secondary pheromone chemicals' act upon their own specialized male sensory cells, perception of these minor components is limited by the primary stimulatory effect caused by the major blend constituent(s) on the various cell types combined within the particular receptor system.

With combinations of receptor cells specific to different (*Z*)-alkenyl acetates separated by not more than two carbon units with respect to chain length or double-bond position (as found in the pheromone receptor systems of many noctuid species), minor components can barely be perceived at a level of 0.1 per cent or less of the major ones. Hence, a receptor system organized such as in *P. pisi* theoretically could accommodate one or two minor pheromone constituents, but only at amounts of 1 per cent or higher relative to the major ones. In the chemical analysis of the *P. pisi* pheromone gland effluents, no *Z*11-16:Ac or *Z*7-12:Ac were detected

(Renou *et al*. 1981), and in field tests the addition of varying amounts of these compounds to the 1 : 1 pheromonal mixture did not reveal any synergistic properties upon trap captures (Priesner 1980b). There are, however, other moth species which do accommodate three pheromone components by using the same four cell types as found in *P. pisi*. Female pine beauty moths, *Panolis flammea*, release a ternary mixture of Z9-14:Ac/Z11-14:Ac/Z11-16:Ac in a ratio of 100 : 1 : 5 (Baker *et al*. 1982), and the *P. flammea* male hair sensilla have been shown to contain specialist receptor cells for these three compounds and the non-pheromonal chemical Z7-12:Ac (Priesner and Schroth 1983; Priesner *et al.* 1978). In wind-tunnel tests, the combination of Z9-14:Ac/Z11-14:Ac induced part of the male behavioural sequence from flight initiation until landing at the source, whereas addition of the other minor component, Z11-16:Ac, was required for inducing male copulatory attempts (Bradshaw *et al*. 1983; Howse *et al*. (Chapter 17)).

Sensory perception of minor pheromone components at levels of 0.1 per cent or even less relative to major blend constituent(s) is achieved by using receptor systems that combine cell types whose 'key compounds' have a far lower (in comparison to the *P. pisi* system) mutual stimulatory effect on the other cell types. In *E. ochrogaster* (mentioned above) landing near the attractant source is induced by the release of a secondary pheromone component, Z9-12:Ac (Palaniswamy *et al*. 1983). This compound is effective behaviourally at a level of 0.25 per cent of the major blend constituent, Z5-12:Ac, in agreement with a thousand times lowered stimulatory effect of the Z5-12:Ac on the Z9-12:Ac receptor cell as determined in single cell recordings. A similar pattern applies to male moths of *R. baja* which, in field trapping tests, were maximally attracted to a ternary Z5-12:Ac/Z7-12:Ac/Z9-14:Ac mixture in a 0.3 : 10 : 100 ratio (Priesner 1984a). This attraction pattern again conforms to single-cell data obtained from *R. baja* male antennae, indicating an effect of the Z9-14:Ac on the Z5-12:Ac receptor cell of a thousandfold below the 'key' stimulus (Priesner 1984a).

Another system permitting the perception of extremely low amounts of minor blend constituents is provided by the combination of receptor cells tuned to compounds having different functional groups (acetate, alcohol, aldehyde), as mentioned above for the armyworm *M. unipuncta*. In field trapping tests, responses of male *M. unipuncta* to the primary pheromone component Z11-16:Ac were strongly enhanced by the addition of 0.2 per cent of the Z11-16:OH and 0.5 per cent of the Z11-16:Ald (Steck *et al*. 1980; Steck *et al*. 1982), indicating a combination of two 'trace synergists' with a single primary component.

With all 'trace co-attractants' studied more closely, the 'synergistic' effect upon trap captures was found to be limited to a narrow range of stimulus amount. Higher doses of the same compound strongly 'inhibited' the attraction of males. Thus, Z9-12:Ac in *E. ochrogaster,* Z5-12:Ac in *R. baja*, or Z11-16:OH in *M. unipuncta*, were synergistic at levels of 0.2

to 0.3 per cent relative to the major blend constituent, but were all 'inhibitory' at the 1 per cent level (Palaniswamy *et al.* 1983; Priesner 1984*a*; Steck *et al.* 1980, 1982) (Similar findings were reported for sawflies by Löfquist (Chapter 13)). Examination of receptor cell activity showed that these compounds, in their proper blend ratio, elicited nerve impulse responses from the respective receptor cell type of a level just slightly above that produced by the primary blend component(s) alone. In fact, in *R. baja* the 'synergistic' range of the *Z*5-12:Ac had been inferred from receptor data (Priesner 1984a), and would probably have been missed by empirical field screening alone. The study of across-fibre cell type activation in lepidopterous pheromone receptor systems has proved highly valuable for the detection of 'trace synergists' and the prediction of their effective ranges (including chemicals currently classified as 'attraction-inhibitors').

26.7. Experimental limitations

In this brief illustration of the author's approach used for correlating sensory and behavioural data in the responses of male Lepidoptera to pheromonal stimuli, a number of simplifying assumptions had to be made. No reference was made to possible differences in the relative abundance among cell types, the different response thresholds and intensity ranges as found for individual cells in the electrophysiological recordings, the compound-specific temporal patterns of nerve impulse discharge as reported for pheromone-sensitive receptor cells in some moth species, or the potential selective adaptation of cell types from the receptor system during the process of male upwind orientation into higher plume concentrations. Although these various parameters may contribute to the coding process (see Bartell and Rumbo (Chapter 19); Boeckh (Chapter 34); Kaissling (Chapter 22); O'Connell (Chapter 25); Van Der Pers and Löfstedt (Chapter 27), there was no evidence for a crucial role of any of these in the species examples referred to here.

Nerve impulse activity patterns were related in Table 26.2 to trap captures, used as a measure for the behavioural outcome. With all compound combinations listed that revealed a maximum capture rate, an orientation sequence by the responding males similar to that shown towards the natural (pheromonal) stimulus is likely to have occurred. However, a lowered capture rate (given by ++ or + in Table 26.2) may have been due to a smaller number of initial upwind flight responses elicited from the male population by some test stimuli, but a modified close-range behaviour for others. This indicates the need for using better defined behavioural data. Current work is concerned with correlating electrophysiologically-determined patterns of nerve impulse activity with successive stages of the male orientation sequence, using species in which quantitative behavioural data are available for sets of test stimuli with defined chemical composition.

27 Signal–response relationship in sex pheromone communication

J. N. C. Van Der Pers and C. Löfstedt

27.1. Introduction

Under natural conditions insects are surrounded by a great variety of air-borne chemicals, only a limited number of which are of biological significance. The ability of insects to detect and identify minute amounts of compounds of behavioural importance is particularly well developed for moth sex pheromones. Sex pheromones of harmful insects are being analysed for practical purposes, and attention is mainly focused on compounds that cause attraction of conspecifics (Kydonieus and Beroza 1982; Minks 1984). Lack of information about chemicals that induce other types of sexual behaviour may have resulted in distortion of our knowledge about sex pheromones.

Full understanding of sex pheromone communication should include a complete knowledge of the characteristics of both the sender (the production of the pheromone gland) and the receiver (the olfactory receptor system) in related and non-related species. Such knowledge would provide information not only about the degree of matching between the composition of the sex pheromone and the sensitivity spectrum of the olfactory receptors, but also about the insect's ability to recognize signals from other species or sources.

Löfstedt *et al*. (1982) analysed the contents of sex pheromone glands of the turnip moth *Agrotis segetum* (Schiff.) (Lepidoptera: Noctuidae) and identified 13 aliphatic acetates and alcohols with possible biological activity. A blend of four compounds was proposed, on the basis of behavioural experiments, to be a good approximation of the natural female sex pheromone. The present paper describes the properties of the male olfactory receptors with respect to these and some other compounds. The male sensitivity spectrum has been measured by electrophysiological methods. The data from the Swedish population is compared with those obtained from French and English populations of *A. segetum* moths. In addition, similar data from four closely related species of the genus *Yponomeuta* (Lepidoptera: Yponomeutidae) are presented.

27.2. Materials and methods

Swedish *A. segetum* moths were obtained from the culture used by Löfstedt *et al*. (1982). French insects were provided by Dr R. Bues, Station de

Zoologie, Montavet, France, and the English moths were obtained from Dr P. L. Sherlock, Rothamsted Experimental Station, Harpenden, England. Insects were kept under standard conditions as described by Löfstedt *et al.* (1982). Pupae and moths of *Yponomeuta cagnagellus*, *Y.padellus*, *Y.evonymellus* and *Y.vigintipunctatus* were provided by Dr W. M. Herrebout, Leiden, The Netherlands, or collected in the field in southern Sweden.

Responses from male olfactory receptors were recorded using electroantennogram (EAG) and single-sensillum techniques combined with stimulation methods as described by Van Der Pers and Den Otter (1978) and Van Der Pers (1982). In all experiments, stimulation with synthetic test chemicals was performed using an amount of 1 μg of the odour sources.

27.3. Results and discussion

Averaged EAG responses evoked during stimulation of Swedish *A.segetum* males with 16 different test compounds are presented in Fig. 27.1. The underlined compounds constitute a blend that induces male behaviour comparable to that evoked by the female sex pheromone (Löfstedt *et al.* 1982). Two of these, (*Z*)-5-decenyl acetate (*Z*5-10:Ac) and (*Z*)-7-dodecenyl acetate (*Z*7-12:Ac) evoke a high response, whereas (*Z*)-9-tetradecenyl acetate (*Z*9-14:Ac) evokes almost no response. It has been found, however, that the latter compound is absolutely necessary for good attraction and is produced by the female gland in relatively large amounts (Löfstedt *et al.* 1982). It is also remarkable that the response to

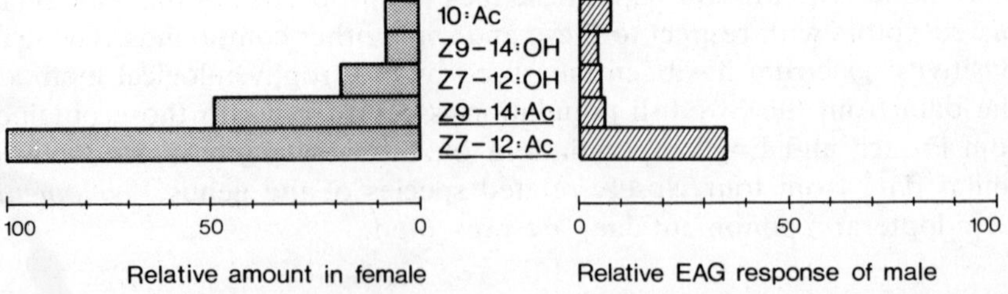

Fig. 27.1. Relative amount of aliphatic acetates and alcohols found in sex pheromone glands of females of *A. segetum* (after Löfstedt *et al.* 1982) (left side of figure) compared with the relative EAG response evoked by these compounds in conspecific males (right side of figure) (underlined compounds: see text).

*Z*5-10:Ac is very high, whereas this compound is found only in small quantities in the female gland. There is obviously no simple relationship between the amounts produced by the female and the response of the male antenna.

A more detailed impression of the male receptor response can be obtained from recordings of single olfactory sensilla on the male antenna. The results of these experiments are presented in Fig. 27.2. Recordings were made from 100 randomly chosen male sensilla trichodea containing two or three receptor cells (Hallberg 1981), two of which were active and recognizable by their difference in spike amplitudes. During the experiments sensilla were not selected and all recordings were taken into account (only two sensilla did not show any activation). To avoid misinterpretation of the results, responses to the 16 test stimuli are presented as individual reactions from the receptor cells in the 100 sensilla and not as the average of these responses. The individual responses are expressed as spikes per second of stimulation.

The high EAG response (Fig. 27.1) to *Z*5-10:Ac seems to be based on the large number of sensilla containing a receptor sensitive to this compound. The lower EAG response evoked by *Z*7-12:Ac (Fig. 27.1) can be explained by the same argument, that is, by the lower number of cells activated by this compound. The EAG response associated with (*Z*)-5-decenyl alcohol (*Z*5-10:OH), however, can not be attributed to the number of activated receptor cells. Based on the single sensillum measurements (Fig. 27.2) a higher EAG response would be expected than has been recorded (Fig. 27.1). This may be explained by the small spike amplitude of the receptor cells (B type cell, Fig. 27.2), which, in the recordings, was associated with relatively small receptor potentials. Consequently, the EAG generated by these receptor cells, as a result of their receptor potentials, is lower.

Only two sensilla contain a cell showing a good response to *Z*9-14:Ac. Although the sensitivity of these cells is not low, their activity is hardly expressed in the EAG (Fig. 27.1) because of their low number (2 per cent). This compound would not have been proposed as a possible pheromone component on its EAG response alone.

On the other hand, two compounds not found to be a component of the female sex pheromone of *A. segetum*, (*Z*)-8-dodecenyl acetate (*Z*8-12:Ac) and *Z*5-10:OH, activate a relatively large number of receptor cells (Fig. 27.2). Arn *et al.* (1980) reported the presence of *Z*8-12:Ac in *A. segetum* female gland extracts, but found that it was an inhibitor for male attraction. When tested in the field *Z*5-10:OH also seemed to reduce attraction when added to an attractive blend (Löfstedt *et al.*, 1985).

The compound (*Z*)-7-dodecenyl alcohol (*Z*7-12:OH) was identified from female gland extracts by Löfstedt *et al.* (1982) and has been reported to activate receptors on male *A. segetum* antennae (Priesner, cited in Arn *et al.*, 1980). In our experiments, however, this compound did not elicit a

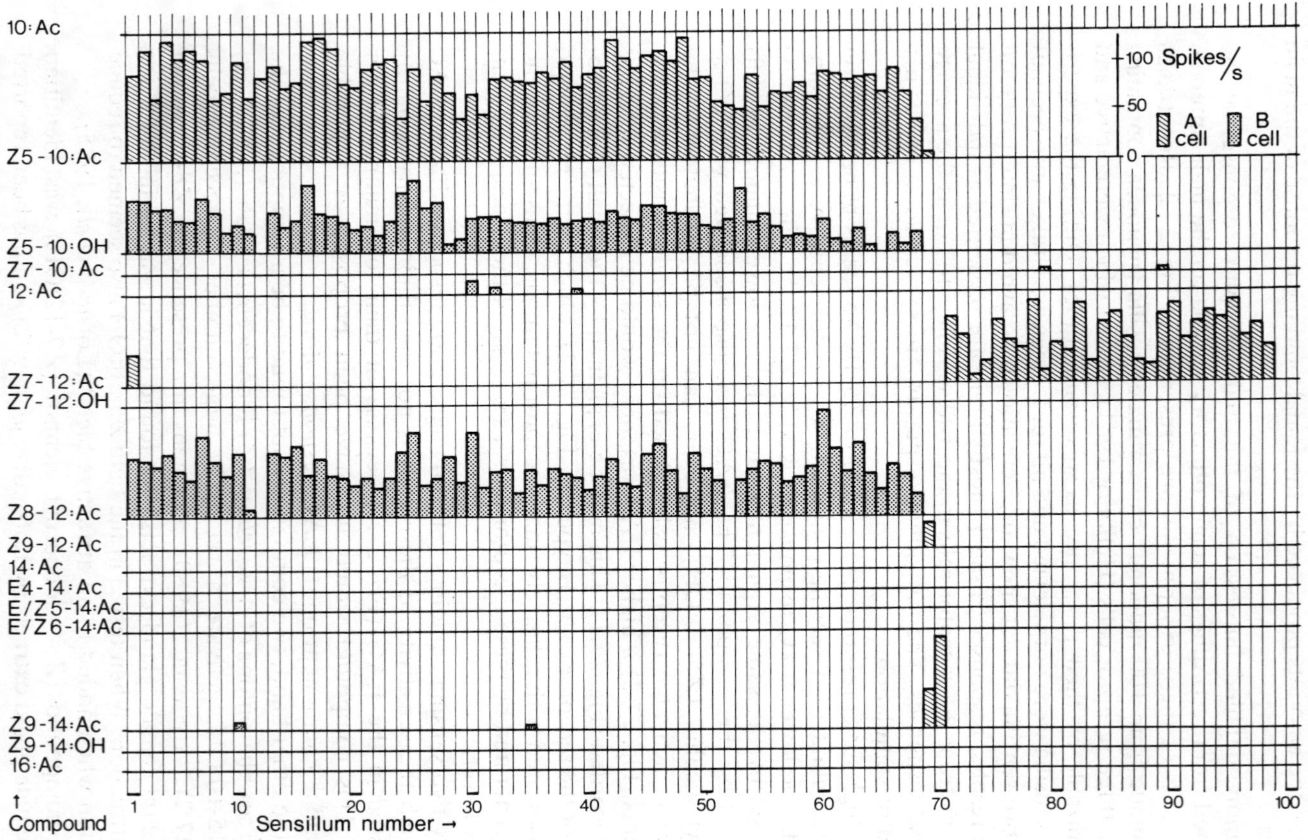

Fig. 27.2. Responses evoked by stimulation with 16 compounds from single antennal receptor cells measured from 100 male antennal sensilla trichodea of *A. segetum*. Cell A is characterized by large spike amplitudes, cell B by small spike amplitudes. Response scale value refers to all responses.

significant EAG response (Fig. 27.1), nor did we find a receptor activated by this compound (Fig. 27.2)). The possibility cannot be excluded that such a receptor cell would be found in more extensive studies.

It is possible that the difference between the results of Arn *et al*. (1980) and our findings are due to variation between geographical races of the species. By extensive field experiments on locations in Denmark, Switzerland, France, and Hungary Arn *et al*. (1983) found significant differences in attraction of males depending on the composition of a three-component blend. It would be interesting to establish whether such differences also are reflected in variations of the olfactory receptor system among males of different populations.

Preliminary results of such a comparison are shown in Fig. 27.3. Single sensillum activity to a set of 12 test substances were recorded on antennae of male *A. segetum* moths from locations in France and England, and the results compared with those from the Swedish insects. It is speculative to draw conclusions from the limited number of experiments that could be performed, but there is an indication that significant differences do indeed occur.

The proportion of sensilla containing receptors sensitive to *Z*5-10:Ac is highest in the French population. This corresponds with the result of field attraction studies and chemical investigations. Arn *et al*. (1983) found that of four European populations tested only the French one responded to *Z*5-10:Ac presented as a single compound in the field. Löfstedt *et al*. (in press) found that the relative abundance of this compound is much higher in the French females than in females of other European populations analysed. Thus there seems to be an interesting correspondence between

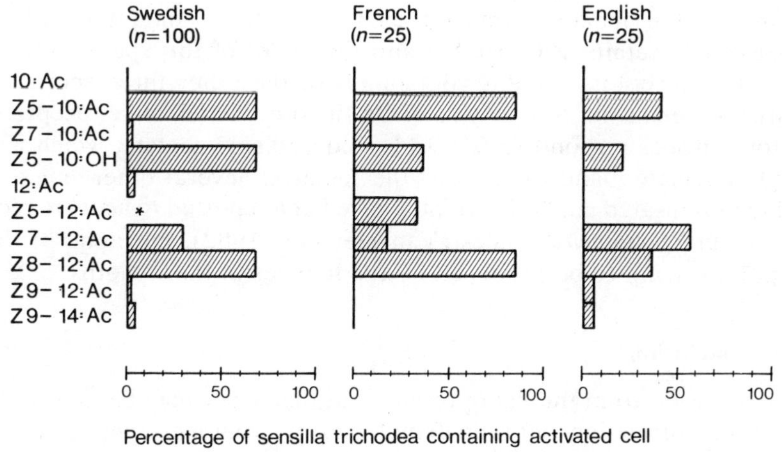

Fig. 27.3. Comparison of male antennal olfactory receptor responses among three geographically different populations of *A. segetum* moths during stimulation with 12 different compounds (*: not measured).

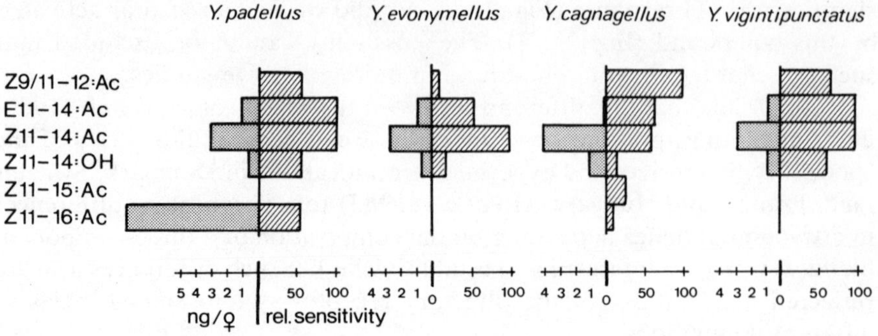

Fig. 27.4. Comparison between amount of six compounds produced by female *Yponomeuta* moths of four closely related species (left side in each diagram; nanogram per female), and the male antennal receptor response evoked by these compounds in conspecific males (right side in each diagram).

behaviour, electrophysiology and female production. In all these respects *Z*5-10:Ac seems to be more important in the French population than in the others.

Identification of the sex pheromones of four species of small ermine moths of the genus *Yponomeuta* (Löfstedt and Van Der Pers, 1985) has recently permitted a similar comparison between the female sex pheromone signal and male olfactory receptor response. The electrophysiological characteristics of the male antennal receptors were described by Van Der Pers (1982). Females of the four species studied so far (Fig. 27.4) mainly produce *E*- and *Z*11-14:Ac and *Z*11-14:OH. Receptors specialized for these compounds occur on the antennae of males of all four species, but their response profiles differ (Van Der Pers 1982). In *Y. padellus* the female attractant contains a high amount of (*Z*)-11-hexadecenyl acetate (*Z*11-16:Ac) and the males of this species are provided with receptors for this compound. In the other three species this compound seems not to play a role. On the other hand, male receptors of the four species respond to (*Z*)-9,11- dodecadienyl acetate, which is not found in female gland extracts of the females. Several other single and double unsaturated aliphatic acetates have been reported to activate receptor cells in *Yponomeuta* males (Van Der Pers 1982), but none of these compounds seem to be produced by the females of these moths.

27.4. Conclusion

The conclusion from the data presented in this paper may be that there is no simple correlation between female sex pheromone composition and male olfactory receptor system. To a certain degree, male receptors are tuned to conspecific female sex pheromone components, but they are also able to detect non-specific substances. This ability might enhance the male's discrimination of the specific sex pheromone from the complicated 'background' consisting of non-specific pheromone compounds and volatiles of other origin. Geographical variations in sex pheromone composition occur between populations of the same species, and they may be reflected in variations of the conspecific male olfactory sensitivity spectrum. The electrophysiological data obtained from males of *A. segetum* (Fig. 27.1 and 27.2) show that the predictive value of such data for sex pheromone identification is rather limited, especially with respect to EAG measurements. Furthermore, our investigation of 100 sensilla of males of *A. segetum* (Fig. 27.2) show that receptors sensitive to certain sex pheromone components might occur in very low numbers. It is also clear that the sensitivity of the individual receptor cells is highly variable. Recordings from 1000 rather than from 100 sensilla are necessary to find receptors specialized to trace components and to improve our knowledge about sex pheromone communication in insects.

28 Peripheral chemoreceptors and regulation of insect behaviour

E. E. Davis

28.1. Introduction

The behaviour of an organism is generally accepted to be controlled by processes in the central nervous system (CNS), whereas the role of the peripheral sensory system is to provide the organism with information about its environment on which the CNS can act (Kandel *et al*. 1979), that is, the peripheral sensory system serves to modulate or affect the intensity of a behavioural response. The literature is replete with examples of just such modulatory activity. Several excellent examples were presented earlier in this volume.

The theory assumes that the response characteristics of the peripheral sensory organs are relatively unchanging and that they continuously provide the CNS with sensory information relevant to all behaviours in the insect's repertoire; however, because of some unspecified central—that is, CNS—inhibitory (or excitatory) gating mechanism, the information is not always acted upon. Alternatively, centrifugal efferent activity from the CNS may prevent the transmission of information from a peripheral sensory neurone by direct inhibition of its output. In neither case are the response properties of the sensory receptors being modified. It is also presumed that any switching between behavioural modes is due solely to control mechanisms in the CNS.

The purpose of this chapter is to challenge this view and to offer an additional and important new role for the peripheral sensory organs—specifically, that the peripheral sensory organs may also act to determine which behavioural mode the insect will display in addition to the traditional modulatory role stated above.

I will present evidence to demonstrate that the initial onset of host-seeking behaviour following adult emergence is linked to the development of high receptor sensitivity to the host-attractant, lactic acid (LA), and that the hormonally related inhibition of host-seeking behaviour following a blood meal is linked to a depression of LA sensitivity due to a haemolymph-borne factor associated with oöcyte development. The correlations between LA-receptor sensitivity and host-seeking behaviour are 1 : 1 in both the above cases. This strongly suggests a causal relationship between the sensitivity of a peripheral receptor and the mediation of a behaviour whose intensity is also modulated by the receptor. The organism, the yellow fever mosquito, *Aedes aegypti*, from which these examples are taken, may not be representative of Insecta as a class. However, there

TABLE 28.1

Comparison of electrophysiological reactions of LA-excited neurones to LA from host-seeking and non-host-seeking female mosquitoes

Behavioural response level	Age of female mosquitoes (hours post-emergence)	No. of females tested	No. of sensilla responded/ tested*	Electrophysiological reactions				
				Spikes absent	Spikes present			
					No response†	Non-specific‡	Low sensitivity§	High sensitivity‖
Host-seeking behaviour absent	0–24	4	11/11	11¶				
	25–48	2	8/8	1	3	3	1	
	49–72	2	7/7	1	1	3	2	
	73–96	2	5/5		3		2	
Totals		10	31/31	13	7	6	5	0
% Sensilla Responding**				41.9	22.6	19.4	16.1	

Host-seeking	18–24	1	3/3		1			2
behaviour	25–48	6	11/13	1				10
present:	49–72	6	20/23		3	2		15
Ages	73–100	3	11/12		3		2	6
<100 h PE								
Totals		16	45/51	1	7	2	2	33
% Sensilla Responding				2.2	15.6	4.5	4.5	73.3
Host-seeking	106–120	1	3/4					3
behavior	142–148	1	2/2					2
present:	169–216	1	7/7				2	5
ages	336–360	2	8/9				1	7
>100 h PE	(14 to 15 d)							
Totals		5	20/22	0	0	0	3	17
% Sensilla Responding							15.0	85.0

* Number of sensilla responded/no. tested: a small number of sensilla either were damaged on penetration ($n = 4$) or the responses of the neurones were mixed—i.e. both LA-excited and LA-inhibited nerve cells were present—such that it was not possible to quantitate responsiveness ($n = 4$).

† No response: spikes present but no chemical or physical stimuli elicited a change in spike frequency.

‡ Non-specific: spikes present and all chemical and physical stimuli elicited changes in spike frequency.

§ Low sensitivity: spikes present and response relatively specific for LA. However, stimulus intensity/response function either too weak to obtain or had thresholds $\geqslant 10\times$ less sensitive to LA than high sensitivity cells.

‖High sensitivity: spikes present and relatively specific for LA. Stimulus intensity/response functions had thresholds at least $10\times$ more sensitive to LA than low-sensitivity cells.

¶ Number of sensilla responding in each category in each age group.

** Percent of sensilla responding in each of the three behavioural categories.

Reprinted with permission from the *Journal of Insect Physiology*, **30**. E. E. Davis: Development of lactic acid-receptor sensitivity and host-seeking behaviour in newly emerged female *Aedes aegypti* mosquitoes (1984).

are several examples of situations where similar causal relationships may be found in a variety of insect orders, which suggests that peripheral sensory mediation of behaviour may be more widespread than we might at first suspect. A few of these cases will be presented below.

28.2. The development of receptor sensitivity and the initiation of a behaviour

The first point to be considered deals with the potential role of the peripheral sensory organs in the mediation of a behaviour that has not been observed previously.

Adult female mosquitoes will not engage in host-seeking behaviour until 30 to 100 h post-emergence (Fig. 28.1; Davis 1984a). This appears to be due to an insensitivity of the antennal olfactory receptors for the host-attractant, lactic acid. It is not until between 30 and 100 h post-adult emergence that these receptors achieve a level of activity observed in 5- and 15-day-old avid host-seeking female mosquitoes (Table 28.1; Davis 1984a). During the transition period of a population of newly emerged females, one can see an age-dependent progression in the development of LA sensitivity, from the complete absence of spikes—both spontaneous and evoked—in non-host responsive females, to neurones exhibiting spontaneous spike activity but responding in a non-specific or insensitive manner to a variety of chemical stimuli, and, finally, to neurones that are highly sensitive to LA. High LA sensitivity is observed *only* in host-seeking females. The time (age)-dependent coincidence of LA-receptor sensitivity correlates 1:1 with host-seeking behaviour. This correlation strongly suggests that the development of sensitivity and specificity in a chemosen-

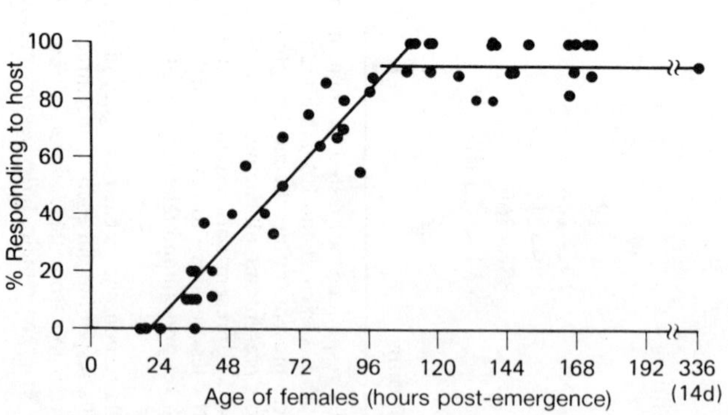

Fig. 28.1. Onset of host-seeking behaviour following emergence. Each point is at the mid-point of the age range of individual trials, with 7 to 10 females per trial. Linear regression line for the data for $x \leq 97$ hr post-emergence is $y = 1.11x - 22.7$, $r^2 = 0.91$. Reprinted with permission from the *Journal of Insect Physiology*, **30**, E. E. Davis: Development of lactic acid-receptor sensitivity and host-seeking behavior in newly emerged female *Aedes aegypti* mosquitoes (1984).

sory neurone is necessary before the behaviour that it modulates may be expressed.

Several attempts have been made to determine whether a hormone—specifically juvenile hormone (JH)—is involved in the onset of host-seeking behaviour. Allatectomized female *Culex pipiens* and *C. quinquefasciatus* failed to develop biting behaviour by seven days after adult emergence (Meola and Petralia 1980). Treatment of the allatectomized females with JH-I or the reimplantation of corpora allata restored biting behaviour. JH deprivation in *Aedes aegypti* by allatectomy prior to eclosion did not prevent the appearance of biting behaviour on day 7 (Meola personal communication). However, these investigators were testing biting behaviour, *not* host-seeking behaviour. It may be that biting behaviour can be induced in the absence of host seeking (Mitchell 1983). Female *A. aegypti* allatectomized at 1 h post-emergence were found to develop host-seeking behaviour by day 3 (Klowden 1983). However, it has been suggested (Feinsod and Spielman 1980) that JH release occurs before adult emergence, and hence it would have already been present when the female *A. aegypti* were allatectomized at 1 h post-emergence. Thus, the role of JH in the development of host-seeking behaviour *per se* is unresolved.

28.3. Regulation of receptor sensitivity and the control of behaviour

The second point to be considered involves the potential role of the peripheral sensory organs in switching the behaviour of female mosquito from host seeking to oviposition.

Once a female mosquito has located a host and taken a blood meal, further host-seeking behaviour is prevented in one of two ways. First, abdominal distention resulting from engorgement inhibits host seeking via a neural pathway involving the ventral nerve cord (Klowden and Lea 1979b). This inhibition is transient; host-seeking behaviour returns upon relief of the abdominal distention. A second phase of host-seeking inhibition begins to appear about 30 h after the blood meal and lasts until the female oviposits. This delayed inhibition appears to be mediated by a haemolymph-borne factor, because it can be transferred by injecting haemolymph from a blood-fed female into a sucrose-fed female (Klowden and Lea 1979a). Although the precise identity and source of this factor are not yet known, vitellogenic ovaries are required to trigger its release (Klowden 1981).

Experiments conducted in my laboratory suggest that the probable cause of oöcyte-induced inhibition of host-seeking behaviour is a reduction of the sensitivity of the peripheral receptor for LA. The time-course for the reduced LA sensitivity is coincident with that for the inhibition of host-seeking behaviour. The reduction in LA sensitivity is seen as early as 24 h post blood meal (PBM), reaches a maximum about 48 h PBM, and recovers to pre-blood-meal levels following oviposition (Davis 1984b;

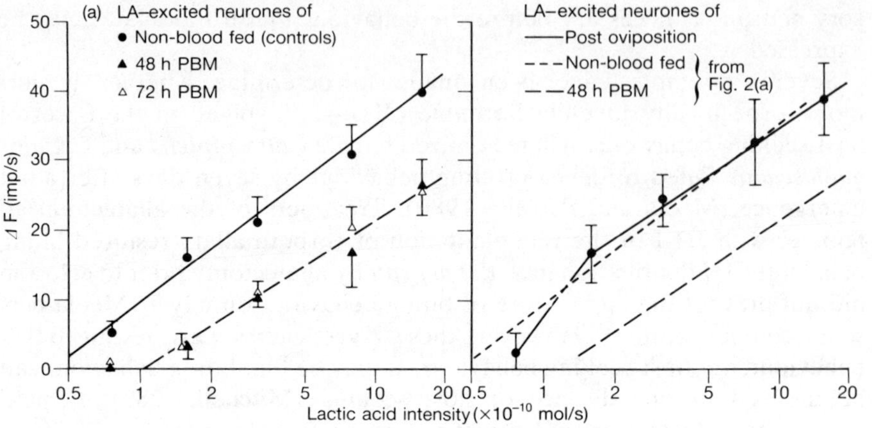

Fig. 28.2. Mean response functions for the LA-excited neurones (a) non-blood-fed (●—●), and 48 (▲), and 72 (△) h PBM (---), and (b) 24 to 72 h following oviposition (●---●). In (b) the dotted line is the non-blood-fed curve and the dashed line is the post-blood-feeding curve, taken from (a) (the vertical lines indicate the SEM). Reprinted with permission from the *Journal of Insect Physiology*, **30**, E. E. Davis: Regulation of sensitivity in the peripheral chemoreceptor system for host-seeking behavior by a hemolymph-borne factor in *Aedes aegypti* (1984).

TABLE 28.2

Behavioural responses and LA sensitivity with respect to ovarian development following a blood meal

	Follicular stage	
	⩽2	⩾3
Behavioural response	(no. of females):	
Responders	4*	0
Non-responders	7†	11
LA-sensitivity	(no. of sensilla):	
High	14‡	0
Low	0	16§

* Three of 4 were ⩾28 h post-blood meal (PBM) (1 at 18 h PBM) and were probably 'doubly anautogenous'.
† Behavioural inhibition may also be attributed to abdominal distention (4 were ⩽6 h PBM; 3 were ⩽30 h PBM (still engorged-looking).
‡ No difference in LA-sensitivity of those of oöcyte stage ⩽2 regardless of the length of time since blood meal. *All* had high LA sensitivity.
§ In those females having oöcytes ⩾3, *none* had high LA sensitivity.

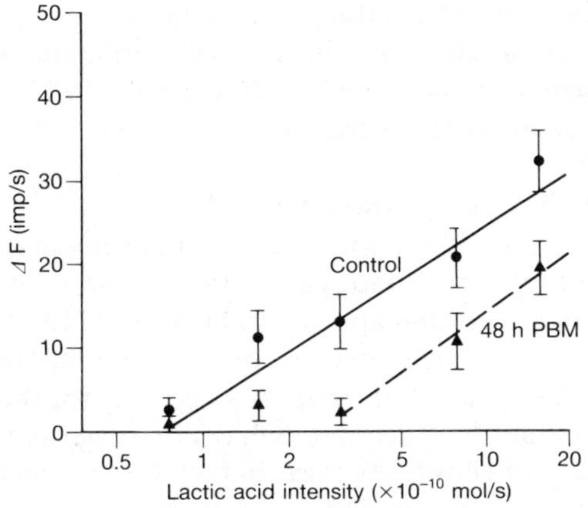

Fig. 28.3. Stimulus intensity vs. response functions of the LA-excited neurones to LA obtained from female mosquitoes that were fed only on sucrose and were injected with 0.5 μl of haemolymph drawn from non-blood-fed females (control) and blood-fed females (48 h PBM). Reprinted with permission from the *Journal of Insect Physiology*, **30**. E. E. Davis: Regulation of sensitivity in the peripheral chemoreceptor system for host-seeking behavior by a hemolymph-borne factor in *Aedes aegypti* (1984).

Fig 28.2a,b). Furthermore, that the reduction of LA-receptor sensitivity is due to a haemolymph-borne factor can be demonstrated by transfusing haemolymph from blood-fed females into sugar-fed females, with a subsequent depression of the non-blood-fed recipients' LA sensitivity (Davis 1984b; Fig. 28.3). The requirement for the presence of vitellogenic ovaries in the depression of LA sensitivity and coincident loss of host-seeking behaviour can be seen in Table 28.2. Blood-fed females whose ovaries were previtellogenic (i.e. follicular stage ≤2) always possessed receptors having high LA sensitivity. (Because these females had been fed to repletion, 7 of the 11 exhibited distention inhibition of host-seeking behaviour.) In contrast, LA-receptors of all females having vitellogenic ovaries were observed to have low sensitivity to LA and would not seek a host. The correlation between low LA-receptor sensitivity and the absence of host-seeking behaviour was 1:1, strongly supportive of a hypothesis of peripheral control of host-seeking behaviour.

Recovery of LA sensitivity and host-seeking behaviour follows oviposition (Klowden and Lea 1979b). Approximately 48 to 72 h PBM, there is an increase in JH titre that may be involved with the reactivation of LA-receptor sensitivity and host-seeking behaviour (Feinsod and Spielman 1980). This increase in JH level coincides with the development of oviposition behaviour and with a possible increase in the sensitivity of the receptors for oviposition site attractants (Davis and Takahashi 1980). This notion has not yet been tested but would follow the same logic, involving the possible role of JH in the initial development of LA sensitivity and host-seeking behaviour after adult emergence. Also, in mosquitoes that diapause as adults, an increase in JH will terminate diapause with a reactivation of host-seeking behaviour (Case *et al*. 1977; Spielman 1974). It has also been demonstrated that an intact, innervated ovary, with the retention of not more than two mature ova, is necessary for the resumption of host-seeking behaviour (Klowden 1981).

We have demonstrated that both the initial development of a behaviour and its subsequent inhibition and reactivation following blood feeding and oviposition are linked, in what appears to be a causal relationship, with the sensitivity of a peripheral olfactory receptor that modulates the behaviour. The necessity for a high level of receptor sensitivity for the host-attractant, LA, in order to obtain host-seeking behaviour suggests that the humoral depression of receptor sensitivity may, in fact, be the mechanism by which host-seeking behaviour is inhibited, that is, the peripheral receptor mediates changes between behaviours as well as modulating a given behaviour. The identity of the haemolymph-borne factor(s) responsible for regulating the sensitivity of the LA-receptor is not yet known, but the most popular candidate is JH. The important point here is that previously most hormones were thought to exert their behavioural effects through a direct action on the CNS. Here, we have presented a case in which their action appears to be directly on a peripheral sensory neurone.

28.4. Other systems in which peripheral receptors may mediate changes in behaviour

Palaniswamy *et al*. (1979) have suggested that topical application of a JH analogue (Altosid^{TM}) depresses pheromone sensitivity in the spruce budworm moth. Bernays (1980) and Bernays and Chapman (1972) have shown that a hormone from the corpora cardiaca is responsible for postprandial rest in locusts and acts to reduce peripheral receptor responsiveness. More recently, Klun and co-workers (personal communication) indicated that a transfusible haemolymph-borne factor underlies the cyclic pheromone production that accompanies the repeated behavioural switches between mating and oviposition in *Heliothis zea* females. Might this haemolymph-borne factor also act on the antennal olfactory receptors for pheromone and/or host–plant oviposition site and thus mediate the changes in behaviour? JH has been shown to terminate diapause in some species of mosquitoes. Could JH reactivate host-seeking behaviour by triggering an increased LA sensitivity? How many other systems are there in which a change in the level of activity in a peripheral receptor neurone might result in a change in the behaviour pattern of an organism?

We have demonstrated that it may occur in at least one case—the mosquito—and on this basis have postulated an important new role for the peripheral sensory system—that is, the mediation of changes in behavioural mode as well as the modulation of a behaviour. We have also demonstrated that the activity of a peripheral sensory neurone may be regulated by endogenous humoral substances. It should be noted, however, that these data do not preclude a role of the CNS in the selection and expression of host-seeking behaviour in female *A. aegypti* mosquitoes.

29 Specificity in perception of pheromones and host odours in Coleoptera

J. C. Dickens

29.1. Introduction

The 'sensory windows' (Griffin 1958) through which organisms view their environment (= *Umwelt* Von Uexküll 1909) have been carefully contoured through evolutionary processes. Olfactory receptor systems of insects have been shaped through evolutionary time to decipher chemical messages in an ever-changing environment. Thus while insects may possess receptor systems responsive to a wide array of odorants, individual receptors may respond best to specific chemicals, such as those important in host finding and species propagation.

Although other sensory modalities are also involved, olfactory perception of semiochemicals plays an important role both in mating and in host selection strategies of coleopterous insects. Orientation of beetles to conspecifics for mating purposes may involve a single component sex pheromone, as in the Japanese beetle, *Popillia japonica* Newman (Scarabaeidae) (Tumlinson *et al*. 1977). But in most instances, beetle aggregation for mating and/or host colonization is mediated by a mixture of several odorants which may be produced by either the insect or its host. For example, male boll weevils, *Anthonomus grandis* Boh. (Curculionidae) release in their frass an aggregation pheromone consisting of four components which attracts both sexes for mating (Hardee *et al* 1969; Tumlinson *et al*. 1969) (Table 29.1). Although some reports purport primary attraction of the boll weevil to its host plant, cotton (*Gossypium hirsutum* L.) (Mistric and Mitchell 1966; Mitchell and Taft 1966; Parencia *et al*. 1964; Smith *et al*. 1965), other reports indicate host selection by random flight (Cross 1973, 1983; Hardee *et al*. 1969; Hunter and Pierce 1912).

In bark beetles (Scolytidae) a multicomponent pheromonal bouquet produced by 'pioneering' individuals, characteristically a particular sex for a given species, is released in the frass and aggregates individuals of both sexes (see Borden 1982a). Host odours may synergize response to the insect-produced pheromone and/or elicit primary attraction. For example, pioneering female Douglas-fir beetles, *Dendroctonus pseudotsugae* Hopk. (Scolytidae), are attracted to potential hosts by certain monoterpenes (Rudinsky 1966) (Table 29.1). Upon locating a host, females release a multicomponent aggregation pheromone which, along with host resin odours, attracts both sexes for mass attack and host colonization (Ryker *et al*. 1979).

TABLE 29.1

Chemical communication systems in the Douglas-fir beetle, D. pseudotsugae* *and the boll weevil,* A. grandis†

Family Species	Scolytidae *Dendroctonus pseudotsugae*	Curculionidae *Anthonomus grandis*
Sex producing aggregation pheromone	Primarily females	Males
Sex responding	Primarily males to initial attractant produced by pioneering females	Both sexes in spring and fall, primarily females in summer
Major components of attractant pheromone	3,2-MCHone, 3-methylcyclo-hex-2-n-1-one	*Compound I*, (+) *cis*-2-isopropenyl-1-methylcyclo-butaneethanol
	3,2-MCHol, 3-methylcyclo-hex-2-n-1-ol	*Compound II, cis*-3,3-dimethyl-$\Delta^{1,\beta}$ cyclohexaneethanol
	frontalin, 1,5-dimethyl-6,8-dioxabicyclo [3.2.1] octane	*Compound III, cis*-3,3-dimethyl-$\Delta^{1,\alpha}$ cyclohexaneacetaldehyde
		Compound IV, trans-3,3-dimethyl-$\Delta^{1,\alpha}$-cyclohexane-acetaldehyde
Host plant attractant synergists	Douglas fir oleoresin; limonene, camphene and α-pinene	Cotton essential oil in field tests; β-bisabolol indicated in laboratory bioassays
Insect-produced synergists	1,2-MCHol, *trans*-verbenol, verbenone	None known

* See references in Rudinsky and Ryker 1977.
† See references in Dickens 1984.

These examples clearly show that chemical signals are involved in beetle communication, consist of one or several components, and may be either insect-produced compounds (i.e. pheromones) or host odours. However, chemical communication systems in Coleoptera are further complicated by the chirality of semiochemicals involved and apparent multifunctionality of some odorants. For example, the attractant pheromones of *P. japonica* and

Gnathotrichus sulcatus Le Conte (Scolytidae) each consist of one compound (Borden *et al*. 1976; Tumlinson *et al*. 1977). In the case of *P. japonica*, one optical isomer of the pheromone attracts, while its optical antipode is inhibitory. In contrast, *G. sulcatus* is attracted by a mixture of pheromonal enantiomers, while neither enantiomer is attractive alone. In many cases only one enantiomer of a coleopterous pheromone has an effect, while its optical antipode is inactive (Wood *et al*. 1976), less active (Payne *et al*. 1982), or elicits behavioural responses only at higher dosages (Dickens *et al*. 1985).

Pheromonal parsimony occurs in the use of a multifunctional compound in the aggregation behaviour of the Douglas-fir beetle (Rudinsky 1973). Pioneering *D. pseudotsugae* females release a multicomponent attractant pheromone which includes low concentrations of 3,2-MCHone (Table 29.1). Upon orientation to a potential host tree and location of the female's gallery by the male, sonic interactions between the sexes stimulate changes in both the quality and quantity of pheromone released (see Rudinsky and Ryker 1977). Increased release by both sexes of 3,2-MCHone leads to inhibition of further aggregation. Thus, 3,2-MCHone in low concentrations may be a part of the attractive bouquet, but at higher concentrations it inhibits response to a previously attractive mixture.

Since the insect is likely to encounter ever-changing mixtures of chemical messengers in various concentrations in its environment, both the sensitivity and specificity of the olfactory receptor system are of paramount importance in deciphering molecular-bound information. The purpose of this chapter is to discuss recent work on the sensitivity and specificty of olfactory receptor systems for multicomponent pheromones and host odours in two coleopterous species: the Douglas-fir beetle, *D. pseudotsugae*, and the boll weevil, *A. grandis* (Table 29.1). Due to limited space, relevant findings in other species will be discussed only where especially pertinent.

29.2. Specificity at the olfactory receptor cell level

29.2.1. Receptor cells for major pheromone components

The attractant pheromones of both *D. pseudotsugae* and *A. grandis* consist of several major components (Table 29.1). Dosage–response curves constructed from electroantennograms (EAGs) from both species in response to their individual pheromone components indicate separate receptors for each (Dickens 1984; Dickens *et al*. 1983). In general, the responding sex is more sensitive to one or more of the major pheromone components than is the sex primarily responsible for pheromone production. Thus, *D. pseudotsugae* males are more sensitive to 3,2-MCHol and 3,2-MCHone than are females, while *A. grandis* females have a lower threshold for compound I than males have. However, both sexes of each species appear to utilize similar receptor mechanisms since the shapes of dosage–response curves for each odorant are much the same. Therefore differences in thresholds and responses near saturation in each sex probably reflect differences in the size of receptor site (acceptor) populations for each odorant for each sex. Thus, for a larger population of responsive receptor sites, the probability of an odorant interacting with a responsive receptor site would be greater.

Single-cell studies on *D. pseudotsugae* males and females further clarify mechanisms suggested by EAG experiments (Dickens *et al.* 1984). At least four olfactory cell types are present in each sex of *D. pseudotsugae*. Three of these cell types are most sensitive to one of the three major aggregation pheromone components, that is, 3,2-MCHone, 3,2-MCHol or frontalin. However, each of these cell types may be activated by other odorants. A fourth cell type responds to insect-produced synergists and host attractants/synergists. This cell type also responds to major pheromone components at higher dosages. Thus, while individual cells respond best to a single component of the attractant pheromone, neural activity in each cell type may be elicited by other odorants at higher dosages. This is in contrast to findings for at least two other bark beetle species, *Ips paraconfusus* (Le Conte) and *I. pini* (Say), which only receive pheromones and host odours by separate receptor populations (Mustaparta 1984).

Both chirality and multifunctionality may play important roles in chemically-mediated behaviour in *Dendroctonus*. For example, frontalin is the major aggregation pheromone component of *D. frontalis* Zimm. and is produced as an 85:15 mixture of (+)- and (−)-enantiomorphs (Kinzer *et al.* 1969; Stewart *et al.* 1977). Both frontalin enantiomers attract beetles, however (−)-frontalin is significantly more active than its optical antipode (Payne *et al.* 1982). EAGs indicate the presence of antennal receptors responsive to each enantiomer (Payne *et al.* 1982). Single-unit recordings further reveal individual cells to be responsive to both enantiomorphs, but more responsive to the (−)-enantiomer. Responses to the racemic mixture are greater than responses to either enantiomer alone. Thus, receptor sites for each enantiomer occur on the same olfactory cell. This phenomenon was further substantiated in *D. pseudotsugae* (Dickens *et al.* 1985). In contrast, *I. paraconfusus* and *I. pini* have separate receptor cells specific for each pheromonal enantiomer (Mustaparta *et al.* 1980).

Perception of the multifunctional pheromone, 3,2-MCHone, by *D. pseudotsugae* involves a multiplicity of olfactory mechanisms (Dickens *et al.* 1984). Olfactory cells which respond with increased nerve impulse frequency to 3,2-MCHone in males and females are some of the most sensitive and specific cells on the antenna. Responses of this cell type increase over a wide range of concentrations. However, while increasing concentrations of 3,2-MCHone lead to increasing activation of this cell type, there is a concomitant decrease in spontaneous activity in other cell types. Thus the attractant response to this pheromone component at low concentrations is probably due to activation of the cell type most sensitive to it. However, while the observed inhibitory effect of 3,2-MCHone at higher concentrations may be partially a product of increased activation of this cell type, it is likely that attractant suppression may also involve inhibition of neural activity in other cell types.

It appears that perception and orientation in Coleoptera involve the simultaneous activation of different cell types responsive to attractant pheromone components. However, perception of ratios of individual components and changing concentrations could be facilitated by across-fibre patterning of information transmitted via other cell types (O'Connell 1975; Pfaffmann 1941) and 'opposing' receptors (Rees 1970), for example for 3,2-MCHone, a multifunctional compound.

29.2.2. Receptor cells for synergists and plant attractants

Both the boll weevil and the Douglas-fir beetle are attracted to volatiles of their prospective host plants (Minyard *et al.* 1969; Rudinsky 1966; Thompson *et al.* 1970). EAGs from both sexes of each species reveal a slightly lower threshold to plant volatiles for the sex responsible for host selection and pheromone production (Dickens *et al.* 1983; Dickens 1984). For example, female Douglas-fir beetles have a 10 times lower threshold than males for the host terpene, limonene, while male boll weevils are at least 10 times more sensitive than females to β-bisabolol (a sesquiterpene and the major volatile of cotton) (Thompson *et al.* 1971) and nerol (an oxygenated monoterpene and possible pheromone precursor) (Thompson and Mitlin 1979). Although *D. pseudotsugae* males and females are 10 to 1000 times more sensitive to certain pheromone components than to plant odours, *A. grandis* males are as sensitive to β-bisabolol and heptanal, both cotton volatiles, as to any pheromone component.

Olfactory-mediated behaviour in *D. pseudotsugae* involves synergistic effects of both host odours and insect-produced compounds with its three major aggregation pheromone components (Rudinsky *et al.* 1974; Rudinsky and Ryker 1977). Cells occur on the antenna which respond to host odours, insect-produced synergists and, at higher dosages, some of the major pheromone components (Dickens *et al.* 1984). For example, a cell maximally responsive to the host odour, limonene, may be activated by higher dosages of *trans*-verbenol, and vice versa. Therefore both insect-produced and host odour synergists activate the same cells.

Specificity of olfactory receptors for enantiomers of dissymmetric plant odours has not yet been investigated for the Douglas-fir beetle. However, electrophysiological responses of antennal olfactory receptors in male and female boll weevils reveal no significant differences in acceptor populations for several optically active monoterpenes (Dickens 1984). Another scolytid beetle, *Ips typographus* L., has receptor sites specific for enantiomers of the host terpene and pheromone precursor, α-pinene (Dickens 1978, 1981). In *Monachamus notatus* (Drury) (Cerambycidae) specific receptor sites for different terpenes occur on the same cell (Dyer and Seabrook 1978).

29.3. Maturation and modulation of receptor responsiveness

Investigations into the effects of age on antennal olfactory responsiveness in two lepidopterous species (Palaniswamy *et al.* 1979; Seabrook *et al.* 1979) and one hymenopterous species (Masson and Arnold 1984) show increases in EAGs to selected odorants from the day of emergence through the first few days of adult life. Since the magnitude of the EAG is considered to be a measure of the relative number of responding receptor sites (Dickens and Payne 1977; Payne 1975), receptor site populations for the odorants tested were thought to increase with age (Seabrook *et al.* 1979).

Little is known about the effects of age on peripheral responsiveness to odours in Coleoptera. However, preliminary results of an experiment to determine the effects of a juvenile hormone analogue on antennal olfactory responsiveness in the boll weevil showed that EAGs to pheromones and host odours increased at least through to day 3 of adulthood (Dickens unpublished).

Modulation of olfactory receptor responsiveness by hormones and hormonal analogues occurs in female mosquitoes, *Aedes aegypti* (L.) (Davis 1984b; see Davis (Chapter 28); Davis and Takahashi 1980) and the spruce budworm, *Choristoneura fumiferana* (Clem.) (Palaniswamy *et al.* 1979). In the mosquito, olfactory receptor cells responsive to lactic acid, a host attractant, become less sensitive 24–48 h after feeding, while cells sensitive to oviposition attractants exhibit a concomitant increase in sensitivity. Antennal olfactory responses of spruce budworm moths are significantly reduced within 10–15 h after topical application of a juvenile hormone analogue.

Results similar to those for the spruce budworm were recently found for the boll weevil (Dickens unpublished). Topical application of 10 μg of a juvenile hormone analogue (ZR515 = Altosid, Zoecon Corp.) to newly emerged female boll weevils significantly reduces antennal olfactory responses to 1-hexanol within 24–30 h post-treatment ($P < 0.05$) (Fig. 29.1). Similar reductions in antennal responses also occur for β-bisabolol and components of the aggregation pheromone. Thus, while the juvenile hormone analogue reduces overall antennal responsiveness, responses to pheromone components and plant odours are equally affected.

29.4. Significance of receptor cell groups for behaviour

The orientation of beetles to pheromones and/or plant odours involves the activation of neural pathways descending from first-order sensory neurones to the deutocerebrum. In the case of a beetle which utilizes a

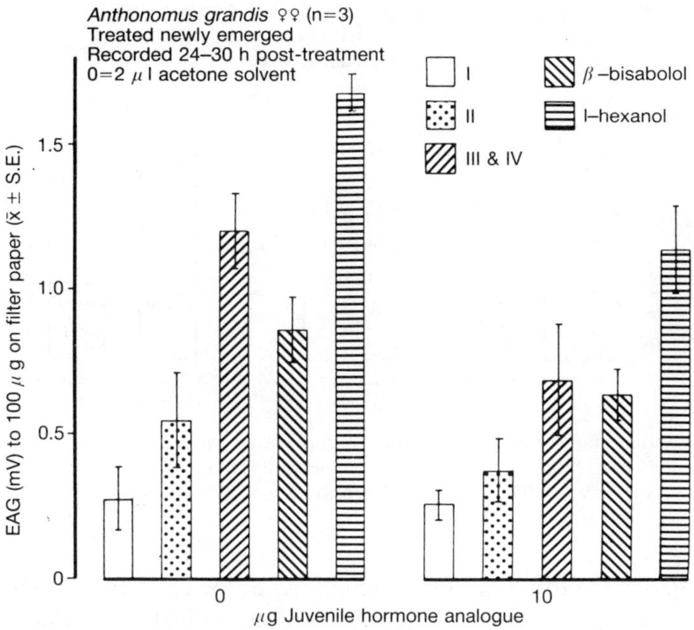

Fig. 29.1. Mean EAGs (mV) of *A. grandis* females to aggregation pheromone components and host plant odours (i.e. β-bisabolol and 1-hexanol) (100 μg on filter paper) 24–30 h post-treatment with solvent alone and 10 μg of a juvenile hormone analogue (Altosid).

multicomponent attractant pheromone, orientation must require simultaneous stimulation of receptor cell groups required for the behaviour. For example, orientation of the Douglas-fir beetle to conspecifics necessitates activation of three receptor cell types responsive to 3,2-MCHone, 3,2-MCHol, or frontalin (Dickens *et al*. 1984; Rudinsky *et al*. 1974). This attraction may be augmented by concurrent stimulation of the fourth receptor type responsive to synergists (Dickens *et al*. 1984; Rudinsky and Ryker 1977).

A similar, but slightly different situation occurs with the boll weevil. The boll weevil has two olfactory receptor populations with different thresholds and saturation levels: one receptor population is primarily responsive to pheromone components and related compounds; the other is primarily responsive to plant odours, for example β-bisabolol (Dickens 1984). Laboratory experiments utilizing a recently developed dual choice olfactometer have yielded interesting results with regard to activation of the two receptor populations (Dickens 1985). Male boll weevils respond positively to either grandlure or β-bisabolol when given a choice between either one and the solvent control (Fig. 29.2a, b). However, when male weevils are allowed to chose between grandlure and β-bisabolol, grandlure is preferred and response to β-bisabolol falls to a level similar to that for the

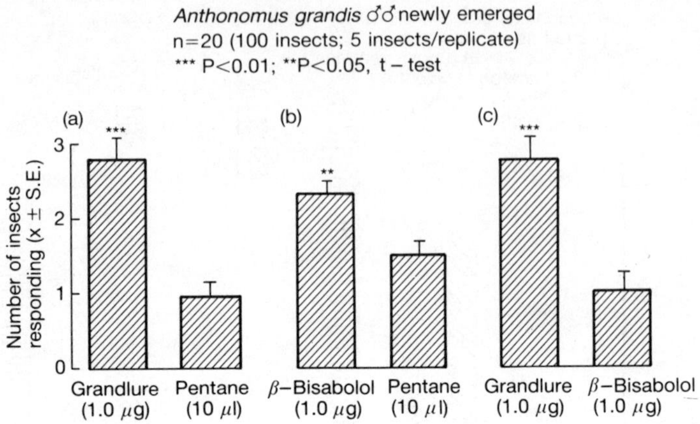

Fig. 29.2. Orientation of *A. grandis* males in dual-choice laboratory olfactometer (Dickens 1985) in response to its aggregation pheromone (grandlure) (a), β-bisabolol (b), and in a choice situation between its aggregation pheromone and β-bisabolol (c).

pentane control versus grandlure (Fig. 29.2a, c). Therefore, interaction of odorants with either receptor population activates neural pathways responsible for an attractant response. However, simultaneous stimulation of both receptor populations, as in a choice situation, leads to accentuation of response to the pheromone, with concomitant inhibition of response to the plant attractant.

29.5. Concluding remarks

The perception of pheromones and host odours by both the boll weevil and the Douglas-fir beetle involve the activation of specific populations of receptor cells responsive to each (Dickens 1984; Dickens *et al.* 1983, 1984). However, 'cross-talk' does occur in which various pheromones and plant odours at high concentrations elicit activity in cells other than those specialized for their perception. This phenomenon may be adaptive, since sensory neurones with different thresholds for an odorant could facilitate perception of varying concentrations of a multicomponent pheromone. The activation by a pheromone of several different cell types with differing sensitivities was recently reported for the cabbage looper, *Trichoplusia ni* Hübner (Lepidoptera: Noctuidae) (O'Connell *et al.* 1983).

Perception of pheromonal chirality in *Dendroctonus* is through enantiomer-specific receptor sites on individual cells (Dickens *et al.* 1985; Payne *et al.* 1982). Both *D. pseudotsugae* and *D. frontalis* have receptor cells responsive to both (+)- and (−)-frontalin (Dickens *et al.* 1985; Payne *et al.* 1982). Furthermore, in *D. pseudotsugae*, even the less specialized synergist cells respond differentially to individual enantiomers (Dickens *et al.* 1985). Thus perception of frontalin enantiomers in *Dendroctonus* occurs via across fibre patterning (O'Connell 1975; Pfaffmann 1941) elicited by the interaction of individual enantiomers with chiral acceptors on individual cells. Although several other coleopterous species have been reported to have single cells responsive to both optical antipodes of a pheromone (Hansen 1983; Wadhams *et al.* 1982), *Ips* bark beetles have acceptors for only one enantiomer on individual cells (Mustaparta *et al.* 1980).

Interestingly, a multiplicity of olfactory mechanisms are involved in perception of a multifunctional pheromone by the Douglas-fir beetle (Dickens *et al.* 1984). The activation of specific 3,2-MCHone cells with concurrent inhibition of activity in other cell types could greatly enhance perception of concentration of this odorant, and, in this instance, explain the observed pheromonal parsimony.

In the classification of olfactory receptor cell types, threshold and shape of dosage–response curves for various odorants are frequently used parameters (Kaissling 1971). In view of recent reports of the effects of age (Dickens unpublished; Seabrook *et al.* 1979) and hormones (Davis 1984b; Davis and Takahashi 1980; Dickens unpublished; Palaniswamy *et al.* 1979) on the responsiveness of peripheral receptors, future experiments on olfaction in Coleoptera and other insects should be conducted on physiologically well-defined animals.

30 Allelochemical effects of pheromones: receptor responses

H. Mustaparta

30.1. Introduction

Pheromones are defined as signals which act intraspecifically (Karlson and Lüscher 1959), and their primary function is to increase the chances for the species to reproduce. However, the same compounds, acting as pheromones in one species, may also influence the behaviour of another species, in which case the pheromones act as allelochemics (Whitaker and Feeny 1971). Depending upon whether the advantage is on the side of the emitting, the perceiving, or both species, interspecifically acting compounds have been termed allomones, kairomones and synomones, respectively (Brown *et al.* 1970; Nordlund 1981). It is well established that closely related species which are sympatric may be influenced by pheromones of each other (Lanier and Wood 1975; see also Birch 1984). Thus, a pheromone compound of one species may interrupt the attraction of the other, and in that manner contribute to the isolation between the two species. One well-studied example of interspecific interruption is that of the sympatric bark beetle species, *Ips paraconfusus* and *I. pini* in California. Here, a mutual interruption is clearly demonstrated (Birch and Wood 1975; Birch *et al.* 1977, 1980; Light and Birch 1979).

We have recently suggested another interspecific effect of pheromones in bark beetles of different genera, where the effect is synergistic (Tømmerås *et al.* 1984). The European bark beetles, *I. typographus* and *Dendroctonus micans*, which are sympatric in some areas, live on the same species of host trees, but are not competing. *D. micans* attacks alive trees and bores into the lowest part of the trunk where the bark is thickest. This species shows no aggregation and the trees usually survive (Gregoire 1983). In contrast, *I. typographus* primarily attacks weakened trees, bores into the bark above the level of *D. micans* and initiates aggregation, which usually results in the death of the trees.

A third type of interspecific effect of pheromones concerns beetles more distantly related. Many field studies have demonstrated that the predators of bark beetles are attracted by pheromones of their prey (Bedard *et al.* 1980; Payne *et al.* 1984; Stephen and Dahlsten 1976; Vité and Williamson 1970; Wood *et al.* 1968). The clerid beetle, *Thanasimus formicarius*, which feeds on many bark beetle species, was the first beetle observed to enter the pheromone traps of *I. typographus* early in the season (Bakke and Kvamme 1981). Here, the predator obviously has evolved sensory cells capable of detecting the odour of its prey. This development has, in all probability, taken place independently of the bark beetle's receptor cells, by evolutionary 'backfire' (Blum 1974). This communication concerns olfactory receptor responses to compounds acting interspecifically as interruptants, synergists, and food volatiles.

30.2. Olfactory receptor responses

The specificities of olfactory receptor cells in bark beetles and in a clerid beetle were determined by screening the cells for sensitivities to various bark beetle pheromones and host volatiles. Different concentrations of each substance were used, and dose–response curves were plotted. Such studies of olfactory cells in several bark beetle species of *Ips* and *Scolytus* have previously shown that each receptor cell is keyed to one pheromone compound and that mixtures of pheromone synergists show no increase of responses above that of the single pheromones (Mustaparta 1984). In these species pheromone mixtures seem to be perceived via separate receptor cells activated in a certain ratio to each other which is also suggested for lepidopteran species (Kaissling 1979; Priesner 1979a). In contrast, it has been reported in three American bark beetle species of *Dendroctonus* that more than one pheromone compound activates the same cells (Dickens and Payne 1977; Dickens *et al*. 1984; Payne 1975).

30.3 Interspecific responses

30.3.1. Interspecific interruption

The reception of compounds acting as interspecific interruptants by blocking the attraction to pheromones has been reported in the two sympatric species *I. paraconfusus* and *I. pini* (Mustaparta 1979; Mustaparta *et al*. 1977). It was shown that these compounds are received by activation of separate receptor cells (Fig. 30.1a). Mixtures of pheromones and an interruptant did not elicit a weaker response than the pheromone alone. Therefore, it was suggested that the interruption of behavioural responses is caused by blocking pheromone pathways in the central nervous system (CNS), rather than pheromone receptors.

30.3.2. Interspecific synergistic effects of pheromones

More recent studies of olfactory cells in *I. typographus* and *D.micans* revealed cells keyed to pheromones which are characteristic of the reciprocal genus (Tømmerås *et al*. 1984). Extensive chemical analyses of pheromones in various species of both genera have shown that *exo-/endo-*brevicomin and frontalin are characteristic of the genus *Dendroctonus* and that ipsenol and ipsdienol are characteristic of the genus *Ips* (Borden 1974, Silverstein 1979). More recently the production of *exo-/endo-*brevicomin

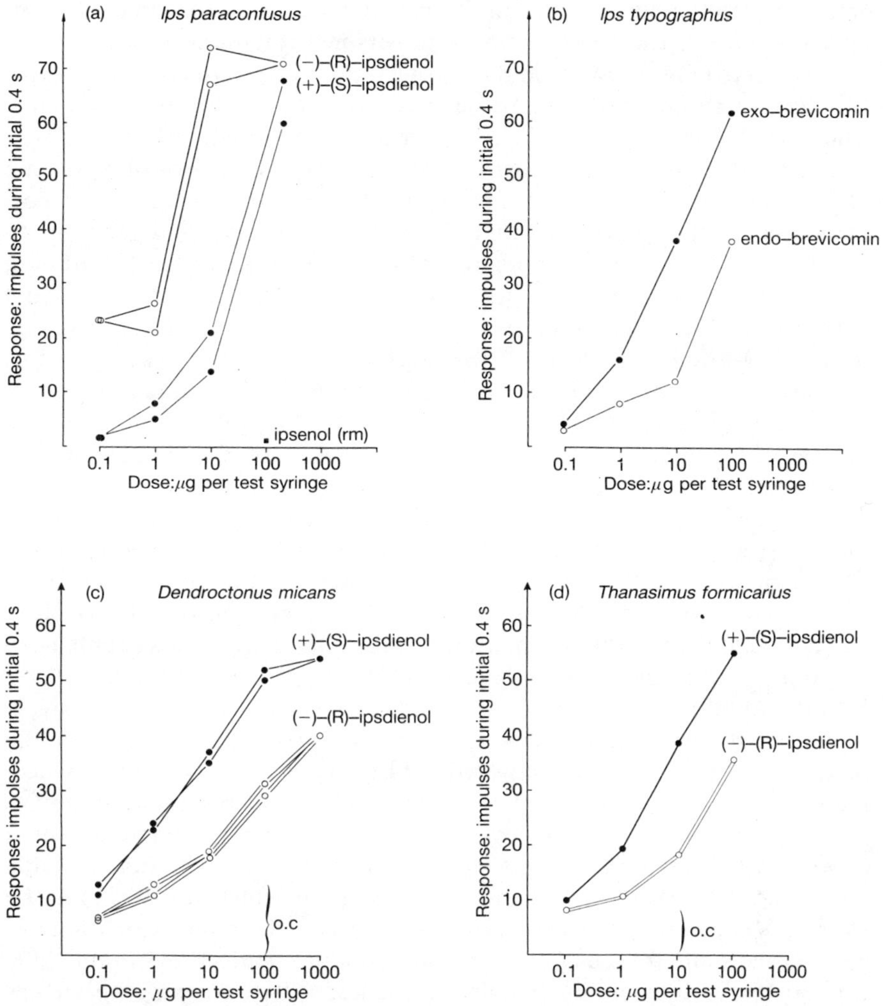

Fig. 30.1. Dose–response curves of receptor cells specialized (a) to an interspecific interruptant, (−)-(*R*)-ipsdienol, in *Ips paraconfusus*; (b) to an interspecific synergist, exo-brevicomin, in *Ips typographus*; (c) to a possible interspecific cue, (+)-(*S*)-ipsdienol in *Dendroctonus micans*; and (d) to a food attractant, (+)-(*S*)-ipsdienol, for the predator *Thanasimus formicarius*, the compound being a pheromone in the prey (d:from Tømmerås 1983). The lower effect of the opposite enantiomer in a, c, and d can be explained by optical impurities. Enantiomers of brevicomin were not available and the effect of chemical and optical impurities can not be stated in this case (b) (*o.c.* = other compounds.)

has been found in *Dryocoetus* species (Kohnle and Vité 1984; Schurig *et al.* 1983). In *I. typographus* the three-component mixture, ipslure, consisting of ipsdienol, *cis*-verbenol and methyl-butenol, acts as a pheromone attractant (Bakke *et al.* 1977). The pheromone of *D. micans* is so far not known.

However, preliminary studies of *D. micans* have indicated production of and responses to *exo*-brevicomin (Vité personal communication).

Our electrophysiological studies showed that *I. typographus* possesses a relatively high number of cells specifically responsive to *exo*-brevicomin. The specificities of these cells, determined by single-cell recordings, showed high response to *exo*-brevicomin, low response to *endo*-brevicomin and no response to other test compounds (Fig. 30.1b). The sensitivity corresponded to that of the same type of cells in *D. micans* and generally to that of other types of pheromone receptor cells in bark beetles. In addition, *D. micans* was found to have cells specialized for (+)-ipsdienol (Fig. 30.1c), the sensitivities and specificities of which corresponded well to the (+)-ipsdienol cell in *I. typographus* and in other species of *Ips*. Electroantennogram recordings showed a relatively high stimulatory effect of both *exo*-brevicomin and (+)-ipsdienol on the antennae of the two species. This indicates that these two species possess a relatively high number of both cell types.

Behavioural responses to *exo*-brevicomin were recorded only in *I. typographus*. Tests in the laboratory (Tømmerås *et al*. 1984) as well as in the field (Tømmeråas and Mustaparta 1984) showed that *exo*-brevicomin alone had no attractive effect on *I. typographus*. However, when *exo*-brevicomin was added to the pheromone mixture (ipslure) the attraction was enhanced by 60 per cent in the laboratory tests and by 40 per cent in the field tests. Furthermore, in the laboratory tests this enhancement was due exclusively to increased attraction of males. However, this was not the situation in the field tests, where both sexes contributed to the *exo*-brevicomin enhancement. The results of these electrophysiological and behavioural studies suggest that *exo*-brevicomin acts as an interspecific synergist and that *I. typographus* uses this compound as a chemical cue in finding suitable host materials. The source of *exo*-brevicomin may in this case be *D. micans*, or perhaps *Dryocoetus autographus*. The latter, which is common on the same trees as *I. typographus*, produces *endo*-brevicomin along with some *exo*-brevicomin (Kohlne and Vité, 1984). *D. autographus* has been considered as a secondary species on these trees. However, early in the season *D. autographus* and *I. typographus* are flying at the same time, and it might be that the interaction is advantageous to both species.

30.3.3. Interspecific effect of pheromones acting as food volatiles

In contrast to the high specialization of pheromone receptor cells, it has generally been found that cells responding to food and host volatiles possess a lower degree of specialization by responding to several constituents. In the interaction between *T. formicarius* and its prey, the compounds which act as pheromones in the prey are food volatiles for the predator. Electrophysiological recordings from olfactory cells of the predator showed, however, that cells responding to bark beetle pheromones possessed the same high degree of specialization as those of bark beetles

(Tømmerås 1983). Also, in the predator, the cells could be classified according to one key compound. Several different groups of cells, each keyed to one pheromone compound, were obtained, indicating that mixtures of bark beetle pheromones are also perceived in the predator via separate cells, one type for each compound. This also applies to optical isomers. Thus, the predator had separate cells specialized to (+)- and (−)-ipsdienol, respectively, with the same sensitivities and specificities as those of *Ips* species (Fig. 30.1d). Furthermore, like in bark beetles, it was found that all cells responding to ipsenol in the predator were keyed to (−)-ipsenol, the only ipsenol enantiomer isolated from species of *Ips*. Hence, the single-cell recordings from *T. formicarius* demonstrated that the predator and its prey have evolved olfactory cells for the same compounds which are specialized in the same manner, indicating the presence of types of membrane receptors. This has occurred in spite of the fact that they have developed as pheromone receptors in the prey and as food odour receptors in the predator. The predator thus seems to be equipped with sensory cells which can as easily detect and discriminate between pheromones of various species of bark beetles as can the receptors of the bark beetles themselves. Predator–prey specificity has also been shown by electroantennogram recordings combined with behavioural tests, for *T. dubius* in response to pheromonal enantiomers of *D. frontalis* (Payne *et al*. 1984).

30.4. Concluding remarks

The sensory aspects involved in the three types of interspecific relationship are based on reception of interspecific chemical cues by separate, specialist types of receptor cells. Cells keyed to the same compound show no difference when involved in pheromone perception, interspecific interruption, interspecific synergism, or food attraction. The presence of specific receptor cells for these chemical cues indicate that the interaction is of selective advantage for the perceiving species. This is obvious in the case of a predator–prey relationship. Here, the interaction is disadvantageous to the emitting species, and the bark beetle pheromones thus act as kairomones.

In the case of interspecific interruption, compounds have previously been termed allomones (Birch *et al*. 1980). The term allomone is used differently by various authors, sometimes including compounds acting with mutual advantage. However, the latter have also been termed synomones (Nordlund 1981). Interspecific interruptants are apparently of advantage also for the perceiving species and may therefore be termed synomones. The interspecific effect of *exo*-brevicomin, acting synergistically on pheromone attraction for *I. typographus*, is obviously of advantage for the perceiving species. The emitting, not competing, species may be unaffected by, or may benefit from this attraction. Thus *exo*-brevicomin either acts here as a kairomone or as a synomone.

Acknowledgements

Studies by the author *et al*. referred to in this review were supported by grants from the Norwegian Research Council for Sciences and Humanities and from Borregaard Industries Ltd. Norway.

31 Computer analysis of chemosensory signals

F. E. Hanson, S. Kogge, and C. Cearley

31.1. Introduction

Our understanding of the basic mechanism of the perception of external chemicals has been greatly accelerated by the use of electrophysiology as a tool. This technique provides us with a window into the nervous system to view responses elicited upon chemosensory stimulations. Understanding how this translates into behaviour will ultimately depend on understanding how the nervous system works. To achieve this goal, we are destined to increase our reliance on electrophysiological techniques and on the methods to analyse the data they provide.

One of the advantages of using electrophysiological techniques is that large amounts of data can be obtained very rapidly. Viewed from the outside, this would appear ideal; however, in many cases it is only the silver lining of a dark cloud. Analyses and interpretation of this large mass of data are often more difficult than obtaining the data originally. Thus the full potential of electrophysiology is not likely to be realized until analysis can be made less burdensome.

31.2. Methods

The search for better methods of analysis has led to the development of a wide array of equipment and techniques. These range from inexpensive analogue integration devices to a complete computer installation. Thus analysis methods exist for each level of sophistication that a particular research problem requires.

Historically, the method of data analysis has been to film the action potential with an oscilloscope camera and manually count the action potentials as a function of time. Although this is still the method of choice for small projects, the drudgery involved has probably inhibited more experiments in chemosensory electrophysiology than all other factors combined. If extended projects are envisioned, something more automated is almost a necessity.

A simple data reduction system might consist of a variable threshold device and a monopulse generator (Schmitt trigger) connected to an integrator and chart recorder to display the number of suprathreshold action potentials versus time. Other hardward modules can provide cumulative counts, rates, or histograms.

These hardware analysis modules are useful but are dedicated to specific tasks. A large increase in flexibility is obtained if the threshold device or other analogue preprocessor (Rumbo 1980) is interfaced to a computer that possesses a clock. Each action potential is then stored in the computer as an event at a particular time, data which can subsequently be manipulated for analysis or graphing.

A more sophisticated interface employs an analogue-to-digital converter between the amplifier and the computer to provide digital values of high-speed samples of the input voltage. Information about the peak amplitude of each spike can then be stored along with the time of event, or the entire action potential can be stored for subsequent reconstruction and scrutiny.

31.3. Utility of computer analysis

31.3.1. Quantitation

It is common knowledge that the best experiments are those from which the results are an obvious yes or no. However, many experiments are more equivocal. Investigating a single problem in depth rather than skimming the cream off many problems inevitably leads to experiments requiring quantitation.

For example, consider the definitive dose–response curve using a series of related compounds on many sensilla of enough animals to overcome the problems of high intra- and inter-animal variability. This might involve recording from five olfactory sensilla on each of five animals stimulated by five concentrations of five compounds using five repetitions on each, for a total of 5^5 (3125) trials. If responses averaged about 100 action potentials each, the total number of spikes catalogued for this study would exceed 300 000. An investigator executing such experiments would soon be driven to seek help, from either a computer expert before the experiment or a therapist after.

Fig. 31.1 shows such an experiment using the tarsal gustatory chemosensors of the blowfly. Here each thin-lined curve represents the average responses of 8–10 sensilla of 1 animal to 8 concentrations of RbCl, with 2–3 repetitions on each sensillum. This approach documents the variability inherent in recording chemosensory data: the standard errors and the scatter among the curves illustrate the intra- and inter-animal variability respectively. Each thin-lined curve in Fig. 31.1 extracts information from about 18 000 spikes. The thick-lined curve represents the average of nine such curves, thereby summarizing over 150 000 spikes. A study this comprehensive would clearly have been impractical without computer analysis.

31.3.2. Display

Perhaps one of the most useful aspects of computer assistance in electrophysiology is the immediate display of the data. Humans are highly visual

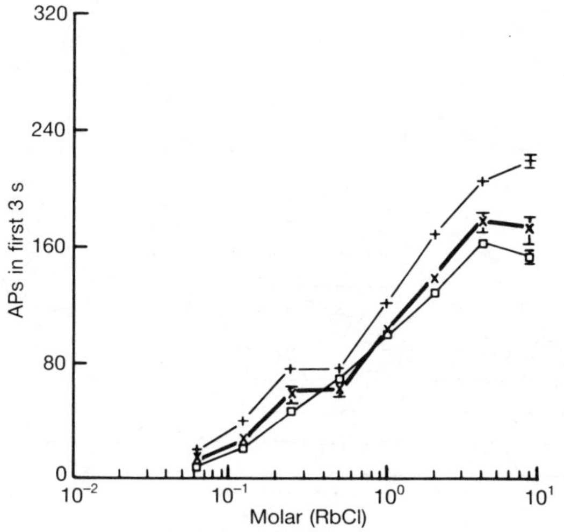

Fig. 31.1. Dose–response curve of the blowfly tarsal salt cell to RbCl. Ordinate is the cumulative number of action potentials in the first 3 s. Thin lines represent data from individual animals, with each point an average of 20–25 trials (2–3 trials on each of 8–10 sensilla of the largest class). Thick line represents similar data averaged from 9 animals.

organisms, and we tend to search for visual concepts and metaphors. Thus an olfactory test series is best understood by visualization of a plotted dose–response curve.

Viewing a high-density summary of large amounts of data in one glimpse often does more to provide an intuitive 'feel' for the data than looking at many individually filmed responses. For example, Fig. 31.2 is a hidden line plot of spike amplitude histograms of responses from one sensillum in a dose–response experiment. Each line plots the response to one concentration of RbCl. A single glance shows the incrementing response of the large spike with increased concentration. Near the left side of each trace, a smaller sized spike is seen responding at low frequencies.

In addition to the enhancement of appreciation of the relationships among the data, displays are important for procedural reasons. Computer analysis can generate artifacts at very high rates if not monitored at every step of the process. The programs we have written make extensive use of displays, as exemplified in the following discussions.

31.3.3. Separation of spikes

If more than one cell is active during a response, it is often desirable to analyse activities of each cell separately. Such a separation can be done by the computer on the basis of the size or shape of the recorded action potential. Usually, different cells will have action potentials which differ in height, width, or time constant of the return to the baseline by the negative phase. The parameter most often used to discriminate between spikes is

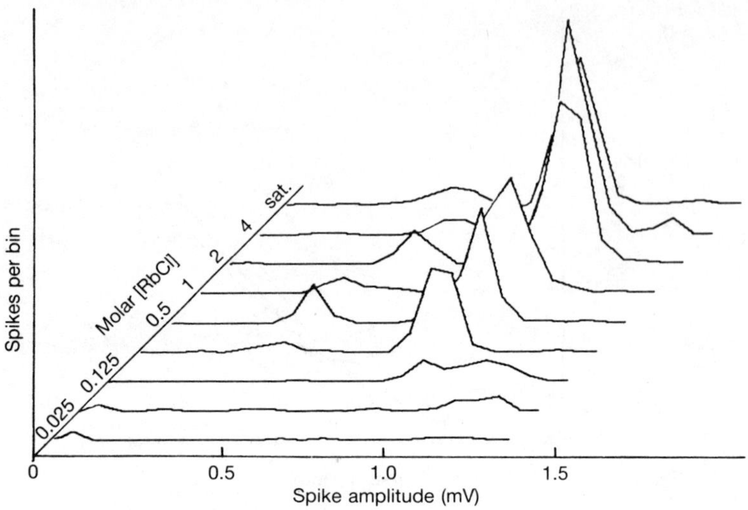

Fig. 31.2.. Amplitude histograms of raw data. The amplitude of action potentials of one trial at each concentration was histogrammed into 20 bins and graphed as a hidden line plot. All data are from the same tarsal sensillum (largest class).

the total amplitude (peak-to-trough height). For example, Figs. 31.3a and 3c show histograms of action potential amplitudes of responses of the medial styloconica of the maxilla of the tobacco hornworm. This organ is both gustatory and olfactory (Städler and Hanson 1975). The bimodal distributions displayed in Fig. 31.3a suggest that two cells are active; separation of these two distributions into 'classes' is easily accomplished by manual insertion of limits or by a computer program. Each spike is then assigned to a class by the computer.

Similar separations can be obtained by plotting the standard deviation of the voltage samples comprising a digitized spike (Figs. 31.3b and 3d). This parameter is somewhat more sensitive to the overall shape than just the height alone, and has been used by O'Connell *et al*. (1973) in separating action potentials of the two antennal olfactory cells.

The data graphed in Figs. 31.3a and 3b are nearly ideal: spike sizes are sufficiently different to be easily separated by nearly any method. Often this is not the case, as illustrated in Figs. 31.3c and 3d; the spikes may be of similar sizes or have overlapping distributions, making separation difficult. In such cases, the entire shape of the spike must be used as a discriminant.

One such algorithm compares on a point-by-point basis the outline of the digitized spike to that of a template, using the sum of the squared differences as the discriminant (Gerstein and Clark 1964; Millechia and McIntyre 1978). This is basically the same as the differences in areas between the template and each spike. Templates are selected automatically, and separately from each trial. If the spike is similar to the first template within operator-designated criteria, the spike is assigned to class 1. If not, it is set

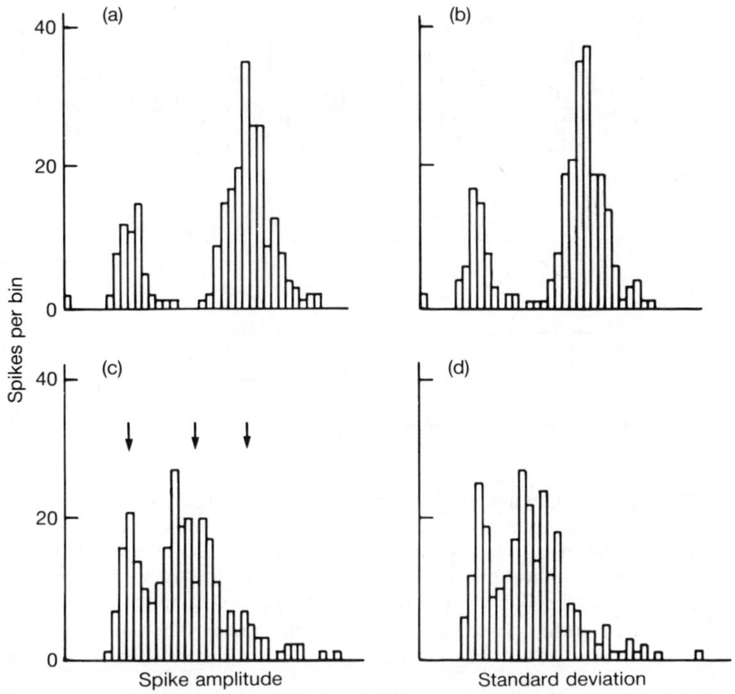

Fig. 31.3. Histograms of spike amplitudes and standard deviations. Plots are the responses of the medial styloconica of the maxilla of the tobacco hornworm to leaves of the unacceptable plant, *Canna generalis* (a and b) and the host-plant, tomato (c and d). Arrows in c denote average spike amplitude of the three classes of spike found by template separation techniques. Note that the intermediate-sized spike in c is absent in a.

aside for a second iteration with a second template. This procedure is repeated until all spikes are classified. An example of such a separation is shown in Fig. 31.4, in which event markers denote the time of occurrence of each spike matching the templates at left. This algorithm is more discriminating than those based on height or standard deviation alone. For example, we have used this method to separate two spikes having nearly equal amplitudes and forming a unimodal height distribution, but having slight differences in shape because the negative phase of one spike type was slightly wider than the other.

The critical aspect is deciding when the separation has been successful. If not, the program can be re-run with different discrimination criteria. To assist the operator in this decision, a variety of displays have been developed. At the individual spike level, Fig. 31.4 shows graphically the class to which each spike is assigned. For an overview, plotting one shape parameter against another as a scattergram is often useful. Fig. 31.5, for example, shows by the clustering of points that the data of Fig. 31.4 have

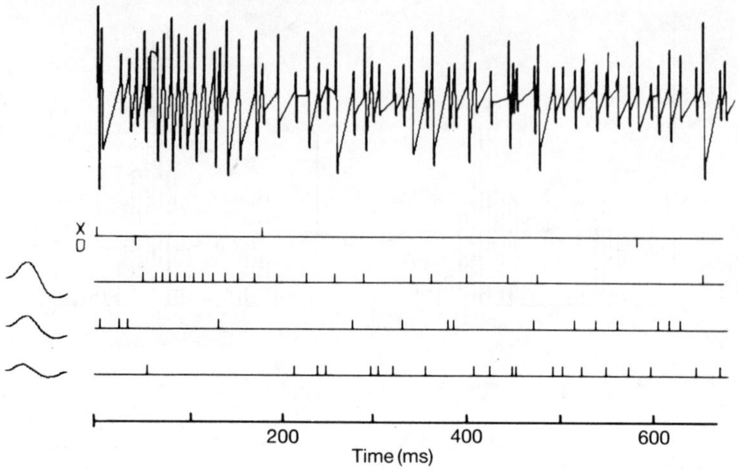

Fig. 31.4. Classification of individual action potentials: first portion of response of the medial styloconica to tomato (same data as in Figure 31.3 c and d). (Top trace: computer reconstruction of analogue signal using rapid plotting techniques (five points per spike). Subsequent traces: marks denote classification of each spike seen in the top trace.) The program found three primary classes of spikes, and generated the average templates shown at left (X: misshapen spike, not further classified, D: double spike, not further classified).

been adequately separated. Another scatter plot (Fig. 31.6) graphs one shape parameter (e.g., amplitude) versus time, which is probably one of the most useful and intuitively meaningful displays. In addition to providing information on the quality of the separation, this display provides an overview of the time-related information about the spike train. The homo-

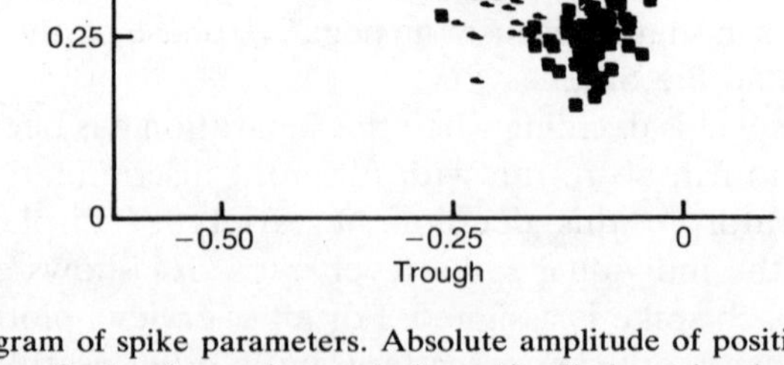

Fig. 31.5. Scattergram of spike parameters. Absolute amplitude of positive peak is plotted against absolute amplitude of negative peak (trough). Circles, triangles, and squares are respectively classes 1, 2 and 3 of Fig. 31.4.

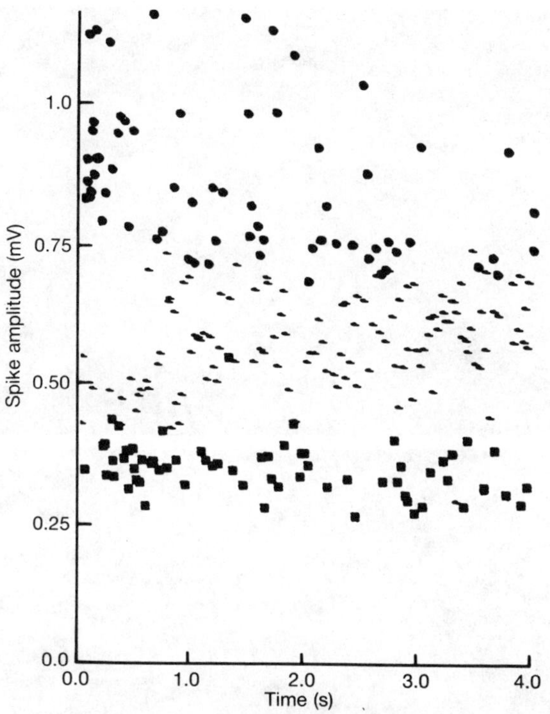

Fig. 31.6. Scattergram of spike parameter vs. time. Peak-to-trough amplitude of each spike in a 4 s trial is plotted at its time of occurrence (details and data as in Fig. 31.5).

geneity of each class can be ascertained by displaying all of the spike waveforms in each class, 20 at a time on the screen (Fig. 31.7).

31.3.4. Averaging

The goal of the chemosensory physiologist is to describe the 'species response' elicited by a chemical stimulant. To do so requires averaging across many animals to overcome the problems of individual variability. The usual method is to select for publication 'representative' records, a process which is a form of mental averaging by the investigator. More quantitative methods of averaging have long been sought that lessen the potential for bias. Numerical averaging across many trials serves this function by smoothing fluctuations inherent in low-level responses and permits the extraction of the mean response. For example, Fig. 31.8a shows the adaptation curve of a single response of one chemosensory cell. When two responses from each of eight such cells are averaged, the smooth curve of Fig. 31.8b is generated. Because the latter contains more information, more confidence can be placed in its shape and in the equations derived from curve-fitting procedures.

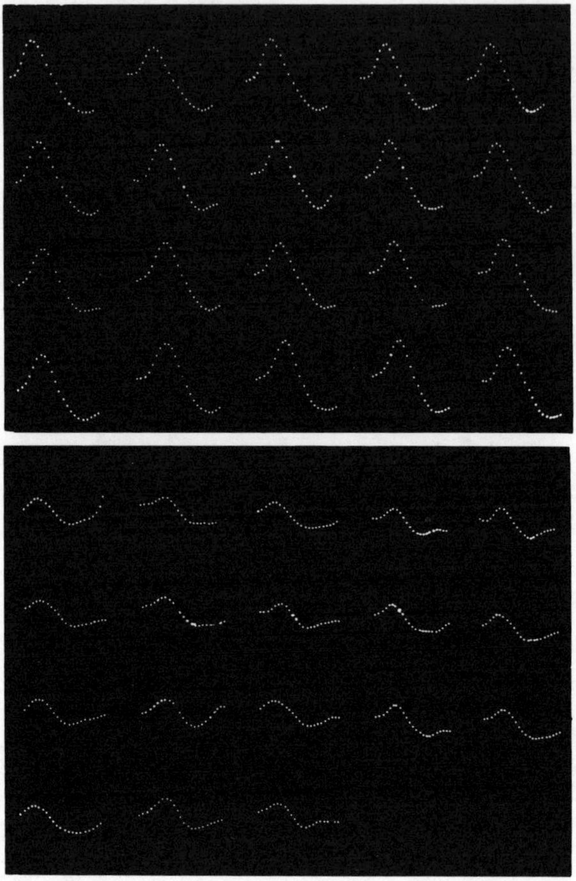

Fig. 31.7. Computer screen display of classified spikes showing homogeneity of each class. Data are the same as in Fig. 31.3 a and b.

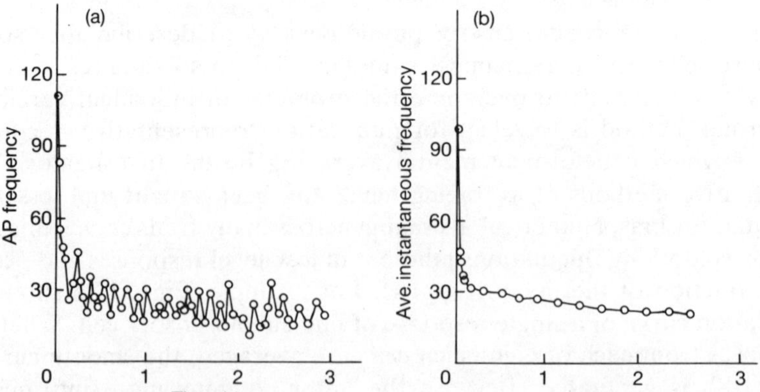

Fig. 31.8. Averaging: (a) Instantaneous frequency of a single response trial of a single cell (i.e. one class of spikes) of a blowfly tarsal sensillum; (b) same, but average of two trials from each of 10 cells using the average post-stimulus time histogram method.

31.3.5. Analysis

Once the classed data are in a computer file, the sensory coding problem becomes more amenable to analysis. The question may arise, for example, whether there is any information in the spike train other than instantaneous frequency (Fig. 31.8). There may be synchronous cross-channel information from several cells in complex responses such as that shown in Fig. 31.4. Cross-correlation analysis would detect this. Or, perhaps certain stimuli elicit continuous firing with constant intervals, whereas others evoke irregular activity. This situation is illustrated in Fig. 31.9: The interspike intervals of the responses of the medial gustatory sensilla of the tobacco hornworm in two different plants show different distributions (Fig. 31.9, large cell and possibly also for the small cell). These results and

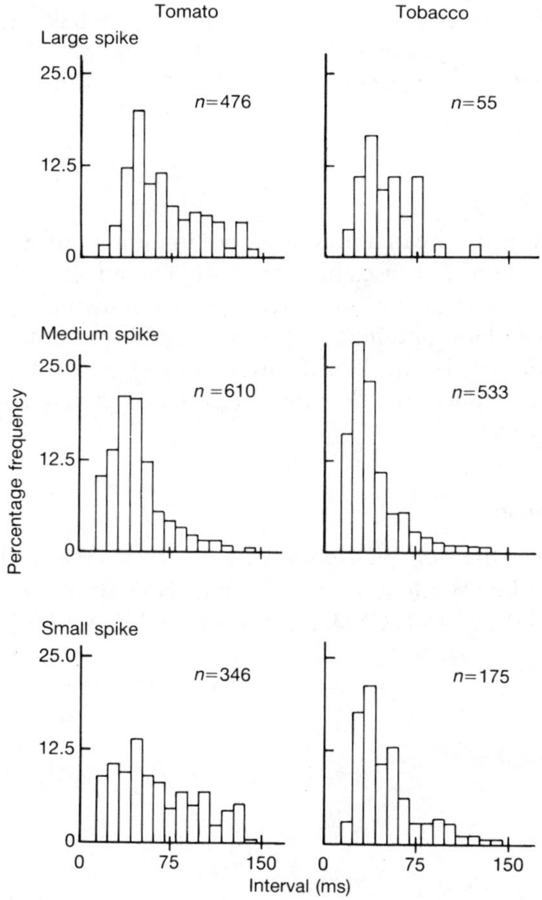

Fig. 31.9. Interspike intervals of three classes of action potentials from the medial styloconica of the maxilla of the tobacco hornworm in response to two host plants. Data were obtained from two trials on each of three animals. Note different interval distributions of the large spike, and possibly also of the small spike.

those of Dethier and Crnjar (1982) suggest that there is some complex information in the spike trains of chemosensory organs that merits further investigation.

The sensory coding problem can be further investigated by multivariate analyses such as cluster analysis (van der Molen *et al*. 1978), factorial analysis, and multidimensional scaling. These techniques are useful in determining cell 'types' by extracting similarities and differences in responses of many cells to a large battery of stimuli. Correlation of these data with appropriate measures of the behaviour of the animal will provide some insight into the coding for that stimulus.

The ideal example of coding is the case of receptor cells for a specific pheromone; all cells of this type respond to a particular compound and none other. The behaviour they activate is also specific, and therefore understanding coding for these compounds is simple. Most chemosensitive cells are more broadly tuned, however, and the behaviour they elicit is often more equivocal. These more complicated responses will require more complex analyses and large behavioural data bases with which they are to be correlated.

31.4. Conclusions

Electrophysiology has been a powerful tool in aiding our understanding of the role of the chemical senses in behaviour. The advent of the laboratory computer promises to extend this usefulness by allowing us to analyse the large masses of data obtained. This will make possible the extensive characterization of cells, the identification of cell types, comparisons across species and races, and provide more confidence in our solutions to the sensory coding problem.

Acknowledgements

We acknowledge the help of Neil Wiley and Jack Glaser. The work was supported by the Whitehall Foundation, NSF BSN-7914043, USDA CRGO 59010410902460, USDA CANUSA 075, and UMBC.

32 Functional organization of olfactory pathways in the central nervous system of *Manduca sexta*

J. G. Hildebrand and R. A. Montague

32.1. Introduction

In insects, olfactory stimuli such as pheromones and kairomones stimulate receptor cells associated with particular classes of sensilla on the antennae. Physiological activity in these sensory neurones, in the form of patterned firing of action potentials, transmits information about the biologically meaningful odours along the axons of the receptor cells to the antennal lobes (ALs) of the brain. In the central nervous system (CNS) this olfactory information passes through several stages of synaptic processing in the ALs and higher centres in the protocerebrum (PR), where it is integrated with physiological information in other sensory and central channels to generate 'descending' premotor activity and ultimately behaviour. Recently there has been growing interest in central olfactory mechanisms in insects, and as a result we are beginning to learn much about the layout and operation of the antennal sensory pathways in the CNS.

Building upon earlier work (e.g. Boeckh and Boeckh 1979; Kaissling 1971; Masson 1972; Schneider 1969; Suzuki and Tateda 1974; Waldow 1975, 1977; Yamada 1971), and paralleling the current research of others using different insect species (e.g. Boeckh *et al*. 1984; Burrows *et al*. 1982; Ernst and Boeckh 1983; Prillinger 1981; Rospars 1983; Stocker *et al*. 1983), we study the anatomy, physiology, neurochemistry, development, and behavioural roles of the olfactory pathways in the CNS of the experimentally favourable sphinx moth *Manduca sexta*. Because other aspects of our work have been reviewed recently elsewhere (Hildebrand 1985; Hildebrand and Maxwell 1980; Hildebrand *et al*. 1982), and owing to severe space limitations, this brief review focuses on some of our findings and speculations about the functional organization of olfactory pathways in the brain of *Manduca*.

32.2. Antenna

The antenna of adult *Manduca* comprises two basal segments (scape and pedicel), containing mechanosensory organs, and a long sexually dimorphic flagellum divided into about 80 annuli bearing numerous sensilla of several types, the great majority of which are olfactory (Hildebrand *et al*. 1980; Sanes and Hildebrand 1976a). A male flagellum has about 2.6×10^5 sensory neurones associated with about 10^5 recognized sensilla. About 40 per cent of these are male-specific sensilla trichodea each of which houses the dendrites of two male-specific olfactory neurones finely specialized to detect sex pheromones (bombykal and other, unidentified substances) released by sexually receptive, calling female moths (Sanes and Hildebrand 1976a,b; Schweitzer *et al*. 1976; Starratt *et al*. 1979; Hildebrand and Kaissling, unpublished data). Antennal flagella of both male and female *Manduca* (whose larvae feed, and adult females oviposit, exclusivly on plants of the family *Solanaceae*) have many receptors that respond to volatiles given off by plants, such as aldehydes, alcohols, and esters (Schweitzer *et al*. 1976; Harrow, Kovelman and Hildebrand, unpublished data).

32.3. Antennal afferent projections

Axons of antennal receptor cells run through the antennal nerve (AN) to enter the brain at the level of the ipsilateral AL of the deutocerebrum (DEU), as schematized in Fig. 32.1. Axons of olfactory receptors and possibly also fibres of other modalities (such as humidity receptors and perhaps certain flagellar mechanoreceptors) project from the flagellum to targets in the AL, but axons from mechanosensory neurones in the basal segments of the antenna bypass the AL and project instead to a deutocerebral 'antennal mechanosensory and motor centre' in the DEU, ventral to the AL (with respect to the body axis of the animal); some project further to the subesophageal ganglion and prothoracic ganglion (Camazine *et al*. 1979; Hildebrand *et al*. 1979, 1980; Matsumoto and Hildebrand 1981; Sanes *et al*. 1977; also Camazine, Harrow, Kent and Hildebrand, unpublished data).

32.4. Antennal lobes

32.4.1. Glomerular organization

The AL (Figs. 1b and 2) has a central zone of coarse neuropil (largely neurites of AL neurones) surrounded by a quite reproducible but not invariant array of about 60 spheroidal glomeruli, which is bordered by three clusters of neuronal somata—lateral (LC), medial (MC), and anterior (AC)—probably totalling about 1000 cells (Hildebrand *et al*. 1979, 1980; Matsumoto and Hildebrand 1981; Sanes *et al*. 1977; Schneiderman *et al*. 1983). Sensory axons projecting into the AL terminate within the glomeruli (Camazine and Hildebrand 1979; Tolbert and Hildebrand, 1981; also Camazine *et al*. unpublished data) where they make chemical synapses with dendritic processes of AL neurones (Matsumoto and Hildebrand 1981; Tolbert and Hildebrand 1981). The glomeruli, which are condensed knots of neuropil 50–100 μm in diameter, comprise

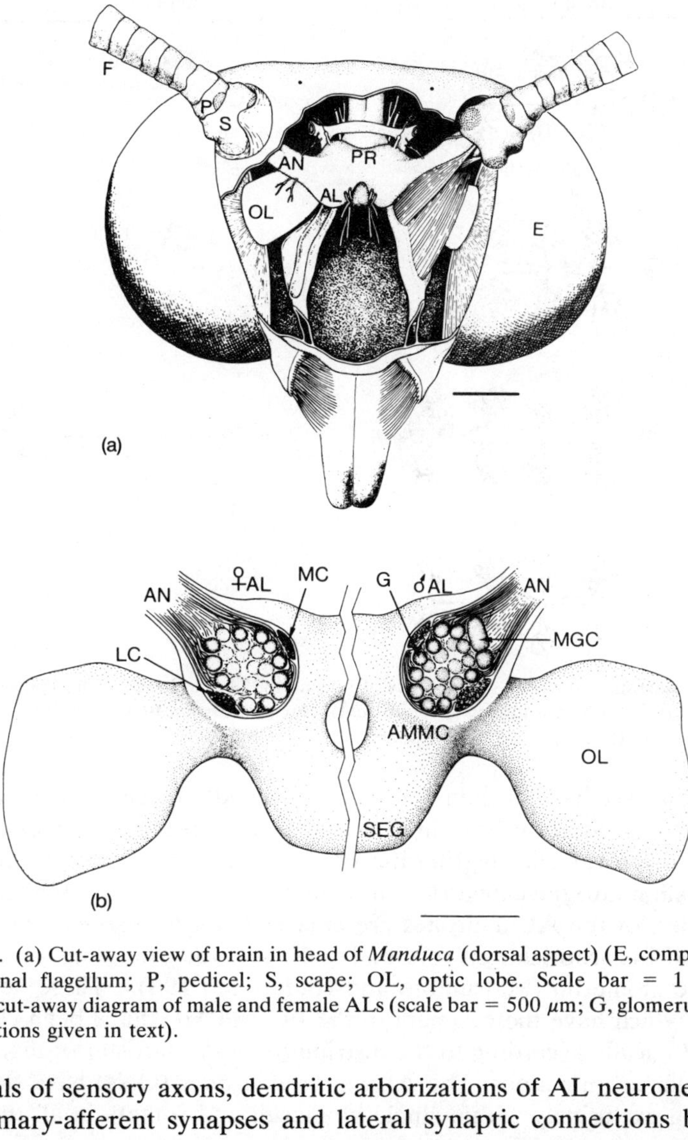

Fig. 32.1. (a) Cut-away view of brain in head of *Manduca* (dorsal aspect) (E, compound eye; F, antennal flagellum; P, pedicel; S, scape; OL, optic lobe. Scale bar = 1 mm); (b) Frontal, cut-away diagram of male and female ALs (scale bar = 500 μm; G, glomerulus; other abbreviations given in text).

terminals of sensory axons, dendritic arborizations of AL neurones, all of the primary-afferent synapses and lateral synaptic connections between AL neurones, and possibly other neural elements, and are nearly surrounded by an investment of glial processes (Hildebrand *et al*. 1979; Matsumoto and Hildebrand 1981; Tolbert and Hildebrand 1981; Tolbert *et al*. 1983).

32.4.2. AL neurones

Using intracellular recording and staining techniques, we have begun to explore the neurones of the AL (Christensen and Hildebrand 1984; Harrow and Hildebrand 1982; Matsumoto and Hildebrand 1981; Montague

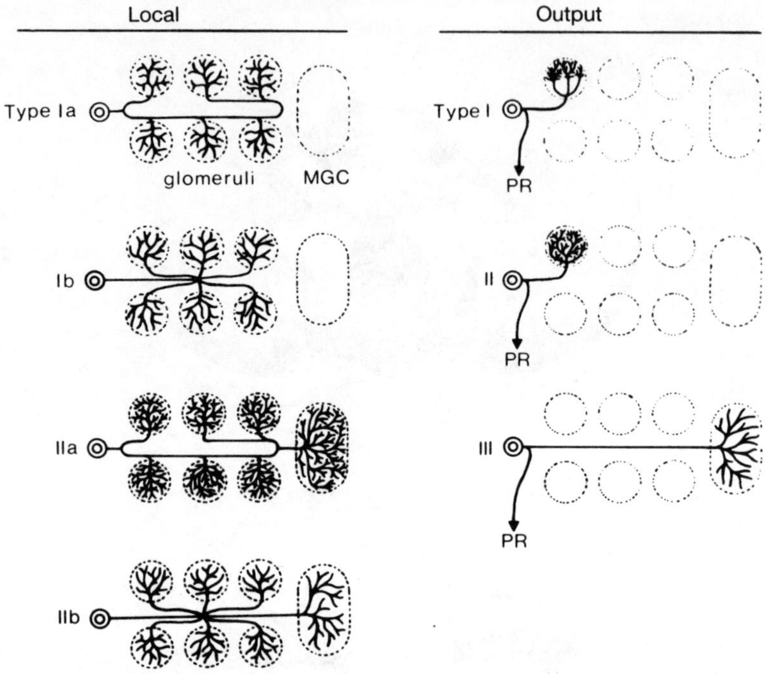

Fig. 32.2. Schematic diagram of the types of LNs and ONs that have been recognized in the AL in *Manduca* (abbreviations as given in text). (From Matsumoto and Hildebrand (1981) *Proc. Roy. Soc.* **B213**, 249–277, with permission.)

et al. 1983). We have examined several thousand AL cells to date. All have had their cell bodies in the three AL neuronal groups, and most, on the basis of the form and distribution of their dendritic arborizations, have clearly fallen into the categories outlined in Fig. 32.2. Many, and probably a majority, of the AL neurones are amacrine, *local interneurones* (LNs), with dendritic arborizations confined to the AL and extending into most or all of the glomeruli (Matsumoto and Hildebrand 1981). These neurones, most of which have their somata in the LC and MC, have been classified morphologically according to the distribution and character of their multiglomerular arborizations. The AL *output neurones* (ONs) have dendrites ramifying exclusively within single glomeruli and somata in all three AL cell groups.

32.4.3. AL sexual dimorphism

Like the antennae, the ALs are sexually dimorphic. Both male and female ALs have similar arrays of spheroidal 'ordinary' glomeruli, which receive the axon terminals of antennal olfactory neurones responsive to plant-associated odours (Camazine *et al*. unpublished), but only male ALs or experimental female ALs innervated by male antennal sensory axons (Schneiderman *et al*. 1982) exhibit a large macroglomerular complex

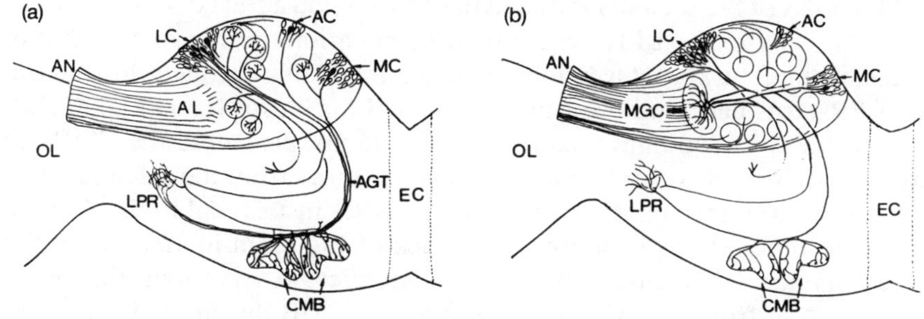

Fig. 32.3. Diagram of types of ON projections observed in the PR in *Manduca* (EC, esophageal canal; other abbreviations as given in text) (a) ONs (in males and females) with dendritic arborizations in ordinary glomeruli; (b) ONs (in males) with dendrites confined to MGC.

(MGC), as schematized in Figs. 32.1b and 32.3. The MGC is the target of the axons of the pheromone-detecting antennal sensory neurones in the male antennal flagellum (Camazine and Hildebrand 1979; Camazine, *et al.* unpublished) and a site of dendritic arborization of all recognized male-specific AL neurones (Figs. 32.2 and 32.3)—including all LNs and ONs that have been observed to date to respond postsynaptically when the antenna is stimulated with female sex pheromones (Christensen and Hildebrand 1984; Matsumoto and Hildebrand 1981).

32.4.4. LPO glomerulus

The MGC is not the only identified glomerulus in the AL. Another distinctive glomerulus in a posteroventral position in the AL has been found to receive bilateral afferent input from the 'labial pit organ' (LPO) in the distalmost segment of each labial palp (Harrow *et al.* 1983). This 'LPO glomerulus' appears to receive no afferent input from the antenna, but probably is a site of arborization of AL neurones. Preliminary evidence suggests that the LPO may be chemosensory—possibly an 'accessory olfactory organ' (Harrow *et al.* 1983).

32.5. AL outputs

The axons of the uniglomerular ONs exit the AL in three separate tracts and exhibit at least three distinct patterns of projection in the PR (Montague *et al.* 1983; also Hishinuma and Hildebrand unpublished data; Montague and Hildebrand unpublished data). Both the ONs from ordinary glomeruli (Fig. 32.3a) and the male-specific MGC ONs (Fig. 32.3b) apparently can have all three types of axonal projections. ON fibres of the first kind run through the antennal–glomerular tract (AGT) to, and send out collaterals that terminate within, the calyces of the mushroom bodies (CMB), and then project further to and terminate within the lateral PR

(LPR). ONs of the second kind send their axons, via a tract distinct from the AGT, directly to the LPR, where their termination 'field' overlaps that of fibres of the first kind. Occasionally an axon running through this tract to the LPR turns back and projects to the CMB (as illustrated in Fig. 32.3a). ONs of the third kind have axons that also do not project to the CMB but instead enter the PR through a distinctive pathway from the AL and, after a 'hairpin' turn, terminate in a region of the PR just caudal to the AL and rostral to the CMB (which are at the back of the brain in *Manduca*). We have the impression, based upon these anatomical observations, that much of the output from the ALs is not sent to the CMB but instead is relayed directly to the LPR and, to a much lesser extent, to other PR centre(s). Moreover, we have encountered ONs that project to the contralateral DEU, but too few of them to permit us to characterize this cell type adequately.

32.6. Some conclusions and speculations

Based on our observations in *Manduca*, as well as findings reported by others from other insect species (see references cited in Section 32.1), we conclude:

(a) In *Manduca*, apparently in contrast with certain other species (e.g. Stocker *et al*. 1983; Strausfeld 1976), antennal sensory afferents have CNS projections that are preponderantly or exclusively ipsilateral.

(b) The CNS projections of antennal afferents in *Manduca* are quite rigorously segregated in separate and distinct mechanosensory and chemosensory regions of DEU neuropil.

(c) Fascicles of sensory fibres (mostly olfactory) in the AN project to individual glomeruli in the AL; the afferents do not branch before entering the one glomerulus that they innervate. All AL synapses occur within glomeruli, and the primary-afferent synapses apparently are cholinergic.

(d) We have recognized no topographic representation of positional information from the antenna in the AL. The central projections of sensory axons appear to be organized on the basis of function, not spatial characteristics.

(e) Each glomerulus probably provides a specific topographic representation of certain odour(s) such that (i) the bundle of afferents projecting into it comes from receptors more or less specific for the odour(s), (ii) integration and synaptic processing of that particular olfactory information occurs largely or entirely within that glomerulus, and (iii) synaptically processed information about the odour(s) is relayed to higher CNS regions from that glomerulus. These notions clearly apply in the case of the sexually dimorphic MGC, which is specialized for processing of sensory information about female sex pheromones.

(f) The multiglomerular LNs predominate among AL neurones, usually exhibit excitatory primary-afferent (in addition to other kinds of) synaptic inputs, often appear to have multiple spike-initiation sites (and thus clearly play complex roles in the synaptic circuitry of the glomeruli), and appear often to contain GABA (Hoskins *et al*. 1984; Kingan and Hildebrand 1982) and to make inhibitory synapses with other AL neurones. Lateral inhibitory synaptic interactions among AL neurones are of major importance in AL function.

(g) The uniglomerular ONs, apparently outnumbered by LNs, send axons through three distinct tracts to three different regions of the PR: the CMB, the LPR, and an unidentified region of PR behind the AL. These PR centres presumably have different functions, each making different use of integrated olfactory information from the AL. Each glomerulus is represented individually and separately in the PR by the ON fibres emanating from it. Many ONs have very divergent projections, relaying outputs from a glomerulus to multiple sites in the CMB and LPR. Preliminary evidence suggests that some type-I ONs projecting directly to the LPR may be cholinergic and that type-II ONs may contain serotonin (Hoskins and Hildebrand 1983).

These and other conclusions that can be drawn from recent investigations of the anatomy and physiology of the antennal sensory pathways must be accommodated as we move toward the day when serious models can be made of central olfactory information processing in insects.

Acknowledgments

We are indebted to all of our co-workers who have contributed to the findings sketched here and whose publications are cited in the text. This research has been supported by a series of grants from the US government, most recently NIH grant RO1 AI-17711, NSF grant BNS-83-12769, and a contract from the Army Research Office (DAAG29-81-K-0091).

33 Central integration of sensory signals: an exploration of processing of pheromonal and multimodal information in lepidopteran brains

D. M. Light

33.1. Introduction

Presently little is known about integration and the coordinated processing of olfactory information in the central nervous system (CNS) of insects. Anatomists have described and modelled only the delineated and aesthetically-organized neuropils of the brain, the deutocerebral antennal lobes (DAL) and the protocerebral optic lobes, mushroom bodies (MB) and central body (Mobbs 1982; Pearson 1971; Strausfeld 1976). Physiologists have described and classified the responses of many randomly probed neurones in various brain regions (e.g. Boeckh and Boeckh 1979; Dingle and Cardwell 1967; Eber 1978; Mimura *et al*. 1970; Olberg 1983b; Suzuki 1975a; Suzuki and Tateda 1974; Suzuki *et al*. 1976; Waldow 1975, 1977; Yamada 1971; Yamada *et al*. 1970). Recently the application of the 'identified neurone' approach has permitted simultaneously both the supravital-staining and the physiological identification of individual interneurones (Burrows *et al*. 1982; Ernst and Boeckh 1983; Homberg and Erber 1979; Matsumoto and Hildebrand 1981; Schildberger 1984; Selzer 1979), thereby enhancing our understanding of the interaction of neural architecture and physiological function. However, at this point only a highly disjointed assemblage of neurones has been described, and the neurones generally defy all but the broadest classifications owing to their individuality in type and pattern of both response and topography. Because thousands of neurones form each cerebral neuropil, no one has succeeded in recording simultaneously from pre- and post-synaptically contiguous olfactory neurones, which would allow for analyses of a circuit of information flow and integration.

The aim of this chapter is to survey the physiological, and, if available, the morphological identities of typical and possibly novel olfactory interneurones that comprise the ascending and descending olfactory pathway(s). The scope of this chapter will be limited to the perception of female sex pheromones by male insects, especially Lepidoptera. Because of the importance of female sex pheromones in the reproductive biology of insects, their perception would appear to be a model and simple system with which to begin the study of sensory integration, a process that leads in a seemingly reflexive manner to a highly orchestrated motor output or behaviour. The primary function of the antennal system of males is the detection of female sex pheromone, which is usually multicomponent. In many species (e.g. *Bombyx mori, Manduca sexta, Antheraea* species, etc.) the male antennae are composed of but a few populations of pheromone-responsive receptors, each population being homogeneous and comprised of tens of thousands of 'specialist' receptor-neurones, which are highly sensitive and specific for the chemoreception of a single component of the pheromone (e.g. Kaissling *et al*. 1978). The sensitivity of the specialist receptor cell is so great that in male *Bombyx mori* a single molecule of the sex pheromone component, bombykol, may be sufficient to induce an impulse in the corresponding receptor axon (Kaissling and Priesner 1970). Thus the brain should receive a great amount of coherent receptor input, and this should facilitate the study of olfactory integration.

From ethological studies four processes are known to occur in pheromone perception: (1) a qualitative assessment of an odorant to determine that the pheromonal constituents are present and in the proper blend-ratio, (2) an enhancement of sensitivity, in which behavioural sensitivity to threshold amounts of pheromone seemingly exceeds by an order of magnitude that of the 'average' receptor cell (whose actual response, e.g. a single spike, is apparently obscured from our observation by background discharge 'noise'), (3) a quantitative assessment of the pheromone concentration present that appears to initiate in some species the transitions in behavioural responses that are progressively evoked (i.e. from initial arousal, to orientation and taxis, to alightment, to close-range approach, and onto the various stages of courtship), and (4) a utilization in unison of many forms of environmental sensory information (e.g. olfaction, vision and wind direction) for orientation, taxis, and courtship.

These four processes suggest at least three properties of the CNS 'black-box': (1) quality convergence of inputs from different populations of 'specialist' receptors upon individual interneurones, allowing for the integrative discrimination of the multicomponent pheromone, (2) spatial convergence of inputs from numerous receptors upon individual interneurones (in the range of 100s:1 to 1000s:1), producing a sensitivity amplification and, hypothetically, an evaluation of the intensity of sensation leading to the 'gating' of concentration-dependent progressive behaviours, and (3) modality convergence of inputs from various sensory organs, which could allow for the abstraction of, for example, an odorant in an airflow.

33.2. Physiology of integration and neural response characteristics

Fundamental to the processing of sensory information in the CNS is integration, the process whereby a neurone, under the dictates of its intrinsic properties (excitability level and firing threshold), sums the various excitatory and inhibitory influences converging upon it and synthesizes a new output signal (Huber 1974). Neural integration occurs at the molecular level (not to be considered here), cellular or neuronal level, and at the level least understood, the multicellular or circuit level.

In the majority of neurones (i.e. those that spike), neural output is encoded in the frequency and pattern of the action potentials. There are five major neural response characteristics that, when combined, help in the physiological classification of neurones:

(1) the neurone's background discharge frequency (bg.d.): (reflecting endogenous spontaneous impulse activity and/or exogenous unknown/uncontrolled inputs);
(2) response phasing in relation to stimulus occurrence: 'on' (either transitional or persistent for stimulus duration), 'off' (ms after stimulus cessation), 'on–off', and after-effects: either prolonged activity (up to seconds beyond stimulus cessation) or sustained-activity states (seconds to minutes);
(3) response polarity: excitation (+) (increase in impulse frequency), inhibition (−) (decrease in impulse frequency), no response (0) (no change in bg.d.), and rebound activity: rebound inhibition (often following excitation), and rebound excitation (often following inhibition);
(4) temporal aspects of response: overall magnitude (absolute numbers or frequency of impulses over entire stimulus interval), latency (here used as time period between initiation of receptor response and that of the interneurone; a function of conduction parameters and synaptic circuitry), temporal patterning of response: phasic (ph.) (a short burst of high frequency discharge), tonic (t.) (steady or decremental frequency of impulses often spanning the stimulus duration or beyond), and phasic–tonic (ph.–t.) (a high frequency burst followed by a steady state of lower frequency); and
(5) response-modulating phenomena: variable in its occurrence, either an enhancement of activity (e.g. facilitation, potentiation, sensitization, etc.) or a diminishment of activity (e.g. adaptation, habituation, depression, etc.).

Due to different configurations of synaptic transmission, three basic integrative computations occur in the postsynaptic neurone (Ewert 1980): (1) addition of excitatory inputs (EPSPs) from two or more convergent presynaptic terminals, (2) subtraction of convergent excitatory (EPSP) and postsynaptically inhibitory (IPSP) inputs, and (3) multiplication, e.g. an axo-axonic synapse mediating 'presynaptic inhibition' which lowers the normal excitatory postsynaptic efficiency of the inhibited axon terminal by a specific multiplicative (<1) factor.

33.3. Central olfactory pathway

Tens of thousands of pheromone receptor afferents project via the antennal nerve and invade the ipsilateral DAL. Within the DAL, the pheromone receptor axons terminate immediately in the macroglomerular complex (MGC) of the dorsal DAL (Ernst *et al.* 1977; Koontz unpublished), where they converge in large numbers upon and synapse with various morphological types of DAL interneurones. Other axons from non-pheromonal, olfactory 'generalist' receptors, and afferents from receptors serving other modalities, project to the various spherical ventral glomeruli (VG) of the DAL. A large percentage of the mechanoreceptor axons, especially those from the Johnston's organ and Böhm bristles, circumscribe the DAL and project to and beyond the mechanosensory and motor neuropil of the ventral and posterior deutocerebrum (Camazine and Hildebrane 1979; Koontz and Light unpublished; Suzuki 1975b).

There are about 1100 DAL interneurones in male *Manduca sexta* (see Hildebrand (Chapter 32)), *B. mori* (Koontz unpublished) and *Periplaneta americana* (Ernst and Boeckh 1983; Ernst *et al.* 1977), and these can be divided into two broad morphological classes: local interneurones and output interneurones. The majority (about 88 per cent) of these interneurones in *M. sexta* are intrinsic or local to the DAL, having neurites that arborize in many glomeruli, but having no processes that leave the DAL (Matsumoto and Hildebrand 1981). The other class of DAL interneurones, the extrinsic or output interneurones, ramify and synapse apparently in only single DAL glomeruli and then send ascending axons via either the antennal–glomerular tract (AGT) (Ernst and Boeckh 1983) or other routes to the protocerebrum (Hildebrand (Chapter 32); Montague *et al.* 1983). AGT output neurones send out two collateral axon branches within the protocerebrum, one that diffusely ramifies in the calyces (or input region of the MB) and the other collateral projects to and arborizes in the lateral lobe of the protocerebrum (LLP). This is the extent of the known pheromonal olfactory pathway based on studies of identified neurones, though separate anatomical and physiological studies show that much pheromonal integration and information processing occurs in the protocerebrum.

33.4. Pheromone receptor cells

The type, temporal pattern, and sensitivity of pheromone responses of 'specialist' receptors on the antennae of male *B. mori* and *Antheraea* species are typical of many Lepidoptera. The apparent threshold response concentration (at which the response is significantly greater than the bg.d.) is about 10^{-4} to 10^{-5} μg of pheromone on a filter paper substrate. As concentration increases (e.g. 10^{-1} μg), latency decreases and a typical ph.–t. response pattern occurs that may outlast the stimulation period (e.g. Kaissling and Priesner 1970).

33.5. Deutocerebral local interneurones

Matsumoto and Hildebrand (1981) described three morphological types of identified local interneurones in male *M. sexta* (Fig. 33.1). Their classification was based on the pattern and density of neurite branching in particular neuropils and not upon the odour-spectra of the interneurones; with stimuli limited to a crude pheromone extract, a pheromone component, bombykal (bal.), (*E*)-2-hexenal (*E*-2-h.), and a tobacco leaf extract (t.l.e.).

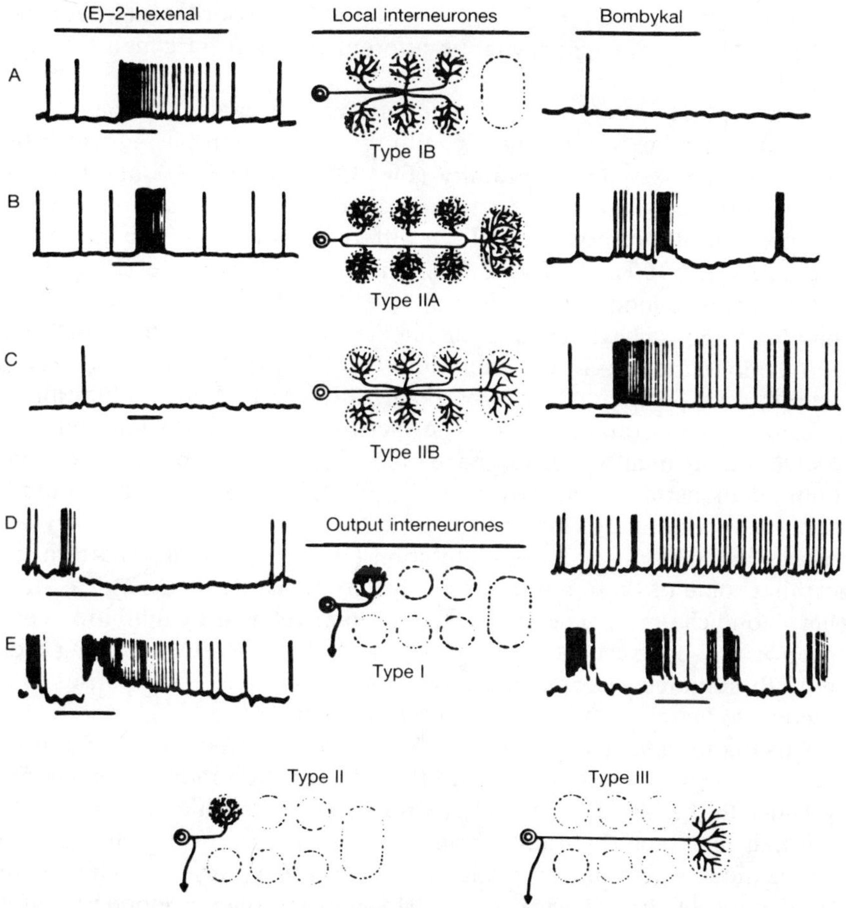

Fig. 33.1. Summary of chemosensory responses and schematic drawings of the arborizations of the various types of local and output DAL interneurones identified through intracellular recordings from male *M. sexta*. The bombykal stimulus dosage ranged from about 1×10^{-4} to 2×10^{-3} μg/cartridge, while the (*E*)-2-hexenal dosage used was cited as 'various concentrations'. Stimulus intervals and spike amplitudes were approximately 600 ms and 60 mV for A–C, and approximately 250 ms and 30 mV for D–E. (Figure has been adapted from that of Matsumoto and Hildebrand 1981.)

Apparently they found no qualitative response differences between bal. and the crude pheromone extract, so reference will be made only to the bal. response. All three types of *M. sexta* local interneurones had a common physiological trait of a low bg.d. rate of about 0–15 Hz.

Type Ib interneurones have radially symmetrical, large arbors that ramify diffusely throughout most glomeruli but not the MGC (Fig. 33.1a). Type Ib interneurones do not respond to pheromone, but do respond to both the plant odour E-2-h. and t.1.e. with a short latency (about 150 ms, reflecting both neural and their stimulation-system delays) and a moderately high ph.–t. response (95:17 Hz, respectively) (Fig. 33.1a). Such a response suggests that type Ib neurones could be first-order interneurones, and possibly receive monosynaptic afferent inputs from generalist olfactory receptors.

Local type IIa interneurones are found only in males and have asymmetrical branching to both the VG and the MGC neuropils, in which fine dendritic processes densely ramify (Fig. 33.1b). Type IIa interneurones have differential, odour-quality responses to *E*-2-h. and bal. In one short-latency, tonically responsive cell the difference was only in magnitude of response (bal. 28 Hz: *E*-2-h. 8 Hz). In the other example (Fig. 33.1b), *E*-2-h. elicited a moderate ph.–t. response (55:30 Hz), while bal. evoked a complex response having a medium latency (270 ms), short 'on' inhibition, then a moderate ph.–t. 'on' excitation (90:21 Hz). This was followed at stimulus cessation by an 'off' hyperpolarization and rebound inhibition for a prolong period (about 1.8 s). The fact that the type IIa interneurones receive odour-quality convergence and respond differentially to such stimuli demonstrates that they possess integrative properties that could be used to discriminate odour quality. Further, the moderate latency and complex inhibition–excitation–inhibition (−, +, −) response pattern suggest that some of these neurones receive polysynaptic inputs at least from pheromone channels. The ph.–t. excitation enveloped by inhibition could enhance the 'contrast' of stimulus 'on' and 'off', thereby encoding more faithfully stimulus onset and duration. Also such (−, +, −) responses could function to 'gate' or time other lateral processing.

Matsumoto and Hildebrand (1981) also identified a second male-specific local interneurone, type IIb (Fig. 33.1c), with radially symmetrical branches to the VG and MGC but with coarser dendritic processes in the neuropils than type IIa interneurones. A high degree of variability in MGC arborizations was found, ranging from only a few large branches to full ramification of the MGC. The type IIb interneurones responded only to bal. and crude pheromone, and were characterized by long latency (about 300 ms), moderate to high frequency ph.–t. (37–170:6–50 Hz) 'on' responses that slowed to a prolonged (4 s) tonic response (about 6 Hz over bg.d.) (Fig. 33.1c). Such prolonged responses suggest either reverberation in presynaptic circuits or postsynaptic feedback self-excitation. Prolonged responses to short pheromone stimuli could elevate the overall activity portions of the pheromone pathway, possibly facilitating threshold responsiveness and 'motivation'.

Local interneurones that respond to pheromones were not morphologically identified in studies of male lepidopteran DAL interneurones of *B. mori* (Olberg 1983b) or *Antheraea* species (Boeckh and Boeckh 1979). However, recently profiles of local interneurones with arborizations in both the VG and MGC were found to be prominent and numerous in anatomical Golgi preparations of the DAL of male *B. mori* (Light unpublished), and a few local interneurones have been 'identified' in physiological studies (Light unpublished). In the cockroach, *P. americana*, pheromone-responsive local interneurones have yet to be found. However, a single food-odour-responsive local interneurone without macoglomerular arbors has been described (Selzer 1979), and a few local multimodal interneurones were identified that had dense arborizations in the numerous ordinary glomeruli and just a single or few arbors in the MGC, and that respond to mechanical brushing of the ipsilateral antenna and to general food odours (Ernst and Boeckh 1983).

33.6. Deutocerebral ascending or output interneurones

In all insect species studied, the DAL output interneurones, which send ascending axons via the AGT to the protocerebral MB and LLP, make only uniglomerular arborizations in the DAL, in either single ordinary glomeruli or just the MGC. Matsumoto and Hildebrand (1981) described three morphological types of output neurones (Fig. 33.1), though only their type I output neurones, with dense hemiglomerular arbors in single VG, were identified from recordings. Type I output interneurones have high bg.d. rates (<50 Hz), and the cells that respond to *E*-2-h. (33 per cent) did so with either long-lasting 'on' inhibition (about 1 s) (Fig. 33.1d) or long-lasting ph.–t. (140:6 Hz above bg.d.) 'on' excitation (Fig. 33.1e). Of the DAL interneurones identified by Matsumoto and Hildebrand (1981) 95 per cent had somata in the lateral cluster, while recently Christensen and Hildebrand (personal communication) have identified their type III output interneurones, with somata predominately in the medial cluster and only MGC arborizations, which have low bg.d. (<10 Hz) and respond to pheromone with either inhibition or ph.–t. excitation of various durations.

In two *Antheraea* species, Boeckh and Boeckh (1979) found DAL cells with medial cluster somata and widely branching arborizations only in the MGC, but no output axons leaving the DAL were resolved (perhaps due to incomplete cobalt migration). These MGC neurones are thought to be true output interneurones. They responded only to the pheromone components and only to both components and not specifically single components, as do their antennal receptor 'specialists'. However, the responses of these MGC neurones to the two pheromone components were differential, with each species having higher sensitivity (> three orders of magnitude) to their major component than to their minor component. Thus, in each cell a complete odour-quality convergence of the two pheromone components occurs, although quantitatively differential, so component discrimination could still be encoded within and across such fibres. These MGC neurones had very low bg.d. rates (about 1 Hz) and responded to the pheromone components with low to moderate frequency responses that were mostly tonic for the minor component, and shifted from tonic to ph.–t. as the concentration increased for the major component.

Furthermore, the MGC neurones of *Antheraea* showed definitive evidence of spatial convergence of numerous pheromone receptor inputs (Boeckh and Boeckh 1979). For their species' major component, they had response thresholds of 10^{-6}–10^{-7} μg/filter paper, which is about two orders of magnitude lower than the average receptor's threshold. At these low threshold concentrations only a small fraction of the receptors react with an impulse above their bg.d. rate, but, probably due to the thousandfold convergence factor, the DAL interneurones are able to respond significantly and detect the pheromone. Similar threshold sensitivities were found in DAL neurones of male *B. mori* (Light unpublished; Olberg 1983b). Thus already in the olfactory pathway these apparently second-order 'output' DAL interneurones have both the CNS properties of quality convergence with specificity discrimination, and spatial convergence with sensitivity amplification.

In contrast to Lepidoptera (e.g. *M. sexta*), output interneurones were more readily identified in *P. americana* (Burrows *et al*. 1982). All of the pheromone-responsive output neurones had arborizations only in MGC and sent axonic collaterals into the ipsilateral MB calyces and the LLP. These output neurones had moderate bg.d. rates ($\bar{x}$ = 27 Hz, range 0–80 Hz) and responded primarily to pheromone, though often differentially to the putative components, and variably to a few other odorants (see for review: Boeckh and Ernst 1983; Boeckh *et al*. 1984). The responses of different cells were: (1) ph.–t. 'on' excitations or ph.–t. 'on' excitation and 'off' rebound–inhibition responses to both pheromone components, (2) prolonged excitatory or inhibitory responses to only one component, (3) ph.–t. 'on' excitation response to pheromone and inhibition to other odours, and the converse. Interestingly, Burrows *et al*. (1982) found that output neurones with specific responses to only one pheromone component arborized specifically in the dorsal–lateral MGC, while neurones responsive to both components arborized over the entire MGC. This suggests that there could be specific afferent projection loci and/or interneuronal integration and processing loci in the MGC, and that parallel processing of single-component labelled-line and multicomponent across-fibre information might possibly occur simultaneously in different DAL output neurones.

33.7. Multimodal deutocerebral interneurones

In *P. americana*, Waldow (1975, 1977; see Boeckh *et al*. 1983) found that a high percentage (about 40 per cent) of odour-responsive DAL interneurones were multimodal, responding also to mechanical stimulation (wind displacement or brushing) of the antennal flagellum (causing receptor responses presumably in the campaniform sensilla). These multimodal mechanical and olfactory neurones (MON) had variable responses to odours, while having uniform responses to mechanical stimuli. However, only 23 per cent of the MON responded to cockroach odours (presumably pheromonal). The others responded to general food odours. MON were found to have significantly shorter latencies (39.6 ms) than the unimodal odour interneurones (57.6 ms) (Waldow 1977) (as has also been found in *B. mori*, Light unpublished). This suggests a possible CNS-coordinating function for the MON. MON have moderate bg.d. rates (about 20 Hz) and respond to odours with ph.–t. 'on', 'off' and 'on–off' reactions of moderate frequency (about ph. 100: t. 50 Hz).

In Lepidoptera, MON responsive to pheromone were rarely observed in local interneurones of the DAL of *M. sexta* (Matsumoto and Hildebrand 1981), but were found quite frequently in both the male *B. mori* DAL (67 per cent) (Light unpublished; Olberg 1983b) and protocerebrum (28 per cent) (Light unpublished). Olberg (1983b) found in male *B. mori* about as many DAL–MON were excited by clean air-puffs as were inhibited. The most prevalent (45 per cent) type of MON recorded was excited by bombykol (bol.) (the attractant pheromone component) compared with types inhibited or not responding to bol. or responding to the secondary component, bal. Thus modality convergence is also a property of the lepidopteran DAL, where pheromone inputs are processed through the MGC and mechanosensory inputs through the mechanosensory centre and the VG. Anatomically, there appear to be both type I and type II local interneurones (designations of Matsumoto and Hildebrand 1981) in the DAL of male *B. mori* (Light unpublished), and certain of the type II could be candidate pheromonal MON types.

33.8. Deutocerebral functions

The MGC is the major site of integration of pheromonal information in the DAL, with all pheromone-responsive neurones (the receptors, local, and output) having arborizations and synapses within its neuropil. As evidenced by the array of complex responses, polysynaptic latencies, convergences, and modalities, a great degree of lateral and tiered interactions occur between local interneurons in the MGC and other DALs, creating truly integrative, novel outputs that encode many stimulus parameters and their context with other modality stimuli. The characteristically high bg.d.

and occurrence of inhibition in the output interneurones suggest that modulation of their spontaneous firing rate may be an important and economical means of transferring olfactory information from the DAL to the protocerebrum. Thus the output neurones of the MGC conduct more than just a filtered and condensed relaying of the receptors' detection of pheromone to higher centres.

33.9. Protocerebrum (PC)

The MB are the most anatomically organized neuropil of insects, especially those of honeybees (see: Mobbs 1982, 1984). The MB are comprised of parallel-aligned intrinsic 'k' (Kenyon) cells, that are 'k' shaped due to their inputs in the calyces, projections through the stalk, and their bifurcation into the 'output' regions, the α- and β-lobes. The MB of Lepidoptera are dissimilar to other insects in having a third γ-lobe (Pearson 1971). In spite of the anatomical analyses, no intrinsic MB interneurones have been identified. Many electrophysiological studies in the vicinity of the MB have found cells with an array of response complexity, though somewhat diminished from that of DAL interneurones. These MB interneurones have bg.d. of 0–30 Hz and mostly (about 60 per cent) have ph.–t. 'on' excitation responses and fewer inhibitory responses (about 15 per cent) (Erber 1978; Suzuki 1975a; Suzuki and Tateda 1974). Surprisingly, there have been no physiological studies published on pheromone-responsive interneurones in any portion of the protocerebrum (PC), in spite of MB having been designated primarily olfactory processing neuropils (see Weiss 1974). It is now apparent that the MB and the surrounding lateral protocerebum (LPC) process many modalities (vision, olfaction, mechanoreception, audition, and gustation), with multimodal cells predominating (about 60 per cent) (Erber 1978; Homberg and Erber 1979; Light unpublished; Mimura *et al.* 1970; Schildberger 1984; Suzuki *et al.* 1976).

Extrinsic MB (EMB) neurones have somata outside the MB calyx cluster, are either presynaptic to intrinsic neurones in the calyces or postsynaptic to 'k' cell axons in the MB lobes, and have processes into various regions of either the LPC or medial PC (see Schildberger 1983, 1984). The physiological responses of EMB neurones are generally multimodal, complex, and individualistic (Erber 1978), while their morphology is just as complex and diverse (see Pearson 1971; Schildberger 1983). Both complexity assessments for the EMB neurones have recently been confirmed by Schildberger (1984) in identified neurone studies on crickets. The advancements in response complexity and integration of many EMB neurones, over other cell types, are their prolonged and often sustained after-effects and their conditional responses based on prior stimuli (Erber 1978; Schildberger 1984).

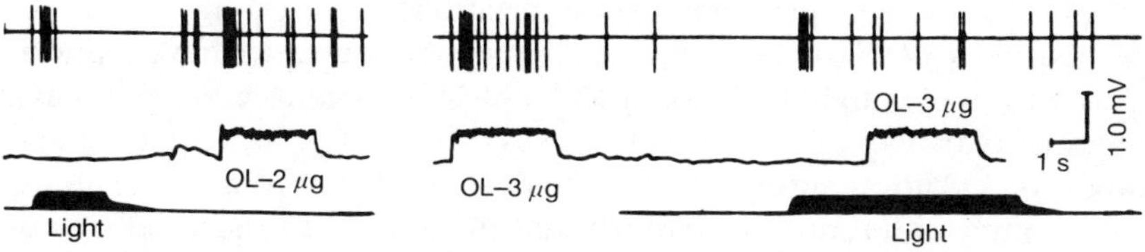

Fig. 33.2. Extracellularly recorded responses of a bimodal LPC interneurone of a male *B. mori* moth that was differentially responsive to modality stimuli, lights 'on' elicited phasic excitation, while bombykol (OL, 1×10^{-3} μg) elicited ph.–t. excitation.

Recently interneurones in the LPC of male *B. mori* were found to have a diverse range in response complexity and differential responsiveness to various presentations of pheromonal, mechanical, and light stimuli (e.g. Fig. 33.2) (Light unpublished). These LPC neurones were predominately (about 60 per cent) multimodal, with responsiveness to pheromones being paired to an equal extent with light or mechanical stimuli. The most frequently found unimodal cells were those responsive to pheromones. The range and degree of complexity in response pattern and phasing of the LPC neurones of *B. mori* was very similar to that seen in EMB neurones of crickets (Schildberger 1984), though, in crickets, mechanical stimuli dominated instead of pheromone stimuli. Fig. 33.3 is an example of differential responsiveness to modalities in an identified LPC neurone that diffusely ramified laterally through the PC and sent a few descending arbors into the tritocerebrum and suboesophageal ganglion. This cell had a low bg.d. (<1 Hz) and responded to lights 'on' and 'off' with ph.–t. (24 : 10 Hz) bursts of spikes, to bal. with two phasic spikes, and to bol. with a ph.–t. (6 : 3 Hz) pattern of spikes. Three other interesting aspects of this cell's physiology are: (1) a transition in a response pattern to bol. from tonic (at 10^{-3} μg) to ph.–t. (at 10^{-2} μg), (2) a frequency enhancement (facilitation?) of response to bol. when presented in context with light, and (3) a threshold-dependent (10^{-2} μg), synergistic response to the combined presentation of bal. and bol. Behaviourally, the major pheromone component, bol., alone elicits wing fanning and anemotaxis in male *B. mori*, while increasing amounts of bal., the minor component (about 10 per cent), dilute and disrupt the effects of the attractant bol. compoent (Kaissling

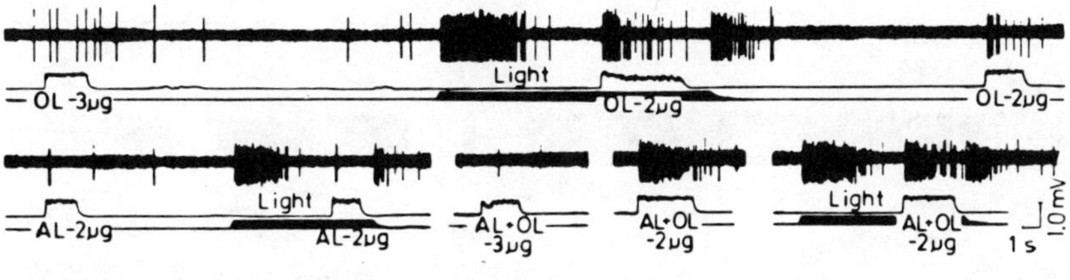

Fig. 33.3. Extracellularly recorded responses of a bimodal LPC interneurone of a male *B. mori* moth that demonstrates many complex modality and odour-quality response features (see text (Stimuli: bombykol (OL) and bombykal (AL)).

et al. 1978; Ludwig personal communication). Interestingly, the DAL interneurones of *B. mori* were predominantly found to have opposite-polarity responses to bol. and bal., with only 13 per cent of the cells having common polarity (+,+ or −,−) responses (Olberg 1983b). Thirty nine per cent of the PC interneurones had common-polarity pheromonal responses, though often differential in pattern and magnitude to these behavioural antagonists (Light unpublished). Sustained, long-term (minutes beyond stimuli) excitation and inhibition are commonly observed in multimodal pheromone-light) LPC interneurones. For example (Fig. 33.4), different LPC interneurones respond with sustained excitation to lights 'on', but to pheromone stimuli the cells differ in having either sustained opposite-polarity responses (e.g. excitation to bol. and inhibition to bal., as in Fig. 33.4a), or sustained common-polarity responses (e.g. inhibition which was stimulus-duration dependent for bol. and sustained for bal., as in Fig. 33.4b).

Thus a new level of pheromonal integration is evident in the PC over the DAL, manifested in three traits. Firstly, modality convergence has expanded to incorporate also illumination levels, creating new multimodal response combinations. Secondly, after-effects are extended for prolonged to sustained periods (possibly due to reverberating EMB-neurone–MB-neurone circuits, Schildberger 1984), which imparts within PC circuits a short-term 'memory' capability that might initiate and then maintain a level of 'command' or 'motivation' in the CNS. Thirdly, conditional responses are evident which are dependent on prior stimuli, making the system plastic. These three PC traits combined should have great bearing on pheromonally-aroused anemotaxis and optomotor responses of insects.

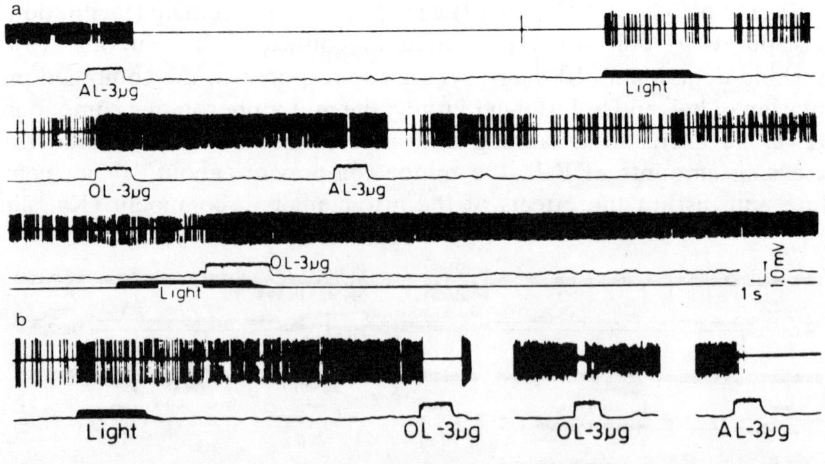

Fig. 33.4. Responses of two (a and b) extracellularly recorded PC interneurones of a male *B. mori* moths that were excited for sustained periods by lights 'on', but differentially excited or inhibited by the two pheromonal components (OL and AL) (see text).

33.10. Pheromonal information descending in the ventral nerve cord (VNC)

Recently studies have been published on descending interneurones that carry sensory information which appears to modulate directly behaviour such as flight (see Bacon and Mohl 1983) and other ambulatory or 'escape' responses (see Strausfeld *et al.* 1984). In most studies these descending interneurones were visually or wind stimulated, but never pheromonally. However, a very intriguing recent study of descending interneurones in the VNC of male *B. mori* has given the first insight into how pheromonal information might command or modulate anemotactic behaviour. Olberg (1983a) recorded from a group of VNC descending interneurones that shared a common sustained output, which switched up and down between long-lasting (>4 min) high (about 45 Hz) and low (about 0–15 Hz) firing states (Fig. 33.5a). These 'flip-flopping' interneurones (FFIs) (named after the properties of such a bistable electronic circuit) were caused to switch from one frequency state to another by either changes in concentration (10 per cent or more) of the attractant bol., or gross changes in light intensity (Fig. 33.5a). At any time, certain FFIs were in high state while others were in low state, and they all switched simultaneously. The high and low frequency states corresponded to two different positions of the antennae,

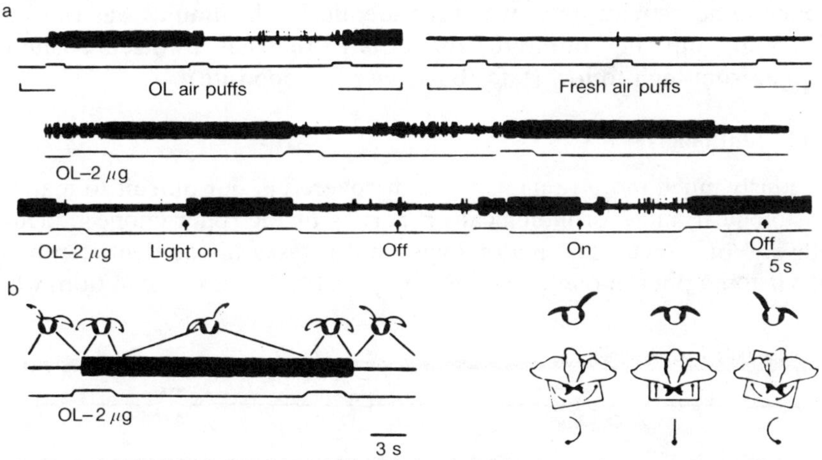

Fig. 33.5. (a) Bimodal responses of a flip-flopping interneurone of a male *B. mori* moth, in which air puffs elicited transitions in frequency states only if they contained bombykol (OL). In addition to bombykol stimuli, lights 'on' elicited transitions in frequency states. Note that phasic response to light-'off' is excitatory during the low state, then inhibitory during the high state (b) Correlation of antennal positions with FFI states. Drawings above traces indicate antennal positions during bracketed periods. Moth mounted ventral side up with head and thorax rigidly fixed. (c) Correlation of antennal positions (top), abdominal ruddering (middle) and animal turning (bottom) (see text). (From Olberg (1983a) Pheromone-triggered flip flopping interneurones in the ventral nerve cord of the silkworm moth, *Bombyx mori*. *J. Comp. Physiol*, **152**, 297–307, with permission.)

dorsal in one, while ventral in the other (Fig. 33.5b). The antennal positions reversed when the FFI switched frequency states. In immobilized animals, free to move their legs but not their heads, the two antennal positions corresponded to right and left turning movements of the legs, and to right and left 'ruddering' of the abdomen (Fig. 33.5c). Olberg (1983a) found that when different concentrations of bol. were presented to the antennae the FFI assumed a high or low firing rate consistent with a walking turn toward the antenna receiving the higher bol. concentration. Altering the levels of bal. (during constant bol.) evoked the opposite effect, that is frequency states consistent with turning away from higher bal. concentration. All of the results support Olberg's hypothesis that the FFIs carry left and right turning instructions for pheromone-directed turns in the male's upwind approach to calling females. The males make this upwind approach by a zigzagging path, which has been hypothesized by Kramer (1975; Chapter 6) to be the vector sum of an upwind turning command and constant turning command, the latter of which changes sign or direction with decreases in bol. concentration. This constant turning command is suggested to be given by the sustained FFI states, with the transitions in states reversing the direction of the turn.

The location of the flip-flop 'generator' probably resides in the LPC, because possible antecedent neurones have been recorded there (Light unpublished). These LPC interneurones switched from sustained high frequency to no-activity states when the identical bol. stimulus was repeated (Fig. 33.6), but they differed from FFIs in that bal. by itself elicited a change, from high to low state (but never the opposite).

33.11. Outlook

Obviously much more remains to be discovered in our pursuit to map the physiological and morphological properties of the 'pheromone olfactory pathway' of insects. The major hypothetical tasks of the deutocerebrum appear to be pheromone detection and possibly species recognition, while

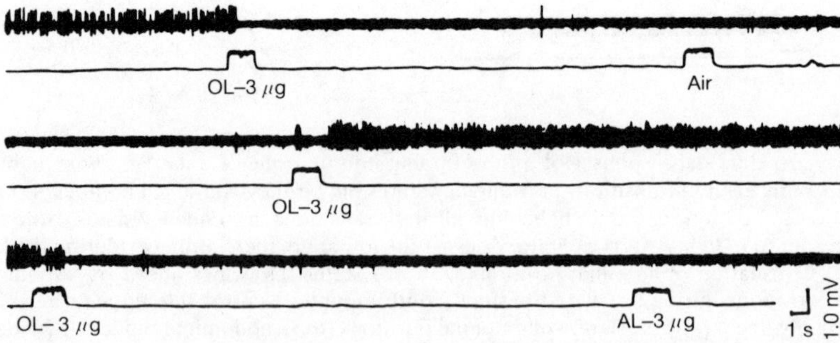

Fig. 33.6. State-transition responses of an extracellularly recorded LPC interneurone of a male *B. mori* that switched frequency states due to consecutive bombykol (OL) stimuli.

the hypothetical tasks of the protocerebrum appear to be: (1) the integration of modalities for the abstraction and reconstruction of real environmental phenomena (e.g. odorant carried in a directional airflow), (2) the 'remembrance' (sustained activities) of prior stimulus events, so that certain circuit elements could possibly generate appropriate novel outputs that encode for sequential multimodal environmental events, and (3) the initiation and possibly direction of motor commands. Thus, there appears to be a definite progression in integration ascending the pathway; however, the underlying circuitry is completely unknown. To advance our understanding of the integrative circuitry, the pursuit to identify interneurones must continue and possibly be coupled with electrical and olfactory stimulations of specific antennal afferents (e.g. Harrow and Hildebrand 1982), so that synaptic communication can be more quantitatively modelled at different levels of the pathway. Also it is necessary that the odour spectra tested be expanded, especially to more biologically relevant odours, and, further, the modality paradigms tested should be aligned to reflect to a greater extent the 'real' sensory inputs encountered in orientation.

Acknowledgements

I am appreciative of the advice and comments of Dr Dietrich Schneider and Dr Robert Olberg, the sharing of unpublished data by Drs M. A. Koontz, R. M. Olberg and D. Ludwig and the typing of the manuscript revisions by Emily Grey and June Ikeda.

Endnote

Original work cited here was performed by the author under a NATO postdoctoral fellowship and a Max-Planck postdoctoral fellowship in the laboratory of Professor Dietrich Schneider, Max-Planck-Institute für Verhaltensphysiologie, Seewiesen, Federal Republic of Germany.

34 Epilogue: physiological mechanisms

J. Boeckh

34.1. Introduction

Perception of insect olfactory signals has attracted workers in several different disciplines not only for the sake of their own curiosity and because of the potential findings which provide a basis for rapid progress in their special fields, but also because insect olfaction represents a problem which can be comprehensively approached only by combination of aspects and methods in a multidisciplinary effort. In this sense one of the most significant features of the seminar that formed the basis for this volume on insect olfaction was that specialists from different fields not only contributed very interesting results and demonstrated methods from their own research, but also shared their approaches and even their attitudes which are specific to their approaches. In this way, the participants learned very much from one another, with one of many consequences being the consideration of formerly neglected parameters of stimuli and response.

Of several interdisciplinary conferences on chemoreception this one was especially successful because of the concentration on a single topic where the participating disciplines have already reached a considerable level of sophistication in their methods, and in a depth of insight as a basis for hypothesizing about orientation mechanisms and the neurological coding of stimuli. Perception of insect olfactory signals might thus be ranked in the series of prominent examples for successful multidisciplinary approaches in sensory biology such as ultrasound–echolocation in bats. The study of such highly evolved and specialized systems has revealed insights into both very general and very basic problems. Since an epilogue dealing with behavioural mechanisms occurs elsewhere in this volume (Cardé (Chapter 20) the following general remarks emphasize the physiological mechanisms.

34.2. Olfactory physiology, state of the art

In the keynote address of the seminar olfactory physiologists were critically reminded of some underdevelopment in their field as compared with other areas of sensory physiology, especially as far as approaches at the cellular and molecular level are concerned (Dethier (Chapter 1)). One reason for this is certainly the notoriously low number of researchers in olfactory physiology and related disciplines. Many students who had started in the field gave up, disappointed by tiny receptor cells and junglelike neuropil, badly defined stimuli, boring work with establishing odour spectra of hundreds of cells, and, not the least, because of a lack of stimulating concepts and perspectives.

This situation seems to be improving slowly but steadily. Critical masses of motivation, talent and expertise are becoming assembled with intense exchange and discussion, all under the influence of constructive competition. Reports and discussions during the seminar and several recent publications show that such development is under way. Modern electrophysiological, neuroanatomical, and cyto-histochemical methods have been developed which are facilitating advances in the field. Direct biochemical studies of receptor membranes are successfully under way. Among all of these approaches, molecular biological methods, especially in immunochemistry and genetics, can be regarded as offering significant potential in the search for understanding the physiological mechanisms of insect olfaction.

34.3. Pheromone receptors, the well-established periphery

Pheromone receptors of insects are especially well suited for many kinds of physiological investigation. There are thousands of cells of a given reaction type with identical chemical specificity on an antenna. They are easily spotted from the outer surface because they are housed in distinct morphological types of sensilla. Recording from most of the sensilla is easy. In several cases only two or three receptor cells are found in one sensillum, and individuals can be discriminated by characteristic time courses and amplitudes of their impulses. Chemically well defined and biologically significant stimuli can be administered and calibrated for their quantity. Thus, the biochemist and the biophysicist have homogeneous and 'pure' material available to study receptor membrane properties and stimulus transduction processes, and to apply the complete arsenal of modern methods in cellular neurophysiology and biochemistry. Former studies about adsorption of odour molecules at the antennal surface and their conduction via cuticular channels to the dendrites are being complemented by investigation of substances in the sensillum liquor which might either serve as transport systems for odour molecules, or play the important role of degrading stimulus molecules after or during their contact with the receptor membrane. The hypothetical model for such events which was present in the seminar demonstrated the potential of a direct approach through modern protein chemistry (Vogt and Riddiford (Chapter 23)). It was pointed out that with the small amount of receptor protein available the 'direct' biochemical approach is limited, and a molecular biological approach using immunebiochemistry might be more promising.

The phenomena of electrogenesis following the primary process have not been dealt with during the seminar, because they seemed not as interesting to an interdisciplinary audience of this composition as is the question of receptor cell spectra or the problem of sensory coding. Especially with the latter, much progress has been achieved by multifold investigation of pheromone receptors, where we know a lot about response patterns of whole populations of cells responding to complex stimuli.

The basic story seemed simple: each of the components of a pheromone affects a separate cell type, and the reaction pattern across the population of participating cells is a measure of the relative amounts of the components and thus of the whole blend. Interaction of such stimuli, for example synergism or antagonism, was believed to take place at the central level. Now there is evidence for overlap in the sensitivity of some pheromone receptors for certain pheromone components for example in the American cockroach, where one type of receptor responds to both of a pair of components. One of these also affects another cell type, which, however, is not sensitive to the other component.

Interesting contributions to this point were reports on synergistic and antagonistic effects between pheromone components taking place at the receptor cells themselves (Priesner (Chapter 26); O'Connell (Chapter 25)). The dose–response characteristics for single compounds and their mixtures which were shown are clearly in favour of the hypothesis (Priesner (Chapter 26)).

As far as the 'complete' reaction pattern is concerned, we are short of detailed information about individual sensitivities within receptor populations, and also about the actual distribution of receptor types. Two very careful investigations presented in the seminar dealt with this question, and we learned that there might indeed be a lot of variation between receptors as far as their intensity functions are concerned (Priesner (Chapter 26); Van Der Pers and Löfstedt (Chapter 27)). Therefore, it remains open for most insects just how many receptor types are firing at which level at a given distance from the pheromone source. Differences in vapour pressure of the compounds and the actual numbers of molecules evaporating during the calling behaviour are critical to the problem. Analyses of the contents in the odour plume received by the male are of utmost interest and there is much need for further microanalytical work.

Moreover, very powerful components might act upon only a few cells on an antenna, so that there are important soloists in the orchestra, which may not appear in the recording of the overall response of an antenna. This gave us one more warning against uncritical use of the electroantennogram (EAG) method that so many laboratories apply as a standard test for sensory effectivity of compounds upon a given antenna (Van Der Pers and Löfstedt (Chapter 27)).

A very important point was brought up in the discussion of the influence of internal state upon sensory input. Electrophysiologists often work with randomly selected animals regardless of their diurnal rhythms, state of satiation, or motivation, etc. There are, however, reports on variations of sensitivity of sensory receptors in parallel, for example to diurnal rhythms. Now we are confronted with such a phenomenon also in insect olfactory receptor cells, which are significantly altered in their responsiveness by internal factors (Davis (Chapter 28); Dickens (Chapter 29)). Not only the phenomenon itself is interesting enough to be studied, but also most intriguing questions arise concerning the nature of the factors and the way in which they exert their influence upon the cell. Even new target organs for endogenus messengers might become accessible for investigation.

As a striking example of physiological experiments inspired by results obtained in behaviour studies, data were presented about 'flicker–fusion' experiments with chemoreceptors (Kaissling (Chapter 22)). When it became apparent that certain moths display regular orientation behaviour only if the spatial/temporal distribution of pheromone resulted in slowly pulsed odour stimulation, electrophysiologists considered again the role of the marked initial (and terminal) phases of impulse responses to changes in odour concentration. When such data were presented 20 years ago, not enough information was available about odour distribution in air currents. Now we know that it is this phase of response which keeps the receptors active in synchrony to repetitive stimulation up to several hertz. This permits an in-phase resolution of time-patterned stimuli, which apparently contributes essential information for the central nervous system (CNS) and enables it to produce motor patterns for correct turning movements. Other quantitative aspects of receptor responses such as the slope of response intensitites in the context of intensity discrimination, have been neglected for a long time. The same is true for the study of osmotropotaxis and its role in pheromone-guided behaviour.

A domaine where physiology can contribute much to our understanding of the evolution of receptors and pheromones is the comparative study of cell specificities. This line of research began 25 years ago with the first investigations of the effectiveness of female pheromone glands in EAG investigations of males of closely related species. Subsequent single-cell recordings revealed many more details about interspecific pheromone action and the corresponding receptor specificities. During the seminar we reviewed, on a greater scale, relationships of this kind within certain groups of Lepidoptera (Priesner (Chapter 26); Van Der Pers and Löfstedt (Chapter 27)). We also learned of an example of the convergent development of receptor cells which are highly specific for the same substances. A bark beetle predator was found to possess olfactory receptors specific for the aggregation pheromone of its prey, and to use that perceptual ability in locating its prey (Mustaparta (Chapter 30)).

A new challenge became apparent for the physiologist when male pheromones of Lepidoptera entered the scene (Schneider (Chapter 21)). There are not only different 'rules' for the roles of such pheromones and striking physiological differences in their effects, but also an interesting phenomenon is found in the sources of the compounds since they appear to be taken up from plants by the insect. A new class of interesting chemicals came into research on olfaction, the pyrrholizidin-alkaloids, which can also serve as chemical defenses against predators. Highly specific receptors for such substances have been detected in some species, and the number of insects found to be 'pharmacophageous' indicates convergent development of such receptors in many insect families. 'Secondary' physiological aspects became apparent with questions on the metabolic pathways of the biosynthesis of the pheromones from the ingested alkaloids, and a very promising and interesting problem dealt with the influence of the ingestion of such substances during larval life upon the development of the male glands and the appendages for the delivery of the pheromone.

34.4. The central olfactory connections, an optimistic start

If one watches the prompt, invariable and very sensitive reaction of some male insects to female pheromone, one is tempted to assume relatively simple if not reflex-like connections within the CNS which mediate the response. This itself might consist, at least in the beginning, of a simple antennal movement, a start of wing and leg muscle activity, and performance of simple right or left turns. Such a system would be ideal to study coding and processing of olfactory information in the CNS in order to come closer to an understanding of nervous mechanisms underlying pheromone-guided behaviour.

The response patterns of receptors perceiving pheromone components are fairly well known in a number of species. Therefore, what would be easier than to follow the central pathway for such stimuli up to 'higher' centres and 'down' to motor pools, using the response to pheromone as an Ariadne's thread in the labyrinth of the neuropil of the insect brain?

The challenge was taken up some years ago, and when this seminar was planned it was an appropriate time for a review of progress in this area (Light (Chapter 33)).

As is often the case with new branches of research, first results are greeted with enthusiasm. It can indeed be regarded as great success that recording from central olfactory neurones and their injection with dye has been accomplished and has revealed significantly new and interesting insights into the functional neuroanatomy of the antennal lobes as the first central station of the olfactory pathway, and some projections of its principal neurones within certain parts of the protocerebrum. This approach to the 'ascending olfactory pathway' by investigating single neurones at subsequent stages is exactly the route D. Hubel and T. Wiesel took for their study of the visual pathway of mammals.

In some moths, the possibility existed to also investigate the 'descending olfactory pathway' by recording from candidate descending neurones in the ventral cord. Neural responses to pheromone have been observed at a level which might represent correlates of certain motor programs, for example turning movements. In this way one possibly can investigate input–output relations for the whole brain. There are only a few data from such studies as yet, but the results are promising.

A host of fascinating, sometimes bewildering data were presented to the audience: at the level of the antennal lobes, a great number of pheromone-sensitive neurones have been investigated, some of them morphologically 'identified' (Light (Chapter 33)). Some grouping into cell types became apparent according to anatomical criterial (Hildebrand and Montague (Chapter 32)). Output neurones innervating a sexual dimorphic glomerulus in the male deutocerebrum with axons to other parts of the brain are distinguishable from axonless local interneurones innervating many glomeruli. Some of these neurones turned out to be extremely sensitive to pheromone components by highly convergent inputs from large numbers of receptor cells. Reaction groups of neurones have not been detected as yet. Many responses which have been recorded have not been reproduced and have to be classified as 'anecdotal'. A mode of coding of pheromone stimuli or clear input–output relationships did not become clear even at this very first stage of the central pathway with its well known sensory input. Very probably we are now in the same situation with the central pathway as we had been after the first recordings from receptors cells 20 years ago: strictly quantitative measurements of the qualitative and quantitative reaction ranges of the cells have to be performed before we can speculate about coding. Moreover, the reactions of neurones must be explained in terms of connectivity. This means that the synaptology has to be dismantled by studying connections between physiologically characterized and anatomically labelled individual neurones and receptor axons.

A major problem at the level of the antennal lobes, and even more, in the protocerebrum or the ventral cord, is the interpretation of the often observed responses of neurones to stimuli of more than one sensory modality. Apart from their odour sensitivity, such neurones are activated by mechanical and/or optical stimuli, and we have to consider that certain combinations of parameters could represent the adequate stimulus for such cells. Thus, we might miss an important hue in the picture if we do not determine the correct or key combination of stimuli. The entire population of neurones must be investigated to determine whether or not they respond to a combination of stimuli, as do neurones in the visual pathways of many animals. For such investigations information from behavioural experiments on the relevance of stimuli other than olfactory is indispensible.

Highly interesting studies were also reported at the seminar about the ontogeny of antennal centres in the brain of a moth, which give insight into the organization of the olfactory system and how it is constructed (Hildebrand and Montague (Chapter 32)). The primary role of the peripheral and central components is clear through the presence of connections and their role as sources of specificity in the perception of odour. Combined with molecular biology and histo-chemistry, the 'developmental approach' to the nervous system has brought about much fascinating and important information on the basic organization of nervous systems. The olfactory system of holometabolous insects seems to be especially suited for such studies because of the ease of access to the structures, the distinct periods for determination and growth, and the possibility to test functionality.

34.5. Can it be done?

There is a profound criticism against use of the single-neurone approach in the study of the central nervous system. One critical question is whether or not (olfactory) neurones in the opened brain of a restrained animal (insect) will react to external and internal stimuli in a manner equivalent to that of the intact and unrestrained animal. A similar question arises when a neurone to be studied is in close connection with other neurones which might have been damaged in preparation of the speciman or by penetration with the electrodes. Such a disturbance can result in a severe impairment of the 'normal' response characteristics.

A second and more basic question is whether or not one can ever explain the modes of function of a nervous system in terms of its cellular constituents. All attempts in this direction have failed thus far, especially with more complex nervous systems.

An answer to the first question might be found with the help of the behavioural physiologist, who could possibly tell us whether or not the motor reactions of the experimental animal resemble the movements of the animal in a behaviour test. Even as neurophysiological preparations, some moths seem to react to pheromone stimuli with flying movements. The second question will be answered when a neuronal network underlying pheromone-guided behaviour is found which is a sufficiently simple description of neuronal responses in terms of how the nervous system of the insect processes olfactory inputs.

Advances in research on the mechanisms in insect olfaction will require the interaction of neurophysiologists, behavioural physiologists, and ethologists. Together they might be able to contribute to the solutions of problems and questions raised by Dethier (Chapter 1) when he pointed out the great significance of evolutionary and comparative approaches to a deeper understanding of a phenomenon such as the perception of insect olfactory signals. Steps towards such combined efforts must include mutual interest, intense discussion and cooperation between the participating disciplines. It was the goal of the seminar and this volume to facilitate such interactions. That goal was successfully reached.

35 Some thoughts on the form of future research

M. C. Birch and L. Barton-Browne

The preceding chapters clearly demonstrate the benefits of communication between workers at the extremes, from single-cell recording to applied field research. They also reveal that while we have a growing understanding of the responses of insects to olfactory signals, some major issues remain unresolved. It is clear, therefore, that olfaction in insects will continue to be a fertile research area for a considerable period. Our brief remarks here are addressed to the question of what general form this research should take and how it might differ in its approach from that which is commonly employed to date.

There are areas of substantial disagreement between workers, particularly regarding orientation mechanisms. One limitation of much of the research on this topic is that a given species has been examined in only one type of laboratory apparatus or under one set of conditions in the field. As a result, it is not possible to assess the extent to which the different views expressed by the various workers reflect differences between the species studied, or differences between conditions under which the observations were made. Such questions could be largely resolved if a species were to be examined in a number of laboratory and field situations, and if different species were examined in the same situations.

There is clearly a continuing need for a multidisciplinary approach to studies on the olfactory responses of insects. A thorough understanding of mechanisms underlying these responses requires, at least, knowledge of (1) the structure of odour plumes emanating from sources to which the insect is responding, (2) the response characteristics of the relevant receptors, and (3) the nature of central processing of olfactory input, and its integration with input representing other modalities and any central programmes involved. The acquisition of such knowledge represents a considerable challenge, which is apparently being met only to a limited extent by the rather fragmentary multidisciplinary approach currently being employed. The challenge is such that it calls for closer collaboration than is now common between workers in different disciplines, such that there is day-to-day contact between them. This requires the creation of multidisciplinary teams dedicated to increased understanding of the mechanisms involved in the responses of insects to olfactory signals.

References

Adler, J. (1969). Chemoreceptors in bacteria. *Science* **166**, 1588–97. [1]

—— (1976). The sensing of chemicals by bacteria. *Sci. Amer.* **234**, 40–7. [1]

Ahmad, S. (ed.) (1983). *Herbivorous insects. Host-seeking behavior and mechanisms.* Academic Press, New York. [2]

Alcock, J. (1982). Natural selection and communication among bark beetles. *Fla. Entomol.* **65**, 17–32. [10]

Altman, J. (1980). The insect brain—problems and progress. *Trends in Neurosci.* **3**, VIII. [1]

Andryszak, N. A., Payne, T. L., Billings, P. M. and Benenati, J. M. (1982). Effect of flight activity on laboratory response of the southern pine beetle to an attractant. *J. Geo. Entomol. Soc.* **17**, 456–60. [10]

Arn, H., Städler, E., Rauscher, S., Buser, H. R., Mustaparta, H., Esbjerg, P., Philipson, H., Zethner, O., Struble, D. L. and Bues, R. (1980). Multicomponent sex pheromone in *Agrotis segetum*: preliminary analysis and field evaluation. *Z. Naturf.* **35c**, 986–9. [27]

——, Esberg, P., Bues, R., Toth, M., Szöcs, G., Guerin, P. and Rauscher, S. (1983). Field attraction of *Agrotis segetum* males in four European countries to mixtures containing three homologous acetates. *J. Chem. Ecol.* **9**, 267–76. [27]

Atkins, M. D. (1959). A study of the flight of the Douglas-fir beetle, *Dendroctonus pseudotsugae* Hopk. (Coleoptera: Scolytidae). I. Flight preparation and response. *Can. Entomol.* **91**, 283–91. [10]

—— (1966a). Behavioural variation among scolytids in relation to their habitat. *Can. Entomol.* **98**, 285–8. [10]

—— (1966b). Laboratory studies on the behaviour of the Douglas-fir beetle, *Dendroctonus pseudotsugae* Hopkins. *Can. Entomol.* **98**, 953–91. [10]

—— (1969). Lipid loss with flight in the Douglas-fir beetle. *Can. Entomol.* **101**, 164–5. [10]

—— (1980). *Introduction to insect behavior.* p. 33. MacMillan, New York. [16]

Aylor, D. E., Parlange, J.-Y. and Grannett, J. (1976). Turbulent dispersion of disparlure in the forest and male gypsy moth response. *Environ. Entomol.* **5**, 1026–32. [3,20]

Bacon, J. and Mohl, B. (1983). The tritocerebral commissure giant (TCG) wind-sensitive interneurone in the locust. I. Its activity in straight flight. *J. Comp. Physiol.* **150**, 439–52. [32]

Bailey, W. J. and Thomson, P. (1977). Acoustic orientation in the cricket *Teleogryllus oceanicus* (Le Guillou). *J. Exp. Biol.* **67**, 61–75. [2]

Baker, R., Bradshaw, J. W. S. and Speed, W. (1982). Methoxymercuration–demercuration and mass spectrometry in the identification of the sex pheromones of *Panolis flammea*, the pine beauty moth. *Experientia* **38**, 233–4. [17,26]

Baker, R. R. (1982). *Migration paths through time and space.* Hodder and Stoughton, London. [2]

Baker, T. C. (1985). Chemical control of behavior. In *Comprehensive insect physiology, biochemistry and pharmocology* (eds. G. A. Kerkut and L. I. Gilbert) Vol. 9, pp. 621–72. Pergamon Press, Oxford. [4]

—— and Cardé, R. T. (1979a). Courtship behavior of the oriental fruit moth (*Grapholitha molesta*). *Ann. Entomol. Soc. Amer.* **72**, 173–88. [21]

—— and Cardé, R. T. (1979b). Analysis of pheromone-mediated behaviors in male *Grapholitha molesta,* the oriental fruit moth (Lepidoptera: Tortricidae). *Environ. Entomol.* **8**, 956–8. [4]

—— and Cardé, R. T. (1979c). Endogenous and exogenous factors affecting periodicities of female calling and male sex pheromone response in *Grapholitha molesta* (Busck). *J. Insect Physiol.* **25**, 943–50. [24]

—— and Kuenen, L. P. S. (1982). Pheromone source location by flying moths: a supplementary non-anemotactic mechanism. *Science* **216**, 424–7. [4,5]

—— and Roelofs W. L. (1976). Electroantennogram responses of the male moth, *Argyrotaenia velutinana,* to mixtures of sex pheromone components of the female. *J. Insect Physiol.* **22**, 1357–64. [4]

—— and Roelofs, W. L. (1981). Initiation and termination of oriental fruit moth male response to pheromone concentrations in the field. *Environ. Entomol.* **10**, 211–18. [4,9]

——, Cardé, R. T. and Roelofs, W. L. (1976). Behavioral responses of male *Argyrotaenia velutinana* (Lepidoptera: Tortricidae) to components of its sex pheromone. *J. Chem. Ecol.* **2**, 333–52. [4,26]

——, Meyer, W. and Roelofs, W. L. (1981). Sex pheromone dosage and blend specificity of response by oriental fruit moth males. *Entomol. Exp. Appl.* **30**, 269–79. [4, 19]

——, Willis, M. A. and Phelan, P. L. (1984). Optomotor anemotaxis polarizes self-steered zigzagging in flying moths. *Physiol. Entomol.* **9**, 365–76. [2,4,5]

Bakke, A. and Kvamme, T. (1981). Kairomone response in *Thanasimus* predators to pheromone components of *Ips typographus. J. Chem. Ecol.* **7**, 305–12. [30]

——, Frøyen, P. and Skattebol, L. (1977). Field response to a new pheromonal compound isolated from *Ips typographus. Naturwissenschaften* **64**, 98. [30]

Barbosa, P., Frongillo, E. A. and Cranshaw, W. (1978). Orientation of field populations of *Brachymeria intermedia* (Hymenoptera: Chalcididae), to host and host habitat cues. *Entomophaga* **23**, 63–7. [16]

Bartell, R. J. (1982). Mechanism of communication disruption by pheromone in the control of Lepidoptera: a review. *Physiol. Entomol.* **7**, 353–64. [12,18]

—— (1985). Pheromone mediated behaviour of male lightbrown apple moth, *Epiphyas postvittana* (Lepidoptera), and correlations with adaptation of pheromone receptors. *Physiol. Entomol.* **10**, 121–6. [19]

—— and Lawrence, L. A. (1973). Reduction in responsiveness of males of *Epiphyas postvittana* (Lepidoptera) to sex pheromone following previous brief pheromonal exposure. *J. Insect Physiol.* **19**, 845–55. [19]

—— and Lawrence, L. A. (1976). Reduction in sexual responsiveness of male light-brown apple moth following previous brief pheromonal exposure is concentration dependent. *J. Aust. Entomol. Soc.* **15**, 236. [19]

—— and Lawrence, L. A. (1977a). Reduction in responsiveness of male apple moths, *Epiphyas postvittana*, to sex pheromone following pulsed pheromonal exposure. *Physiol. Entomol.* **2**, 1–6. [6,22]

—— and Lawrence, L. A. (1977b). Reduction in responsiveness of male light-brown apple moths, *Epiphyas postvittana*, to sex pheromone following pulsed pre-exposure to pheromone components. *Physiol. Entomol.* **2**, 89–95. [19]

—— and Shorey, H. H. (1969). A quantitative bioassay for the sex pheromone of *Epiphyas postvittana* (Lepidoptera) and factors limiting male responsiveness. *J. Insect Physiol.* **15**, 33–40. [19]

Barth, R. (1960). Orgaos odoriferos dos Lepidópteros. Rio de Janeiro: *Parque Nacional do Itatiaia Bol.* **7**, 1–59. [21]

Barynin, J. A. M. and Wilson, M. J. G. (1972). Outdoor experiments on smell. *Atmos. Environ.* **6**, 197–207. [3]

Beck, S. D. (1965). Resistance of plants to insects. *Ann. Rev. Entomol.* **10**, 207–32. [2]

Bedard, W. D., Wood, D. L., Tilden, P. E., Lindhal, K. Q. Jun., Silverstein, R. M. and Rodin, J. O. (1980). Field responses of the western pine beetle and one of its predators to host- and beetle-produced compounds. *J. Chem. Ecol.* **6**, 625–41. [30]

Behrend, K. (1971). Riechen in Wasser und in der Luft bei *Dytiscus marginalis* L., *J. Comp. Physiol.* **75**, 108–22. [21]

Bell, W. J. (1984). Chemo-orientation in walking insects. In *Chemical ecology of insects* (eds. W. J. Bell and R. T. Cardé) pp. 93–109. Chapman and Hall, London and New York. [8]

—— and Cardé, R. T. (1984). *Chemical ecology of insects.* Chapman and Hall, London and New York. [2]

—— and Tobin, T. R. (1982). Chemo-orientation. *Biol. Rev.* **57**, 219–60. [2,4,5,15,20]

——, Tobin, T. R., Vogel, G. and Surber, J. L. (1983). Visual course control of escape orientation in the cockroach *Blaberus craniifer*: role of internal and external information. *Physiol. Entomol.* **8**, 121–32. [15]

Bellas, T. E. and Bartell, R. J. (1983). Dose–response relationship for two components of the sex pheromone of light-brown apple moth, *Epiphyas postvittana* (Lepidoptera: Tortricidae). *J. Chem. Ecol.* **9**, 715–25. [18,19]

——, Bartell, R. J. and Hill, A. (1983). Identification of two components of the sex pheromone of the moth, *Epiphyas postvittana* (Lepidoptera, Tortricidae). *J. Chem. Ecol.* **4**, 503–12. [19]

Benn, M. H., Galbreath, R. A., Holt, V. A., Young, H., Downs, G. and Priesner, E. (1982). The sex pheromone of the silver Y moth, *Chrysodeixis eriosoma* (Doubleday) in New Zealand. *Z. Naturforsch.* **37**, 1130–5. [25,26]

Bennett, R. B. and Borden, J. H. (1971). Flight arrestment of tethered *Dendroctonus pseudotsugae* and *Trypodendron lineatum* (Coleoptera: Scolytidae) in response to olfactory stimuli. *Ann. Entomol. Soc. Amer.* **64**, 1273–86 [10]

Berg, H. C. and Purcell, E. M. (1977). Physics of chemoreception. *Biophys. J.* **20**, 193–219. [23]

Bernays, E. A. (1980). The post-prandial rest in *Locusta migratoria* nymphs and its hormonal regulation. *J. Insect Physiol.* **26**, 119–23. [28]

—— and Chapman, R. F. (1972). The control of changes in peripheral sensilla associated with feeding in *Locusta migratoria. J. Exp. Biol.* **57**, 755–63. [28]

Berryman, A. A. and Ashraf, M. (1970). Effects of *Abies grandis* resin on the attack behavior and brood survival of *Scolytus ventralis* (Coleoptera: Scolytidae). *Can. Entomol.* **102**, 1229–36. [11]

Bierl, B. A., Beroza, M. and Collier, C. W. (1970). Potent sex attractant of the gypsy moth: its isolation, identification, and synthesis. *Science* **170**, 87–9. [7]

Billings, R. F., Gara, R. I. and Hrutfiord, B. F. (1976). Influence of ponderosa pine resin volatiles on the response of *Dendroctonus ponderosae* to synthetic *trans*-verbenol. *Environ. Entomol.* **5**, 171–9. [10]

Birch, M. C. (1970). Pre-courtship use of abdominal brushes by males of the nocturnal moth, *Phlogophora meticulosa* (Lepidoptera: Noctuidae). *Anim. Behav.* **18**, 310–16. [21]

—— (1974a). *Pheromones*. North Holland, Amsterdam. [21]

—— (1974b). Seasonal variation in pheromone-associated behavior and physiology of *Ips pini*. *Ann. Entomol. Soc. Amer.* **67**, 58–60. [10]

—— (1984). Aggregation in bark beetles. In *Chemical ecology of insects* (ed. W. J. Bell and R. T. Cardé) pp. 331–53. Chapman and Hall, London and New York. [2,10,13,30]

—— and Wood, D. L. (1975). Mutual inhibition of the attractant pheromone response by two species of *Ips* (Coleoptera: Scolytidae). *J. Chem. Ecol.* **1**, 101–13. [30]

——, Light, D. L. and Mori, K. (1977). Selective inhibition of response of *Ips pini* to its pheromone by the (*S*)-(−)-enantiomer of ipsenol. *Nature* **270**, 738–9 [30]

——, Svihra, P., Paine, T. D. and Miller, J. C. (1980a). Influence of chemically mediated behavior on host tree colonization by four cohabiting species of bark beetles. *J. Chem. Ecol.* **6**, 395–414. [11]

——, Light, D. M., Wood, D. L., Browne, L. E., Silverstein, R. M., Bergot, B. J., Ohloff, G., West, J. R. and Young, J. C. (1980b). Pheromonal attraction and allomonal interruption of *Ips pini* in California by the two enantiomers of ipsdienol. *J. Chem. Ecol.* **6**, 703–17. [10,30]

Bjostad, L. B. and Roelofs, W. L. (1983). Sex pheromone biosynthesis in *Trichoplusia ni* reveals delta-11 desaturation and chain shortening are keysteps in many Lepidoptera. *Science* **220**, 1387–9. [24,25]

——, Gaston, L. K., Noble, L. L., Moyer, J. H. and Shorey, H. H. (1980). Dodecyl acetate, a second pheromone component of the cabbage looper moth, *Trichoplusia ni*. *J. Chem. Ecol.* **6**, 727–34. [25]

Blanche, C. A., Hodges, J. D., Nebeker, T. E. and Moehring, D. M. (1983). Southern pine beetle: the host dimension. *Miss. Agric. and For. Exp. Stn. Bull.* **917**, 29 pp. [11]

Blondeau, J. (1981). Aerodynamic capabilities of flies, as revealed by a new technique. *J. Exp. Biol.* **92**, 155–63. [7]

Blum, M. S. (1974). Deciphering the communicative rosetta stone. *Bull. Entomol. Soc. Amer.* **20**, 30–5. [30]

Boeckh, J. (1962). Electrophysiologische Untersuchungen an einzelnen Geruchsrezeptoren auf den Antennen des Totengrabers (*Necrophorus,* Coleoptera). *Z. Verg. Physiol.* **46**, 212–48. [21,22]

—— (1967a). Inhibition and excitation of single insect olfactory receptors, and their role as a primary sensory code. In *Olfaction and taste II* (ed. T. Hayashi) pp. 721–35. Pergamon Press, Oxford and New York. [21]

—— (1967b). Reaktionsschwelle, Arbeitsbereich und Spezifität eines Geruchsrezeptors auf der Heuschreckenantenne. *Z. Verg. Physiol.* **55**, 378–406. [21]

—— (1974). Die Reaktionen olfaktorischer Neurone im Deutocerebrum von Insekten im Vergleich zu Antwortmustern der Geruchssinneszelle. *J. Comp. Physiol.* **90**, 183–205. [21]

—— and Boeckh, V. (1979). Threshold and odor specificity of pheromone-sensitive neurons in the deutocerebrum of *Antheraea pernyi* and *A. polyphemus* (Saturniidae). *J. Comp. Physiol.* **132**, 235–42. [21,32,33]

—— and Ernst, K.-D. (1983). Olfactory food and mate recognition. In *Neuroethology and behavioral physiology, roots and growing points* (eds. H. Huber and H. Markl) pp. 78–94. Springer-Verlag, Berlin. [4,21,33]

——, Ernst, K.-E. and Polz, H. (1983). Convergence of inputs from different sense organs upon neurones of the central antennal pathway in insects. In *Multimodal convergences in sensory systems* (ed. Horn) pp. 143–8. Fischer, Stuttgart. [33]

——, Ernst, K. D., Sass, H. and Waldow, U. (1984). Anatomical and physiological characteristics of individual neurones in the central antennal pathway of insects. *J. Insect Physiol.* **30**, 15–26. [4,21,32,33]

Boldt, P. E., Marston, M. and Dickerson, W. A. (1973). Differential parasitism of several species of lepidopteran eggs by two species of *Trichogramma. Environ. Entomol.* **2**, 1121–2. [16]

Boppré, M. (1984). Chemically mediated interactions between butterflies. *Symp. R. Soc. Lond.* **11**, 259–75. [21]

Borden, J. H. (1967). Factors influencing the responses of *Ips confusus* (Coleoptera: Scolytidae) to male attractant. *Can. Entomol.* **99**, 1164–93. [10]

—— (1974). Aggregation pheromones in the Scolytidae. In *Pheromones* (ed. M. C. Birch) pp. 135–60. North Holland, Amsterdam. [10,11,30]

—— (1982a). Aggregation pheromones. In *Bark beetles in North American conifiers* (eds. J. B. Mitton and K. B. Sturgeon) pp. 74–139. Univ. Texas Press, Austin. [10,29]

—— (1982b). Secondary attraction in the Scolytidae: an annotated bibliography. *Simon Fraser Univ. Pest Mgt. Pap. No. 26.* [10]

—— (1984). Semiochemical-mediated aggregation and dispersion in the Coleoptera. In *Insect communication* (ed. T. Lewis). pp. 123–49. Academic Press, New York. [10]

——, Chong, L., McLean, J. A., Slessor, K. N. and Mori, K. (1976). *Gnathotrichus sulcatus*: synergistic response to enantiomers of the aggregation pheromone sulcatol. *Science* **192**, 894–6. [29]

——, Conn, J. E., Friskie, L. M., Scott, B. E., Chong, L. J., Pierce, H. D. Jun. and Oehlschlager, A. C. (1983a). Semio-chemicals for the mountain pine beetle, *Dendroctonus ponderosae* (Coleoptera: Scolytidae), in British Columbia: baited tree studies. *Can. J. For. Res.* **13**, 325–33. [10]

——, Chong, L. J., Pratt, K. E. G. and Gray, D. R. (1983b). The application of behaviour-modifying chemicals to contain infestations of the mountain pine beetle, *Dendroctonus ponderosae. For. Chron.* **59**, 235–9. [10]

Boreham, B. W. and Harvey, J. K. (1984). Dipolar charged particles as markers for dispersion simulation in wind tunnels. *J. Phys. E.* **17**, 994–8 [3]

Bossert, W. H. and Wilson, E. O. (1963). The analysis of olfactory communication among animals. *J. Theor. Biol.* **5**, 443–69. [3,4,6,9,14,20]

Bradshaw, J. W. S., Baker, R. and Howse, P. E. (1979a). Multicomponent alarm pheromones in the mandibular glands of the African weaver ant, *Oecophylla longinoda. Physiol. Entomol.* **4**, 15–25. [17]

——, Baker, R. and Howse, P. E. (1979b). Chemical composition of the poison apparatus secretions of the African weaver ant, *Oecophylla longinoda,* and their role in behaviour. *Physiol. Entomol.* **4**, 39–46. [17]

——, Baker, R. and Lisk, J. C. (1983). Separate orientation and releaser components in a sex pheromone. *Nature* **304**, 265–7. [4,17,25,26]

Brower, L. P., Brower, J. V. Z. and Cranston, F. P. (1965). Courtship behavior of the queen butterfly, *Danaus gilippus berenice* (Cramer). *Zoologica* (New York) **50**, 1–39. [21]

Brown, W. L. Jun., Eisner, T. and Wittaker, R. H. (1970). Allomones and kairomones: trans-specific chemical messengers. *Biosc*. **20**, 21–2. [16,30]

Buda, V. (1980). Spatial structure of multicompound pheromone signal. *Proc. XVI Int. Congr. Entomol*., Kyoto, Japan. [4]

Buddenbrock, W. von (1952). *Vergleichende physiologie*. I. Sinnesphysiologie. Birkhäuser, Basel. [21]

Burrows, M., Boeckh, J. and Esslen, J. (1982). Physiological and morphological properties of interneurones in the deutocerebrum of male cockroaches which respond to female pheromone. *J. Comp. Physiol.* **145**, 447–57. [32,33]

Bursell, E. (1984). Observations on the orientation of tsetse flies (*Glossinia pallidipes*) to wind-borne odours. *Physiol. Entomol.* **9**, 133–7. [2,5]

Butenandt, A., Beckmann, R., Stamm, D. and Hecker, E. (1959). Über den Sexuallockstoff des Seidenspinners *Bombyx mori*. Reindarstellung und Konstitution. *Z. Naturforsch.* **14b**, 283–4. [21]

Byers, J. A. (1982). Male-specific conversion of the host plant compound, myrcene, to the pheromone, (+)-ipsdienol, in the bark beetle, *Dendroctonus brevicomis*. *J. Chem. Ecol.* **8**, 363–71. [10]

——, Lanne, B. S., Lösqvist, J., Schlyter, F. and Bergström, G. (1986). Olfactory recognition of host-tree susceptibility by pine shoot beetles. *Naturwissenschaften*, in press. [11]

Byrne, K. J., Swigar, A. A., Silverstein, R. M., Borden, J. H. and Stokkink, E. (1974). Sulcatol: population aggregation pheromone in the scolytid beetle, *Gnathotrichus sulcatus*. *J. Insect Physiol.* **20**, 1895–900. [10]

Byström, S., Högberg, H.-E. and Norin, T. (1981). Chiral synthesis of (2*S*,3*S*,7*S*)-3,7-dimethylpentadecan-2-yl acetate and propionate, potential sex pheromone components of the pine saw-fly *Neodiprion sertifer* (Geoff.). *Tetrahedron* **37**, 2249–54. [13]

Camazine, S. M. and Hildebrand, J. G. (1979). Central projections of antennal sensory neurones in mature and developing *Manduca sexta*. *Soc. Neurosci. Abstr.* **5**, 155. [32,33]

Cardé, R. T. (1979). Behavioral responses of moths to female produced pheromones and the utilization of attractant-baited traps for population monitoring. In *Movement of highly mobile insects: concepts and methodology in research* (eds. R. L. Rabb and G. G. Kennedy) pp. 286–315. North Carolina State Univ. Press, Raleigh. [18]

—— (1981). Precopulatory sexual behavior of the adult gypsy moth. In *The gypsy moth: research toward integrated pest management* (eds. C. C. Doane and M. L. McManus) pp. 572–87. *US Dept. Agric. Tech. Bull.* **1584**, Washington, DC [8,20]

—— (1984). Chemo-orientation in flying insects. In *Chemical ecology of insects* (eds. W. J. Bell and R. T. Cardé) pp. 111–24. Chapman and Hall, London and New York. [4,8]

—— and Baker, T. C. (1984). Sexual communication with pheromones. In *Chemical ecology of insects* (eds. W. J. Bell and R. T. Cardé) pp. 355–83. Chapman and Hall, London and New York. [20]

—— and Charlton, R. E. (1985). Olfactory sexual communication in Lepidoptera: strategy, sensitivity, and selectivity. *Symp. R. Entomol. Soc. Lond.* **12**, 241–64. [2,20]

—— and Hagaman, T. E. (1979). Behavioral responses of the gypsy moth in a wind-tunnel to air-borne enantiomers of disparlure. *Environ. Entomol.* **8**, 475–84. [4,7,12,20,21]

—— and Hagaman, T. E. (1984). Mate location strategies of gypsy moths in dense populations. *J. Chem. Ecol.* **10**, 25–31. [8,20]

—— and Roelofs, W. L. (1973). Temperature modification of male sex pheromone response and factors affecting female calling in *Holomelina immaculata* (Lepidoptera: Arctiidae). *Can. Entomol.* **105**, 1505–12. [20]

——, Comeau, A., Baker, T. C. and Roelofs, W. L. (1975). Moth mating periodicity: temperature regulates the circadian gate. *Experientia* **31**, 46–8. [24]

——, Dindonis, L. L., Agar, B. and Foss, J. (1984). Apparency of pulsed and continuous pheromone to male gypsy moths. *J. Chem. Ecol.* **10**, 335–48. [6,22,20]

——, Doane, C. C., Balzer, T. C., Iwaki, S. and Marumo, S. (1977). Attractancy of optically active pheromone for male gypsy moths. *Environ. Entomol.* **6**, 768–72. [7]

Case, T. J., Washino, R. K. and Dunn, R. L. (1977). Diapause termination in *Anopheles freeborni* with juvenile hormone mimics. *Entomol. Exp. Appl.* **21**, 155–62. [28]

Castrovillo, P. J. and Cardé, R. T. (1979). Environmental regulation of female calling and mate pheromone response periodicities in the codling moth (*Laspeyresia pomonella*). *J. Insect Physiol.* **25**, 659–67. [24]

Chapman, R. F. (1982). Chemoreception: the significance of receptor numbers. In *Advances in insect physiology* (eds. M. J. Berridge, J. H. Treherne and V. B. Wigglesworth) vol. 16, pp. 247–356. Academic Press, London. [16]

——, Bernays, E. A. and Simpson, S. J. (1981). Attraction and repulsion of the aphid, *Cavariella aegopodii*, by plant odour. *J. Chem. Ecol.* **7**, 881–8. [2]

Charlton, R. L. and Cardé, R. T. (1985a). Walking orientation of male gypsy moths to pheromone sources. Unpublished manuscript. [20]

—— and Cardé, R. T. (1985b). Factors mediating mate recognition and copulatory response in male gypsy moths. Unpublished manuscript. [20]

Choudhury, J. H. and Kennedy, J. S. (1980). Light versus pheromone-bearing wind in the control of flight direction by bark beetles, *Scolytus multistriatus*. *Physiol. Entomol.* **5**, 207–14. [10]

Christensen, T. A. and Hildebrand, J. G. (1984). Functional anatomy and physiology of male-specific pheromone-processing interneurons in the brain of *Manduca sexta*. *Soc. Neurosci. Abstra.* **10**, 862. [32]

Cole, W. E. and Amman, G. D. (1980). Mountain pine beetle dynamics in lodgepole pine forests. Part I: course of an infestation. *USDA Forest Service Gen. Tech. Rep. INT-89.* [10]

Collett, T. S. (1980). Some operating rules for the optomotor system of a hoverfly during voluntary flight. *J. Comp. Physiol.* **138**, 271–82. [5]

—— and Blest, A. D. (1966). Binocular, directionally selective neurones, possibly involved in the optomotor response of insects. *Nature* **212**, 1330–3. [5]

Conn, J. E., Borden, J. H., Scott, B. E., Friskie, L. M., Pierce, H. D. Jun. and

Oehlschlager, A. C. (1983). Semiochemicals for the mountain pine beetle, *Dendroctonus ponderosae* (Coleoptera: Scolytidae) in British Columbia: field trapping studies. *Can. J. For. Res.* **13**, 320–4. [10]

Conner, W. E., Eisner, T., Vander Meer, R. K., Guerrero, A., Ghiringelli, D. and Meinwald, J. (1980). Sex attractant of an arctiid moth (*Utethesia ornatrix*): a pulsed chemical signal. *Behav. Ecol. Sociobiol.* **7**, 55–63. [20]

——, Eisner, T., Vander Meer, R. K., Guerrero, A. and Meiwald, J. (1981). Precopulatory sexual interaction in an arctiid moth (*Utethesia ornatrix*): role of a pheromone derived from dietary alkaloids. *Behav. Ecol. Sociobiol.* **9**, 227–35. [21]

Coppel, H. C., Casida, J. E. and Dauterman, W. C. (1960). Evidence for a potent sex attractant in the introduced pine sawfly, *Diprion similis* (Hymenoptera: Diprionidae). *Ann. Entomol. Soc. Amer.* **53**, 510–12. [13]

Coster, J. E., Payne, T. L., Edson, L. J. and Hart, E. R. (1978). Influence of weather on mass aggregation of southern pine beetles at attractive host trees. *Southwest. Entomol.* **3**, 14–20. [10]

——, Payne, T. L., Hart, E. R. and Edson, L. J. (1977). Aggregation of the southern pine beetle in response to attractive host trees. *Environ. Entomol.* **6**, 725–31. [11]

Coulson, R. N., Hennier, P. B., Flamm, R. O., Rykiel, E. J., Hu, L. and Payne, T. L. (1983). The role of lightning in the epidemiology of the southern pine beetle. *Z. Ang. Entomol.* **96**, 182–93. [11]

Craig, W. (1918). Appetites and aversions as constituents of instincts. *Biol. Bull. Woods Hole* **34**, 91–107. [20,21]

Cross, W. H. (1973). Biology, control and eradication of the boll weevil. *Ann. Rev. Entomol.* **18**, 17–46. [29]

—— (1983). Ecology of cotton insects with special reference to the boll weevil. In *Cotton insect management with special reference to the boll weevil* (eds. R. L. Ridgway, E. P. Lloyd, and W. H. Cross) pp. 53–70. *US Dept. Agric. Handbook* No. **589**. [29]

Culvenor, C. C. J. and Edgar, J. A. (1972). Dihydropyrrolizine secretions associated with coremata of *Utetheisa* moths (Family Arctiidae). *Experientia* **28**, 627–8. [21]

Darwin, C. (1868). *The variation of plants and animals under domestication*. (1st ed.) John Murray, London. [20]

David, C. T. (1978). The relationship between body angle and flight speed in free-flying *Drosophila. Physiol. Entomol.* **3**, 191–5. [5]

—— (1979a). Height control by free-flying *Drosophila. Physiol. Entomol.* **4**, 209–16. [5]

—— (1979b). Optomotor control of speed and height by free-flying *Drosophila. J. Exp. Biol.* **82**, 389–92. [5,21]

—— (1982a). Compensation for height by *Drosophila* in a new 'barber's pole' wind tunnel. *J. Comp. Physiol.* **147**, 485–93. [5]

—— (1982b). Competition between fixed and moving stripes in the control of orientation by flying *Drosophila. Physiol. Entomol.* **7**, 151–6. [5]

—— (1984). The dynamics of height stabilization in *Drosophila. Physiol. Entomol.* **9**, 377–86. [5]

—— (1985). Visual control of the partition of flight force between lift and thrust in free-flying *Drosophila. Nature* **313**, 48–50. [5]

——, Kennedy, J. S. and Ludlow, A. R. (1983). Finding of a sex pheromone source by gypsy moths released in the field. *Nature* **303**, 804–6. [2,4,5,8,9,20,23]

——, Kennedy, J. S., Ludlow, A. R., Perry, J. N. and Wall, C. (1982). A reappraisal of insect flight towards a distant point source of wind-borne odour. *J. Chem. Ecol.* **8**, 1207–15. [5,9,20]

Davis, E. E. (1976). A receptor sensitive to oviposition site attractants on the antennae of the mosquito, *Aedes aegypti. J. Insect Physiol.* **22**, 1371–6. [21]

—— (1977). Response of antennal receptors of the male *Aedes aegypti* mosquito. *J. Insect Physiol.* **23**, 613–17. [21]

—— (1984a). Development of lactic acid-receptor sensitivity and host-seeking behavior in newly emerged female *Aedes aegypti* mosquitoes. *J. Insect Physiol.* **30**, 211–15. [28]

—— (1984b). Regulation of sensitivity in the peripheral chemoreceptor system for host-seeking behavior by a haemolymph-borne factor in *Aedes aegypti. J. Insect Physiol.* **30**, 179–83. [28,29]

—— and Takahashi, F. T. (1980). Humoral alteration of chemoreceptor sensitivity in the mosquito. *Olfaction Taste Proc. Int. Symp. VII*, 138–42. [28,29]

Daykin, P. N. (1967). Orientation of *Aedes aegypti* in vertical air currents. *Can. Entomol.* **99**, 303–8. [5]

de Kramer, J. J. (1983). Transduction in insect olfaction. *Trends In Neurosci.* **6**, 455–8. [1]

Den Otter, C. J. (1974). Electrophysiology of sex pheromone sensitive cells on antennae of the summerfruit tortrix moth, *Adoxophyes orana*. Abstract, *1st ECRO Congr.*, Paris, p. 45. [26]

—— (1977). Single sensillum responses in the male moth, *Adoxophyes orana* (F.v.R) to female sex pheromone components and their geometrical isomers. *J. Comp. Physiol.* **121**, 205–22. [19,26]

Deneubourg, J. L., Pasteels, J. M. and Verhaeghe, J. C. (1983). Probabilistic behaviour in ants: a strategy of errors. *J. Theor. Biol.* **105**, 259–71. [14]

——, Paro, M., Pasteels, J. M., Verhaeghe, J. C. and Champagne, P., (1982). L'exploitation des ressources chez les fourmis: un jeu d'erreurs et d'amplification. *Compte-rendus du Colloque International sur les Insectes Sociaux*, 97–106, Barcelone. [14]

Dethier, V. G. (1982). Mechanisms of host-plant recognition. *Ent. Exp. Appl.* **31**, 49–56. [8]

—— (1984). Analyzing proximate causes of behavior. In *Evolutionary genetics of invertebrate behavior*. Plenum Press, New York, in press. [1]

—— and Crnjar, R. M. (1982). Candidate codes in the gustatory system of caterpillars. *J. Gen. Physiol.* **79**, 549–69. [31]

——, Brown, L. B. and Smith, C. N. (1960). The designation of chemicals in terms of the responses they elicit from insects. *J. Econ. Entomol.* **53**, 134–6. [16]

Dickens, J. C. (1978). Olfactory perception of pheromone and host-odour enantiomers by *Ips typographus* L. (Col.: Scolytidae). 4th Internat. Insect–Host Plant Symposium. *Entomol. Exp. Appl.* **24**, 136–42. [29]

—— (1981). Behavioral and electrophysiological responses of the bark beetle, *Ips typographus*, to potential pheromone components. *Physiol. Entomol.* **6**, 251–61. [29]

—— (1984). Olfaction in the boll weevil, *Anthonomus grandis* Boh. (Coleoptera: Curculionidae): electroantennogram studies. *J. Chem. Ecol.* **10**, 1759–86. [29]

—— (1985). Orientation of the boll weevil, *Anthonomus grandis* Boh. (Coleoptera:Curculionidae), to its pheromone and volatile host compound in the laboratory. *J. Chem. Ecol.* in press. [29]

—— and Payne, T. L. (1977). Bark beetle olfaction: pheromone receptor system in *Dendroctonus frontalis. J. Insect Physiol.* **23**, 481–9. [11,21,29,30]

—— and Payne, T. L. (1978). Olfactory-induced muscle potentials in *Dendroctonus frontalis*: effects of *trans*-verbenol and verbenone. *Experientia* **34**, 463–4. [11]

——, Payne, T. L., Ryker, L. C. and Rudinsky, J. A. (1984). Single cell responses of the Douglas-fir beetle, *Dendroctonus pseudotsugae* Hopkins (Coleoptera: Scolytidae) to pheromones and host odors. *J. Chem. Ecol.* **10**, 583–600. [11,29,30]

——, Payne, T. L., Ryker, L. C. and Rudinsky, J. A. (1985). Multiple acceptors for pheromonal enantiomers on single olfactory cells in the Douglas-fir beetle, *Dendroctonus pseudotsugae* Hopk. (Coleoptera: Scolytidae). *J. Chem. Ecol.* **11**, 1359–70. [29]

——, Gutmann, A., Payne, T. L., Ryker, L. C. and Rudinsky, J. A. (1983). Antennal olfactory responsiveness of the Douglas-fir beetle, *Dendroctonus pseudotsugae* Hopkins (Coleoptera: Scolytidae), to pheromones and host odors. *J. Chem. Ecol.* **9**, 1383–95. [29]

Dindonis, L. L. and Miller, J. R. (1980). Host-finding behavior of onion flies, *Hylemia antiqua. Environ. Entomol.* **9**, 769–72. [8]

Dingle, H. and Cardwell, R. L. (1967). Multimodal interneurones in cockroach cerebrum. *Nature* **215**, 63–4. [33]

Dixon, W. N. and Payne, T. L. (1980). Attraction of entomophagous and associate insects of the southern pine beetle to beetle- and host tree–produced volatiles. *J. Ga. Entomol. Soc.* **15**, 378–9. [11]

Djerassi, C., Shih-Coleman, C. and Dickman, J. (1974). Insect control of the future: operational and policy aspects. *Science* **186**, 596–607. [18]

Doetsch, R. N. (1972). A unified theory of bacterial motile behavior. *J. Theor. Biol.* **35**, 55–66. [1]

Doutt, R. L. (1964). Biological characteristics of entomophagous adults. In *Biological control of insect pests and weeds*. (ed. P. DeBach) pp. 145–67. Reinhold, New York. [16]

Dumpert, K. (1972). Alarmstoffrezeptoren auf der Antenne von *Lasius fuliginosus* (Latr.) (Hymenoptera, Formicidae). *Z. Verg. Physiol.* **76**, 403–25. [21]

Dyar, F. C. and Gould, J. L. (1983). Honey bee navigation. *Amer. Sci.* **71**, 587–97. [8]

Dyar, L. J. and Seabrook, W. D. (1978). Evidence for the presence of acceptor sites for different terpenes on one receptor cell in male *Monochamus notatus* (Drury) (Coleoptera: Cerambycidae). *J. Chem. Ecol.* **4**, 523–9. [29]

Eckert, H. (1982). The vertical–horizontal neurone (VH) in the lobula plate of the blowfly, *Phaenicia. J. Comp. Physiol.* **149**, 195–205. [5]

Eckert, R., Naitoh, Y. and Machemer, H. (1976). Calcium in the bioelectric and motor functions of *Paramecium. Symp. Soc. Exp. Biol.* **30**, 233–55. [1]

Edwards, R. L. (1954). The host finding and oviposition behavior of *Mamoniella vitripennis* (Walker), a parasite of muscoid flies. *Behavior*. **7**, 88–112. [16]

Elkinton, J. S. and Cardé, R. T. (1983). Appetitive flight behavior of male gypsy moths (Lepidoptera: Lymantriidae). *Environ. Entomol.* **12**, 1702–7. [20]

—— and Cardé, R. T. (1984). Odor dispersion. In *Chemical ecology of insects* (eds. W. J. Bell and R. T. Cardé) pp. 73–91. Chapman and Hall, London and New York. [3,4,8,20]

——, Cardé, R. T. and Mason, C. J. (1984). Evaluation of time-average dispersion models for estimating pheromone concentration in a deciduous forest. *J. Chem. Ecol.* **10**, 1081–107. [3,8,20]

——, Schal, C. and Cardé, R. T. (1985). Pheromone puff trajectory and upwind flight of the male gypsy moth. Unpublished manuscript. [2,20]

Erber, J. (1978). Response characteristics and after-effects of multimodal neurones in the mushroom body area of the honeybee. *Physiol. Entomol.* **3**, 77–89. [33]

Ernst, K. D. and Boeckh, J. (1983). A neuroanatomical study of the organization of the central antennal pathways in insects. III. Neuroanatomical characterization of physiologically defined response types of deutocerebral neurones in *Periplaneta americana. Cell Tissue Res.* **229**, 1–22. [32,33]

——, Boeckh, J. and Boeckh, V. (1977). A neuroanatomical study on the organization of the central antennal pathway. II. Deutocerebral connections in *Locusta migratoria* and *Periplaneta americana. Cell Tissue Res.* **176**, 285–308. [33]

Evans, P. D. (1980). Biogenic amines in the insect nervous system. In *Advances in insect physiology* (eds, M. J. Berridge, J. E. Treherne and V. B. Wigglesworth) vol. 15, pp. 317–473. Academic Press, London. [24]

—— (1982). Properties of modulatory octopamine receptors in the locust. In *Neuropharmacology of insects* (eds. D. Everd, M. O'Connor and J. Whelan) pp. 48–69. CIBA Foundation Symposium 88, London. [24]

Evershed, R. P., Morgan, E. D. and Gammaerts, M. C. (1981). 3-ethyl-2,5-dimethylpyrazine, the trail pheromone from the venom gland of eight species of *Myrmica* ants. *Insect Biochem.* **12**, 383–91. [14]

Ewert, J.-P. (1980). *Neuroethology, an introduction to the neurophysiological fundamentals of behaviour*. Springer-Verlag, Berlin. [33]

Fabre, J. H. (1879). *Souvenirs Entomologiques*. I.-X Sér. Ch. Delagrave, Paris. [21]

—— (1913). *Social life in the insect world*. The Century Co., New York. [25]

Fackrell, J. E. (1978). A system for turbulent concentration measurement. *J. Phys. E.* **11**, 1015–22. [3]

Falk, R. and Atidia, J. (1975). Mutation affecting taste perception. *Nature* **254**, 325–16. [1]

Fares, Y., Sharpe, P. J. H. and Magnuson, C. E. (1980). Pheromone dispersion in forests. *J. Theor. Biol.* **84**, 335–59. [3]

Fargo, W. S., Coulson, R. N., Pulley, P. E., Pope, D. N. Kelley, C. L. and Foltz, J. L. (1978). Spatial and temporal patterns of within-tree colonization by *Dendroctonus frontalis* (Coleoptera: Scolytidae). *Can. Entomol.* **110**, 1213–32. [11]

Farkas, S. R. and Shorey, H. H. (1972). Chemical trail-following by flying insects: a mechanism for orientation to a distant odor source. *Science* **178**, 67–8. [4]

——, Shorey, H. H. and Gaston, L. K. (1974). Sex pheromones of Lepidoptera. Influence of pheromone concentration and visual cues on aerial odor-trail following by males of *Pectinophora gossypiella. Ann. Entomol. Soc. Amer.* **67**, 633–8. [4,12]

Fein, B. L., Reissig, W. H. and Roelofs, W. L. (1982). Identification of apple volatiles attractive to the apple maggot, *Rhagoletis pomonella. J. Chem. Ecol.* **8**, 1473–87. [8]

Feinsod, F. M. and Spielman, A. (1980). Independently regulated juvenile hormone activity and vitellogenesis in mosquitoes. *J. Insect Physiol.* **26**, 829–32. [28]

Feldman, R. S. and Quenzer, L. F. (1984). *Fundamentals of neuropsychopharmacology*. Sinauer Ass. Inc., MA. [24]

Feng, A. S., Gerhardt, H. C. and Capranica, R. R. (1976). Sound localization behavior of the green treefrog (*Hyla cinerea*) and the barking treefrog (*H. gratiosa*). *J. Comp. Physiol.* **107**, 241–52. [2]

Finch, S. (1980). Chemical attraction of plant-feeding insects to plants. In *Applied Biology* (ed. T. H. Coaker) pp. 67–143. Academic Press, London. [8]

Fletcher, W. A., Goodman, L. J., Guv, R. G. and Mobbs, P. G. (1984). Horizontal and vertical motion detectors in the ventral nerve cord of the honeybee, *Apis mellifera*. *J. Physiol.* **351**, 158. [5]

Fox, S. W. and Nakashima, T. (1980). The assembly and properties of protobiological structures: the beginning of cellular peptide synthesis. *Biosystems* **12**, 155–66. [1]

Fraenkel, G. S. (1931). Die Mechanik der Orientierung der Tiere im Raum. *Biol. Rev.* **6**, 36–87. [5]

—— and Gunn, D. L. (1940). *The orientation of animals. Kineses, taxes and compass reactions*. Clarendon Press, Oxford. [2,20]

—— and Gunn, D. L. (1961). *The orientation of animals*. Dover Publications, New York. [15]

Frisch, K. von (1921). Über den Sitz des Geruchssinnes bei Insekten. *Zoo. Jhb.* (allg. Zool. und Physiol.) **38**, 449–516. [21]

—— and Lindauer, M. (1955). Über die Fluggeschwindigkeit der Bienen und über ihre Richtungsweisung bei Seitenwind. *Naturwissenschaften* **42**, 377–85. [7]

Furniss, M. M. and Furniss, R. L. (1972). Scolytids (Coleoptera) on snowfields above timberline in Oregon and Washington. *Can. Entomol.* **104**, 1471–8. [10]

Futrelle, R. P. (1984). How molecules get to their detectors; the physics of diffusion of insect pheromones. *Trends in Neurosci.* **7**, 116–20. [23]

Gara, R. I., Vité, J. P. and Cramer, H. H. (1965). Manipulation of *Dendroctonus frontalis* by use of a population aggregating pheromone. *Contrib. Boyce Thompson Inst.* **23**, 55–66. [11]

Gerstein, G. and Clark, A. L. (1964). Stimultaneous studies of firing patterns in several neurons. *Science* **143**, 1325–7. [31]

Gesteland, R. C. (1976). Physiology of olfactory reception. In *Frog neurobiology* (eds. R. Llinas and W. Precht) pp. 234–50. Springer-Verlag, Heidelberg. [22]

Getchell, T. V., Heck, G. L., DeSimone, J. A. and Price, S. (1980). The location of olfactory receptor sites. Inferences from latency measurements. *Biophys. J.* **29**, 397–412. [22]

Gillett, J. D. (1979). Out for blood: flight orientation up-wind in the absence of visual cues. *Mosquito News* **39**, 221–9. [2,5]

Götz, K. G. (1972). Principles of optomotor reactions in insects. *Biblthca. Ophthalmol.* **82**, 252–9. [5]

—— and Buchner, E. (1978). Evidence for one-way movement detection in the visual system of *Drosophila*. *Biol. Cybern.* **31**, 243–8. [5]

Graham, K. (1959). Release by flight exercise of a chemotrophic response from photoresponsive domination in a scolytid beetle. *Nature* **184**, 283–4. [10]

—— (1962). Photic behavior in the ecology of the ambrosia beetle *Trypodendron lineatum*. *Proc. XI Int. Congr. Entomol.* **2**, 226. [10]

Greany, P. D., Tumlinson, J. H., Chambers, D. L. and Boush, G. M. (1977). Chemically mediated host finding by *Biosteres (Opios) longicaudatus*, a parasitoid of tephritid fruitfly larvae. *J. Chem. Ecol.* **3**, 189–95. [16]

Greenway, A. R. (1984). Sex pheromone of the pea moth, *Cydia nigricana* (F.), (Lepidoptera: Olethreutidae). *J. Chem. Ecol.* **10**, 973–82. [9]

Gregoire, J.-Cl. (1983). Host colonization strategies in *Dendroctonus*: larval gregariousness vs. mass attack by adults? In *Population dynamics of forest insects and integrated control of bark beetles* (eds. L. Safranyik and A. A. Berryman) Proc. IUFRO Conf., Banff, Canada. [30]

Gries, G. (1984). Zur Frage der Dispersion des Buchdruckers (*Ips typographus* L.). *Z. Ang. Entomol.* **99**, 12–20. [10]

Griffin, D. R. (1958). *Listening in the dark*. Yale University Press, New Haven, CT. [29]

Gross, H. R. (1981). Employment of kairomones in the management of parasitoids. In *Semiochemicals: their role in pest control.* (eds. D. A. Nordlund, R. L. Jones and W. J. Lewis) pp. 137–50. John Wiley, New York. [16]

Gueldner, R. C., Nordlund, D. A., Lewis, W. J., Thean, J. E. and Wilson, D. M. (1984). Kairomones and their use for management of entomophagous insects. XV. Identification of several acids in scale of *Heliothis zea* moths and comments on their possible role as kairomones for *Trichogramma pretiosum*. *J. Chem. Ecol.* **10**, 245–51. [16]

Hadjitoffi, A. and Wilson, M. J. G. (1979). Fast response measurements of air pollution. *Atmos. Environ.* **13**, 755–60. [3]

Hagen, B. W. and Atkins, M. D. (1975). Between generation variability in fat content and behavior of *Ips paraconfusus* Lanier. *Z. Ang. Entomol.* **79**, 169–72. [10]

Hagstrum, D. W. and Davis, L. R. Jun. (1982). Mate-seeking behavior and reduced mating by *Ephestia cautella* (Walker) in a sex pheromone-permeated atmosphere. *J. Chem. Ecol.* **8**, 507–15. [18]

Hall, C. D. (1975). The simulation of particule motion in the atmosphere by a numerical random walk model. *Quart. J. Royal Meteorol. Soc.* **101**, 235–44. [3]

Hallberg, E. (1979). The fine structure of the antennal sensilla of the pine saw-fly *Neodiprion sertifer* (Insecta: Hymenoptera). *Protoplasma* **101**, 111–26. [13]

—— (1981). Fine-structural characteristics of the antennal sensilla of *Agrotis segetum* (Insecta: Lepidoptera). *Cell and Tissue Res.* **218**, 209–18. [27]

Halliday, T. (1980). *Sexual strategy.* Univ. Chicago, Chicago. [20]

Hansen, K. (1983). Reception of bark beetle pheromone in the predaceous beetle, *Thanasimus formicarius* (Coleoptera: Cleridae). *J. Comp. Physiol.* **150**, 371–8. [21,29]

——, Schneider, D. and Boppré, M. (1982). Chiral pheromone and reproductive isolation between the gypsy- and nun moth. *Naturwissenschaften* **70**, 466–7 [25]

Hantgartner, W. (1967). Spezifität und Inaktivierung des Spurpheromons von *Lasius fuliginosus* Latr. und Orientierung der Arbeiterinnen im Dutfeld. *Z. verg. Physiol.* **62**, 111–20. [14]

—— (1969). Structure and variability of the individual odor trail in *Solenopsis geminata* Fabr. (Hymenoptera, Formicidae). *Z. verg. Physiol.* **75**, 123–42. [14]

Hardee, D. D., Cross, W. H. and Mitchell, E. B. (1969). Male boll weevils are more attractive than cotton plants to boll weevils. *J. Econ. Entomol*. **62**, 165–9. [29]

Harris, M. O. and Miller, J. R. (1983). Color stimuli and oviposition behavior of the onion fly, *Delia antiqua*. *Ann. Entomol. Soc. Amer*. **76**, 766–71. [8]

—— and Miller, J. R. (1984). Foliar form influences oviposition behavior of the onion fly. *Physiol. Entomol*. **9**, 145–55. [8]

Harrow, I. D. and Hildebrand, J. G. (1982). Synaptic interactions in the olfactory lobe of the moth *Manduca sexta*. *Soc. Neurosci. Abstr.* **8**, 528. [32,33]

——, Quartararo, P., Kent, K. S. and Hildebrand, J. G. (1983). Central projections and possible chemosensory function of neurons in a sensory organ on the labial palp of *Manduca sexta*. *Soc. Neurosci. Abstr*. **9**, 216. [32]

Hassell, M. P. and Southwood, T. R. E. (1978). Foraging strategies of insects. *Ann. Rev. Ecol. Syst.* **9**, 75–98. [8]

Hausen, K. (1982a). Motion sensitive interneurones in the optomotor system of the fly. I. The horizontal cells: structure and signals. *Biol. Cybern*. **45**, 143–56. [5]

—— (1982b). Motion sensitive interneurons in the optomotor system of the fly. II. The horizontal cells: receptive field organization and response characteristics. *Biol. Cybern.* **46**, 67–79. [5]

Hauser, G. (1980). Physiologische und histologische Untersuchungen über das Geruchsorgan der Insekten. *Wiss. Zool*. **34**, 367–403. [21]

Havukkala, I. (1980). Klinokinetic and klinotactic humidity reactions of the beetles, *Hylobius abietis* and *Tenebrio molitor*. *Physiol. Entomol.* **5**, 133–140. [2,15]

Havukkala, I. J. and Kennedy, J. S. (1984). A programme of self-steered turns as a humidity response in *Tenebrio*, and the problem of categorizing spatial manoeuvres. *Physiol. Entomol*. **9**, 157–64. [2,15,20]

Hawkes, C. and Coaker, T. H. (1979). Factors affecting the behavioural responses of the adult cabbage root fly, *Delia brassicae*, to host plant odour. *Entomol. Exp. Appl*. **25**, 45–58. [5,8,20]

Heath, R. R., Teal, P. E. A., Tumlinson, J. H. and Mengelkoch, L. J. (1986). Prediction of release ratios of multicomponent pheromone blends from rubber septa: a first approximation using relative vapor pressure calculated from retention indices on a liquid crystal column. *J. Chem. Ecol*. in press. [18]

——, Tumlinson, J. H., Leppla, N. C., McLaughlin, J. R., Dueben, B., Dundulis, E. and Guy, R. H. (1983). Identification of a sex pheromone produced by female velvetbean caterpillar moth. *J. Chem. Ecol.* **9**, 645–56. [18]

Heikkenen, H. J. (1977). Southern pine beetle: a hypothesis regarding its primary attractant. *J. For.* **75**, 412–3. [11]

Hendry, L. B., Greany, P. D. and Gill, R. J. (1973). Kairomone mediated host-finding behavior in the parasitic wasp, *Orgilus lepidus*. *Entomol. Exp. Appl*. **15**, 471–7. [16]

Henson, W. R. (1962). Laboratory studies on the adult behavior of *Conophthorus coniperda* (Schwarz) (Coleoptera: Scolytidae). III. Flight. *Ann. Entomol. Soc. Amer*. **55**, 524–30. [10]

Heran, H. (1955). Versuche über die Windkompensation der Bienen. *Naturwissenschaften Jahrg*. **42**, 132–3. [7]

—— und Lindauer, M. (1963). Windkompensation und Seitenwindkorrektur der Bienen beim Flug über Wasser. *Z. Verg. Physiol.* **47**, 39–55. [7]

Hicks, R. R. Jun. (1980). Climatic, site and stand factors. In *The southern pine beetle*. (eds. R. C. Thatcher, J. L. Searcy, J. E. Coster and G. D. Hertel) pp. 55–68. *USDA FS Tech Bull.* **1631**. [11]

Hildebrand, J. G. (1985). Metamorphosis of the insect nervous system: influences of the periphery on the postembroynic development of the antennal sensory pathways in the brain of *Manduca sexta*. In *Model neural networks and behavior* (ed. A. I. Selverston) pp. 129–48. Plenum, New York. [32]

—— and Maxwell, G. D. (1980). Neurochemical explorations of the central nervous system of the moth *Manduca sexta* and especially of the antennal and visual pathways. In *Insect neurobiology and pesticide action (Neurotox 79)* pp. 101–7. Soc. Chem. Ind., London. [24,32]

——, Hall, L. M. and Osmond, B. C. (1979). Distribution of binding sites for ^{125}I-labeled α-bungarotoxin in normal and deafferented antennal lobes of *Manduca sexta. Proc. Nat. Acad. Sci. USA* **76**, 499–503. [32]

——, Matsumoto, S. G., Tolbert, L. P., Schneiderman, A. M. and Camazine, S. M. (1982). Postembryonic development of the antennal lobes in the moth *Manduca sexta.* In *Neuronal development: cellular approaches in invertebrates* (eds. C. S. Goodman and K. G. Pearson). *Neurosciences Res. Prog. Bull.* **20**, 891–900. [32]

——, Matsumoto, S. G., Camazine, S. M., Tolbert, L. P., Blank, S., Ferguson, H. and Ecker, V. (1980). Organisation and physiology of antennal centres in the brain of the moth *Manduca sexta*. In *Insect neurobiology and pesticide action (Neurotox 79)* pp. 372–82. Soc. Chem. Ind., London. [32]

Hill, A. S., Rings, R. W., Swier, S. R. and Roelofs, W. L. (1979). Sex pheromone of the black cutworm moth, *Agrotis ypsilon. J. Chem. Ecol.* **5**, 439–57. [26]

Hinks, C. F. (1967). Relationship between serotonin and the circadian rhythm in some nocturnal moths. *Nature* **214**, 386–7. [24]

Hodges, J. D. and Pickard, L. S. (1971). Lightning in the ecology of the southern pine beetle, *Dendroctonus frontalis* (Coleoptera: Scolytidae). *Can. Entomol.* **103**, 44–51. [11]

Holt, S. J. (1955). On the foraging activity of the wood ant. *J. Anim. Ecol.* **24**, 1–34. [14]

Homberg, U. and Erber, J. (1979). Response characteristics and identification of extrinsic mushroom body neurones of the bee. *Z. Naturforsch.* **34c**, 612–15. [33]

Hopkins, A. D. (1909). Practical information on scolytid beetles of North American forests. Bark beetles of the *Dendroctonus*. USDA *Entomol. Bull.* **83**, 169 pp. [11]

Hoskins, S. G. and Hildebrand, J. G. (1983). Neurotransmitter histochemistry of neurons in the antennal lobes of *Manduca sexta. Soc. Neurosci. Abstr*. **9**, 216. [32]

——, Kingan, T. G., Christensen, T. A. and Hildebrand, J. G. (1984). Mapping GABA-like immunoreactivity in antennal lobe of the moth *Manduca sexta. Soc. Neurosci. Abstr.* **10**, 688. [32]

Huber, F. (1974). Neural integration (central nervous system). In *The physiology of insecta* (2nd edn.)(ed. M. Rockstein) pp. 3–100. Academic Press, New York. [33]

Hunter, W. D. and Pierce, W. D. (1912). The Mexican cotton boll weevil. *USDA Bur. Entomol. Bull.* **114**, 118pp. [29]

Hynum. B. G. and Berryman, A. A. (1980). *Dendroctonus ponderosae* (Coleoptera: Scolytidae): pre-aggregation landing and gallery initiation on lodgepole pine. *Can. Entomol.* **112**, 185–91. [11]

—— and Berryman, A. A. (1981). *Dendroctonus ponderosae* Hopkins (Coleoptera: Scolytidae): gallery initiation on lodgepole pine during aggregation. *Environ. Entomol.* **10**, 842–6. [10]

Ishikawa, S., Tazima, Y. and Hirao, T. (1963). Responses of the chemoreceptors of maxillary sensory hairs in a 'non-preference' mutant of the silkworm. *J. Seric. Sci.* Tokyo, **32**, 125–9. [1]

Isono, K. and Kikuchi, T. (1974). Autosomal recessive mutation in sugar responses of *Drosophila. Nature* **248**, 243–4. [1]

Jander, R. J. (1975). Ecological aspects of spatial orientation. *Ann. Rev. Ecol. Syst.* **6**, 171–88. [2,15]

Jewett, D. M., Matsumura, F. and Coppel, H. C. (1976). Sex pheromone specificity in the pine sawflies: interchange of acid moities in an ester. *Science* **192**, 51–3. [13]

Jones, C. D. (1977). Ion concentration variations at short distance downwind of continuous and quasi-instantaneous point sources. *Pesticide Science* **8**, 84–95. [3]

—— (1983). On the structure of instantaneous plumes in the atmosphere. *J. Hazardous Materials* **7**, 87–112. [3]

Jones, O. T., Lomer, R. A. and Howse, P. E. (1981). Response of male Mediterranean fruit flies, *Ceratitis capitata*, to trimedlure in a wind tunnel of novel design. *Physiol. Entomol.* **6**, 175–81. [12,17]

Jones, R. L. (1981). Chemistry of semiochemicals involved in parasitoid–host and predator–prey relationships. In *Semiochemicals: their role in pest control.* (eds. D. A. Norlund, R. L. Jones and W. J. Lewis). pp. 239–50. John Wiley, New York. [16]

——, Lewis, W. J., Beroza, M., Bierl, B. A. and Sparks, A. N. (1973). Host seeking stimulants (kairomones) for the egg parasite *Trichogramma evanescens. Environ. Entomol. 2*, 593–6. [16]

——, Lewis, W. J., Bowman, M. C., Beroza, M. and Bierl, B. A. (1971). Host-seeking stimulant for parasite of corn earworm: isolation, identification and synthesis. *Science* **173**, 842–3. [16]

Kafka, W. A. (1970). Molekulare Wechselwirkungen bei der Erregung einzelner Riechzellen. *Z. Verg. Physiol.* **70**, 105–43. [21]

Kaib, M. (1974). Die Fleisch- und Blumenduftrezeptoren auf der Antenne der Schmeißfliege *Calliphora vicina. J. Comp. Physiol.* **95**, 105–21. [21]

Kaissling, K.-E. (1971). Insect olfaction. In *Handbook of sensory physiology* (ed. L. M. Beidler) Vol. **4**, pp. 351–431. Springer-Verlag, Berlin–Heidelberg–New York. [4,21,29,32]

—— (1972). Kinetic studies of transduction in olfactory receptors of *Bombyx mori*. In *Olfaction and taste* (ed. D. Schneider) pp. 207–13. Wiss, Stuttgart. [23]

—— (1974). Sensory transduction in insect olfactory receptors. In *Biochemistry of sensory functions* (ed. L. Jaenicke) pp. 243–73. Springer-Verlag, Berlin-Heidelberg-New York. [21,22,23]

—— (1979). Recognition of pheromones by moths, especially in saturniids and *Bombyx mori*. In *Chemical ecology and odour communications in animals* (ed. F. J. Ritter) pp. 43–56. Elsevier/North Holland, Biomedical Press, Amsterdam. [22,30]

—— and Priesner, E. (1970). Die Riechschwelle des Seidenspinners. *Naturwissenschaften* **57**, 23–8. [6,21,22,23]

—— and Renner, M. (1968). Antennale Rezeptoren für Queen Substanz und Sterzelduft bei Honigbiene. *Z. Verg. Physiol.* **59**, 357–61. [21]

—— and Thorson, J. (1980). Insect olfactory sensilla: structural, chemical and electrical aspects of the functional organization. In *Receptors for neurotransmitters, hormones and pheromones in insects* (eds. D. Sattelle *et al.*) pp. 261–82. Elsevier/North Holland, Biomedical Press, Amsterdam. [23]

——, Kasang, G., Bestmann, H. J., Stransky, W. and Vostrowsky, O. (1978). A new pheromone of the silkworm moth *Bombyx mori*, sensory pathway and behavioural effect. *Naturwissenschaften* **65**, 382–4. [13,22,33]

Kanaujia, S. and Kaissling, K.-E. (1985). Interactions of pheromones with moth antennae: adsorption, desorption and transport. *J. Insect Physiol.* **31**, 71–81. [22]

Kandel, E. R. and Schwartz, J. H. (1982). Molecular biology of learning: modulation of transmitter release. *Science* **218**, 433–43. [24]

——, Krasne, F. B., Strumwasser, F. and Truman, J. W. (1979). Cellular mechanisms in the selection and modulation of behavior. *Neurosci. Res. Program Bull.* **17**, 523–710. [28]

Karieva, P. (1983). Influence of vegetation texture on herbivore populations: resource concentration and herbivore movement. In *Variable plants and herbivores in natural and managed systems* (eds. R. F. Denno and M. S. McClure) pp. 259–89. Academic Press. New York. [8]

Karlson, P. and Lüscher, M. (1959). Pheromone, Ein Nomenklatur-vorschlag für eine Wirkstoffklasse. *Naturwissenschaften* **46**, 63–4. [30]

Kasang, G. (1971). Bombykol reception and metabolism on the antennae of the silkworm *Bombyx mori*. In *Gustation and olfaction* (eds. G. Ohloff and A. F. Thomas) pp. 245–50. Academic Press, New York. [23]

—— and Kaissling, K.-E. (1972). Specificity of primary and secondary olfactory processes in *Bombyx* antennae. In *Olfaction and taste IV* (ed. D. Schneider) pp. 200–6. Verlagsgesellschaft, Stuttgart. [23]

Keil, T. A. (1982). Contacts of pore tubules and sensory dendrites in antennal chemosensilla of a silkmoth: demonstration of a possible pathway for olfactory molecules. *Tiss. Cell* **14**, 451–62. [23]

Kennedy, J. S. (1940). The visual responses of flying mosquitoes. *Proc. Zool. Soc. Lond.* Ser A **109**, 221–42. [4,5,7]

—— (1972). The emergence of behavior (address to closing plenary session, XII Int. Congr. Entomol.). *J. Aust. Entomol. Soc.* **11**, 168–76. [4,20]

—— (1977). Olfactory responses to distant plants and other odor sources. In *Chemical control of insect behavior: theory and application* (eds. H. H. Shorey and J. J. McKelvey Jun) pp. 67–91. John Wiley–Interscience, New York. [2,20]

—— (1978). The concepts of olfactory 'arrestment' and 'attraction'. *Physiol. Entomol.* **3**, 91–8. [2,4,8,15,20]

—— (1982). Mechanism of moth sex attraction: a modified view based on wind-tunnel experiments with flying male *Adoxophyes*. *Les Colloques de l'INRA* **7**, 189–92. [4]

—— (1983). Zigzagging and casting as programmed responses to wind-borne odour: a review. *Physiol. Entomol.* **8**, 109–20. [2,4,5,6]

—— (1985). Migration, behavioral and ecological. In *Migration: mechanisms and adaptive significance* (ed. M. A. Rankin). *Contrib. in Marine Sci.* **27**, Supplement, 5–26 [2]

—— and Booth, C. O. (1963a). Free flight of aphids in the laboratory. *J. Exp. Biol.* **40**, 67–85. [10]

—— and Booth, C. O. (1963b). Co-ordination of successive activities in an aphid. The effect of flight on the settling responses. *J. Exp. Biol.* **40**, 351–69. [10]

—— and Marsh, D. (1974). Pheromone-regulated anemotaxis in flying moths. *Science* **184**, 999–1001. [4,7,21,23]

—— and Thomas, A. A. G. (1974). Behaviour of some low-flying aphids in wind. *Ann. Appl. Biol.* **76**, 143–59. [5]

——, Ludlow, A. R. and Sanders, C. J. (1980). Guidance system used in moth sex attraction. *Nature* **288**, 475–7. [4,5]

——, Ludlow, A. R. and Sanders, C. J. (1981). Guidance of flying male moths by wind-borne sex pheromone. *Physiol. Entomol.* **6**, 395–412. [2,4,5,12,20,21]

Kikukawa, T., Matsumura, F., Kraemer, M., Coppel, H. C. and Tai, A. (1982). Field attractiveness of chirally defined synthetic attractants to males of *Diprion similis* and *Gilpinia frutetorum*. *J. Chem. Ecol.* **8**, 301–14. [13]

——, Matsumura, F., Olaifa, J., Kraemer, M., Coppel, H. C. and Tai, A. (1983). Field evaluation of chiral isomers of the sex pheromone of the European pine sawfly, *Neodiprion sertifer*. *J. Chem. Ecol.* **9**, 673–93. [13]

Kingan, T. G. and Hildebrand, J. G. (1982). GABA in the antennal lobes of metamorphosing and mature *Manduca sexta*. *Soc. Neurosci. Abstr.* **8**, 988. [32]

Kinzer, G. W., Fentiman, A. F. Jun., Page, T. F. Jun., Foltz, R. L., Vité, J. P. and Pitman, G. B. (1969). Bark beetle attractants: identification, synthesis and field bioassay of a new compound isolated from *Dendroctonus*. *Nature* **221**, 477–8. [11,29]

Klimetzek, D. (1984). Grundlagen einer Überwachung und Bekämpfung der Nutzholzborkenkafer (*Trypodendron* spp.) mit Lock- und Ablenkstoffen. *Freiburger Waldschutz-Abhandlungen*, Band 5. Forstzoologisches Institut der Albert-Ludwigs-Universität, Freiburg in Breisgau. [10]

Klowden, M. J. (1981). Initiation and termination of host-seeking inhibition in *Aedes aegypti* during oöcyte maturation. *J. Insect Physiol.* **27**, 799–803. [28]

—— (1983). The physiological control of mosquito host-seeking behavior. In *Current topics in vector research* (ed. K. F. Harris) Vol. **1**, pp. 93–116. Praeger Scientific, New York. [28]

—— and Lea, A. O. (1979a). Humoral inhibition of host-seeking in *Aedes aegypti* during oöcyte maturation. *J. Insect Physiol.* **25**, 231–5. [28]

—— and Lea, A. O. (1979b). Abdominal distention terminates subsequent host-seeking behavior of *Aedes aegypti* following a blood meal. *J. Insect Physiol.* **25**, 583–5. [28]

Klun, J. A. and Robinson, J. F. (1972). Olfactory discrimination in the European corn borer and several pheromonally analogous moths. *Ann. Entomol. Soc. Amer.* **65**, 1337–40. [25]

Knerer, G. and Atwood, C. E. (1973). Diprionid sawflies: polymorphism and speciation. *Science* **179**, 1090–9. [13]

Kohnle, U. and Vité, J. P. (1984). Bicyclic ketals in the chemical communication of European bark beetles. *Naturwissenschaften* **71**, 47–8. [30]

Komarek, E. V. Sen. (1964). The natural history of lightning. *Proc. Third Ann. Tall Timbers Fire Ecol. Conf.* Tallahassee, Fla. pp. 139–86. [11]

Koopowitz, H. and Keenan, L. (1982). The primitive brains of platyhelminthes. *Trends in Neurosci.* **5**, 77–9. [1]

Koshland, D. E. (1983). The bacterium as a model neuron. *Trends In Neurosci.* **6**, 133–7. [1]

van der Kraan, C. and van Deventer, P. (1982). Range of action and interaction of pheromone traps for the summerfruit tortrix moth, *Adoxophyes orana* (F.v.R.). *J. Chem. Ecol.* **8**, 1251–62. [2]

Kraemer, M., Coppel, H. C., Matsumura, F., Kikukawa, T. and Mori, K. (1979). Field responses of the white sawfly, *Neodiprion pinetum*, to optical isomers of sawfly sex pheromones. *Environ. Entomol.* **8**, 519–20. [13]

——, Coppel, H. C., Matsumura, F., Wilkinson, R. C. and Kikukawa, T. (1981). Field and electroantennogram responses of the red-headed pine sawfly, *Neodiprion lecontei* (Fitch), to optical isomers of sawfly sex pheromones. *J. Chem. Ecol.* **7**, 1063–72. [13]

——, Coppel, H. C., Kikukawa, T., Matsumura, F., Thomas, H. A., Thompson, L. C. and Mori, K. (1983). Field and electroantennogram responses to sex pheromone optical isomers by four fall-flying sawfly species (Hymenoptera: Diprionidae, *Neodiprion*). *Environ. Entomol.* **12**, 1592–6. [13]

Kramer, E. (1975). Orientation of the male silkmoth to the sex attractant bombykol. In *Orientation and taste V*. (eds. D. A. Denton and J. P. Coghlan) pp. 329–35. Academic Press, New York. [6,15,21,23,33]

—— (1978). Insect pheromones. In *Taxis and behavior* (ed. G. L. Hazelbauer) pp. 207–29. Chapman and Hall, London and New York. [15]

Kuenen, L. P. S. and Baker, T. C. (1981). Habituation versus sensory adaptation as the cause of reduced attraction following pulsed and constant sex pheromone pre-exposure in *Trichoplusia ni*. *J. Insect Physiol.* **27**, 721–6. [22]

—— and Baker, T. C. (1982a). Optomotor regulation of ground velocity in moths during flight to sex pheromone at different heights. *Physiol. Entomol.* **7**, 193–202. [12]

—— and Baker, T. C. (1982b). The effects of pheromone concentration on the flight behaviour of the oriental fruit moth, *Grapholitha molesta*. *Physiol. Entomol.* **7**, 423–34. [4,12]

—— and Baker, T. C. (1983). A non-anemotactic mechanism used in pheromone source location by flying moths. *Physiol. Entomol.* **8**, 277–89. [2,4,5,9,20]

Kydonieus, A. F. and Beroza, M. (1982). *Insect suppresion with controlled release pheromone systems*. CRC Press Inc., Boca Raton, FL. [27]

Lacher, V. (1964). Elektrophysiologische Untersuchungen an einzelnen Rezeptoren für Geruch, Kohlendioxyd, Luftfeuchtigkeit und Temperatur auf den Antennen der Arbeitsbiene und der Drohne (*Apis mellifica* L.). *Z. Verg. Physiol.* **48**, 587–623. [21]

—— (1967). Elektrophysiologische Untersuchungen an einzelnen Geruchsrezeptoren auf den Antennen weiblicher Moskitos (*Aedes aegypti* L.). *J. Insect Physiol.* **13**, 1461–70. [21]

—— (1971). Arbeitsbereiche von Geruchsrezeptoren auf der Moskitoantenne (*Aedes aegypti*). *J. Insect Physiol.* **17**, 507–17. [21]

—— and Schneider, D. (1963). Elektrophysiologischer Nachweis der Reichfunktion von Porenplatten (Sensilla placodea) auf den Antennen der Drohne und Arbeitsbiene (*Apis mellifica* L.). *Z. Verg. Physiol.* **47**, 274–8. [21]

Lackie, J. M. and Wilkinson, P. C. (1982). *Biology of the chemotactic response*. Cambridge University Press, Cambridge. [1]

Laing, J. (1937). Host finding by insect parasites. I. Observations on the finding of host by *Alysia manducator, Marmoniella vitripennis* and *Trichogramma evanescens*. *J. Anim. Ecol.* **6**, 298–317. [16]

Lancet, D. (1984). Molecular view of olfactory reception. *Trends In Neurosci*. **7**, 35–6. [1]

Lanier, G. M. and Burns, B. W. (1978). Barometric flux. Effects on the responsiveness of bark beetles to aggregation attractants. *J. Chem. Ecol*. **4**, 139–47. [10]

—— and Wood, D. L. (1975). Specificity of response to pheromones in the genus *Ips* (Coleoptera: Scolytidae). *J. Chem. Ecol.* **1**, 9–23. [30]

——, Silverstein, R. M. and Peacock, J. H. (1976). Attractant pheromone of the European elm bark beetle (*Scolytus multistriatus*): isolation, identification, synthesis and utilization studies. In *Perspectives in forest entomology* (eds. J. F. Anderson and K. K. Kaya) pp. 149–75. Academic Press, New York. [10]

——, Birch, M. C., Schmitz, R. F. and Furniss, M. M. (1972). Pheromones of *Ips pini* (Coleoptera: Scolytidae): variation in response among three populations. *Can. Entomol*. **104**, 1917–23. [10]

——, Claesson, A., Stewart, T., Piston, J. J. and Silverstein, R. M. (1980). *Ips pini*: the basis for interpopulational differences in pheromone biology. *J. Chem. Ecol*. **6**, 677–87. [10]

Legg, B. J. (1983). Turbulent dispersion from an elevated line source: Markov chain simulations of concentration and flux profiles. *Quart. J. Royal Meteorol. Soc*. **109**, 645–60. [3]

——, Strange, S. A., Wall, C. and Perry, J. N. (1980). Diffusion of gaseous plumes within crop canopies. *Rep. Rothamsted Exp. Stn. for 1979*, part 1, 162. [9]

van Lenteren, J. C., Nell, H. W. and van der Lelie, L. A. S. (1980). The parasite–host relationship between *Encarsia formosa* (Hymenoptera: Aphelinidae) and *Trialeurodes vaporariorum* (Homoptera: Aleyrodidae). IV. Oviposition behavior of the parasite, with aspects of host selection, host discrimination and host feeding. *Z. Ang. Entomol*. **89**, 442–54. [16]

——, Nell, H. W., van der Lelie, L. A. S. and Woets, J. (1976). The parasite–host relationship between *Encarsia formosa* (Hymenoptera: Aphelinidae) and *Trialeurodes vaporariorum* (Homoptera: Aleyrodidae). I. Host finding by the parasite. *Entomol. Exp. Appl.* **20**, 123–30.

Lewis, T. and Macaulay, E. D. M. (1976). Design and elevation of sex-attractant traps for pea moth, *Cydia nigricana* (Steph.), and the effect of plume shape on catches. *Ecol. Entomol*. **1**, 175–87. [4]

Lewis, W. J., Jones, R. L., Gross, H. R. Jun. and Nordlund, D. A. (1976). The role of kairomones and other behavior chemicals in host finding by parasitic insects. *Behav. Biol*. **16**, 267–89. [16]

——, Beevers, M., Norlund, D. A., Gross, H. T. Jun. and Hagen, K. S. (1979). Kairomones and their use for management of entomophagous insects. IX. Investigation of various kairomone treatment patterns for *Trichogramma* spp. *J. Chem. Ecol.* **5**, 673–80. [16]

——, Norlund, D. A., Gueldner, R. C., Teal, P. E. A. and Tumlinson, J. H. (1982). Kairomones and their use for management of entomophagous insects. XIII. Kairomonal activity for *Trichogramma* spp. of abdominal tips, excreta, and a synthetic sex pheromone blend of *Heliothis zea* (Boddie) moths. *J. Chem. Ecol.* **8**, 1323–31. [16,18]

Light, D. M. and Birch, M. C. (1979). Inhibition of the attractive pheromone response in *Ips paraconfusus* by (R)-(−)-ipsdienol. *Naturwissenschaften* **66**, 159–60. [21,30]

Liljefors, T., Thelin, B. and Van Der Pers, J. N. C. (1984). Structure–activity relationships between stimulus molecules and the response of a pheromone receptor cell in the turnip moth, *Agrotis segetum. J. Chem. Ecol.* **10**, 1661–75. [26]

Lindgren, B. S. (1983). A mulitple funnel trap for scolytid beetles. (Coleoptera). *Can. Entomol.* **115**, 299–302. [10]

Linn, C. E. Jun. and Gaston, L. K. (1981). Behavioral function of the components and the blend of the sex pheromone of the cabbage looper, *Trichoplusia ni. Environ. Entomol.* **10**, 751–5. [4]

—— and Roelofs, W. L. (1981). Modification of sex pheromone blend discrimination in male oriental fruit moths by pre-exposure to (*E*)-8-dodecenyl acetate. *Physiol. Entomol.* **6**, 421–9. [4]

—— and Roelofs, W. L. (1983). The effect of varying proportions of the alcohol component on sex pheromone blend discrimination in male oriental fruit moths. *Physiol. Entomol.* **8**, 291–306. [4,19,24]

—— and Roelofs, W. L. (1984). Sublethal effects of neuroactive compounds on pheromone response thresholds in male oriental fruit moths. *Arch. Insect Biochem. Physiol.* **1**, 331–44. [24]

——, Bjostad, L. B., Du, J. W. and Roelofs, W. L. (1984). Redundancy in a chemical signal: behavior responses of male *Trichoplusia ni* to a six-component sex pheromone blend. *J. Chem. Ecol.* **10**, 1635–58. [4,24,25]

Lofstedt, C. and Van Der Pers, J. N. C. (1985). Sex pheromones and reproductive isolation in four European small ermine moths. *J. Chem. Ecol.* **11**, 649–66. [27]

——, Linn, C. E. Jr. and Löfqvist, J. (1985). Behavioral responses of male turnip moths, *Agrotis segetum*, to sex pheromone in a flight tunnel and in the field. *J. Chem. Ecol.* **11**, 1209–27. [21]

——, Löfqvist, J., Lanne, B. S., Van Der Pers, J. N. C., and Hansson, B. S. Pheromone dialects in European turnip moths *Agrotis segetum. Oikos.* **46** (in press). [27]

——, Van Der Pers, J. N. C., Löfqvist, J., Lanne, B. S., Applegren, M., Bergström, G. and Thelin, B. (1982). Sex pheromone components of the turnip moth *Agrotis segetum*: chemical identification, electrophysiological evaluation and behavioural activity. *J. Chem. Ecol.* **8**, 1305–21. [13,27]

Long, T. F. and Murdock, L. L. (1983). Stimulation of blowfly feeding behavior by octopaminergic drugs. *Proc. Nat. Acad. Sci. USA* **80**, 4159–61. [24]

Longhurst, C., Baker, R. and Mori, K. (1980). Response of the sawfly *Diprion similis* to chiral sex pheromones. *Experientia* **36**, 946–7. [13]

Lorio, P. L. Jun. and Bennett, W. H. (1974). Recurring southern pine beetle infestations near Oakdale, Louisiana. *USDA For. Serv. Res. Pap*. SO–95. 6pp. [11]

Ludlow, A. R. (1984). Application of computer modeller to behavioural coordination. Ph.D. Thesis, University of London. [4,5]

Lythgoe, J. H. (1979). *The ecology of vision.* Clarendon, Oxford. [8]

Magnusson, G. (1977). Pheromone synthesis. Preparation of *erythro*-3,7-dimethylpentadecan-2-ol, the alcohol from pine sawfly sex pheromone (Hymenoptera: Diprionidae). *Tetrahedron Letters* **31**, 2713–6. [13]

—— (1978). Pheromone synthesis-II. Preparation of pine sawfly (Hymenoptera:

Diprionidae) sex attractants and analogues with possible biological activity. *Tetrahedron* **34**, 1385–8. [13]

Markl, H. (1974). Insect behavior: functions and mechanisms. In *The physiology of insects*. (ed. M. Rockstein) **3**, 3–148. Academic Press, New York. [8]

Marsh, D., Kennedy, J. S. and Ludlow, A. R. (1978). An analysis of anemotactic zigzagging flight in male moths stimulated by pheromone. *Physiol. Entomol*. **3**, 221–40. [4,5,7,12]

——, Kennedy, J. S. and Ludlow, A. R. (1981). Analysis of zigzagging flight in moths: a correction. *Physiol. Entomol*. **6**, 225. [4,5,23]

Masson, C. (1972). Le système antennaire chez les fourmis. I. Histologie et ultrastructure du deutocerebron. Etude comparée chex *Camponotus vagus* (Formicinae) et *Mesoponera caffraria* (Ponerinae). *Z. Zellforsch.* **134**, 31–64. [32]

—— and Arnold, G. (1984). Ontogeny, maturation and plasticity of the olfactory system in the worker bee. *J. Insect Physiol*. **30**, 7–14. [29]

Matsumoto, S. G. and Hildebrand, J. G. (1981). Olfactory mechanisms in the moth *Manduca sexta*: response characteristics and morphology of central neurons in the antennal lobes. *Proc. R. Soc. Lond*. B **213**, 249–77. [21,32,33]

Mayer, M. S. (1973). Electrophysiological correlates of attraction in *Trichoplusia ni*. *J. Insect Physiol*. **19**, 1191–8. [25]

McCambridge, W. F. (1971). Temperature limits flight of the mountain pine beetle, *Dendroctonus ponderosae*. *Ann. Entomol. Soc. Amer.* **64**, 534–5. [10]

McLaughlin, J. R. (1979). Disruption of mating communication in peach borers. *US Dept. Agric., Agric. Res. Results* ARR-NE-**6**, 27–34. [18]

—— and Mitchell, E. R. (1981). Practical development of pheromones in *Heliothis* management. In *Proceedings of the international workshop on Heliothis management* (eds. W. Reed and V. Kumble) pp. 309–318. International Crops Research Institute for the Semi-Arid Tropics, Patancherce, AP, India. [18]

——, Brogdon, J. E., Agree, H. R. and Mitchell, E. R. (1975). Effect of trap color on captures of male cabbage loopers and soybean loopers in double-cone pheromone traps. *J. Ga. Entomol. Soc*. **10**, 174–9. [18]

Meola, R. W. and Petralia, R. W. (1980). Juvenile hormone induction of biting behavior in *Culex* mosquitoes. *Science* **209**, 1548–50. [28]

Mercer, A. R. and Menzel, R. (1982). The effects of biogenic amines on conditioned and unconditioned responses to olfactory stimuli in the honeybee *Apis mellifera*. *J. Comp. Physiol*. **145**, 363–8. [24]

Miksad, R. W. and Kittredge, J. (1979). Pheromone aerial dispersion: a filament model. *14th Conf. Agric. and For. Met., Am., Met. Soc.* pp. 283–43. [3,20]

Millechia, R. and McIntyre, T. (1978). Automatic nerve impulse identification and separation. *Computers and Biomed. Res*. **11**, 459–68. [31]

Miller, J. R. and Roelofs, W. L. (1978). Gypsy moth responses to pheromone enantiomers as evaluated in a sustained-flight tunnel. *Environ. Entomol*. **7**, 42–4. [7]

—— and Stickler, K. L. (1984). Finding and accepting host plants. In *Chemical ecology of insects* (eds. W. J. Bell and Cardé, R. T.) pp. 127–57. Chapman and Hall, London and New York. [8,16,20]

Miller, M. C. (1983). Lightning strike simulation for studying southern pine bark and engraver beetle attacks. *USDA For. Serv. Paper* SO-296, 4pp. [11]

Mimura, K., Tateda, H., Morita, H. and Kuwabara, M. (1970). Convergence of antennal and ocellar inputs in the insect brain. *Z. Verg. Physiol*. **68**, 301–10. [33]

Minks, A. K. (1984). *Attractants and pheromones of noxious insects*. SROP/WPRS Bulletin 1984/VII/1, Wageningen, The Netherlands. [27]

Minyard, J. P., Hardee, D. D., Gueldner, R. C., Thompson, A. C., Wiygul, G. and Hedin, P. A. (1969). Constituents of the cotton bud. Compounds attractive to the boll weevil. *Agric. and Food Chem.* **17**, 1093–7. [29]

Mistric, W. J. Jun. and Mitchell, E. R. (1966). Attractiveness of isolated groups of cotton plants to migrating boll weevils. *J. Econ. Entomol.* **59**, 39–41. [29]

Mitchell, C. J. (1983). Differentiation of host-seeking behavior from blood-feeding behavior in overwintering *Culex pipiens* and observations on gonotrophic dissociation. *J. Med. Entomol.* **20**, 157–63. [28]

Mitchell, E. R. and Taft, H. M. (1966). Host-plant selection by migratory boll weevils. *J. Econ. Entomol.* **59**, 390–2. [29]

Mitchell, W. C. and Mau, R. F. U. (1971). Response of female southern green stink bug and its parasite, *Trichopoda pennipes*, to male stink bug pheromones. *J. Econ. Entomol.* **64**, 856–9. [16]

Mittelstaedt, H. (1969). Strategies of cybernetic analysis. In *Biocybernetics of the central nervous system* (ed. M. Proctor) pp. 33–52. Little, Brown and Co., Boston. [6]

Mobbs, P. G. (1982). The brain of the honeybee *Apis mellifera*. I. The connections and spatial organization of the mushroom bodies. *Phil. Trans. R. Soc.* (B) **298**, 309–54. [32]

—— (1984). Neural networks in the mushroom bodies of the honeybee. *J. Insect Physiol.* **30**, 43–58. [33]

Moeck, H. A., Wood, D. L. and Lindahl, K. Q. Jun. (1981). Host selection behavior of bark beetles (Coleoptera: Scolytidae) attacking *Pinus ponderosa* with special emphasis on the western pine beetle *Dendroctonus brevicomis*. *J. Chem. Ecol.* **7**, 49–83. [11]

Van der Molen, J. N., van der Meulen, J. W., de Kramer, J. J. and Pasveer, F. J. (1978). Computerized classification of taste cell responses. *J. Comp. Physiol.* **128**, 1–11. [31]

Montague, R. A., Kent, K. S., Imperato, M. T. and Hildebrand, J. G. (1983). Projections of antennal lobe output neurones in the brain of *Manduca sexta*. *Soc. Neurosci. Abstr.* **9**, 216. [32,33]

Moore, K. R. (1979). Distributions of three species of *Ips* bark beetles within southern pine beetle infestations. MS Thesis. Stephen F. Austin State University, Nacogdoches, TX. [11]

Moore, R. N., Golumbric, C. and Fisher, G. S. (1956). Autoxidation of α-pinene. *J. Amer. Chem. Soc.* **78**, 1173–6. [10]

Moser, J. C. and Dell, T. R. (1979). Predictors of southern pine beetle flight activity. *For. Sci.* **25**, 217–22. [10]

Mudd, A. and Corbet, S. A. (1982). Response of the ichneumonid parasite *Nemeritis canescens* to kairomones from the flour moth, *Ephestia kuehniella*. *J. Chem. Ecol.* **8**, 843–50. [16]

Müller, F. (1877). Uber Haarpinsel, Filzfleckle und ähnliche Gebilde auf den Flugeln männlicher Schmetterlinge. *Jenaische Z. f. Naturwiss.* **5**(11), 99–114. [21]

Murlis, J. and Bettany, B. W. (1977). The night-flight towards a sex pheromone source by male *Spodoptera littoralis* (Boisd.) (Lepidoptera, Noctuidae). *Nature* **268**, 433–4. [5]

—— and Jones, C. D. (1981). Fine-scale structure of odour plumes in relation to

insect orientation to distant pheromone and other attractant sources. *Physiol. Entomol*. **6**, 71–86. [2,3,4,6,20,22]
——, Bettany, B. W., Kelley, J. and Martin, L. (1982). The analysis of flight paths of male Egyptian cotton leafworm moths, *Spodoptera littoralis*, to a sex pheromone source in the field. *Physiol. Entomol*. **7**, 435–41. [5,12]
Mustaparta, H. (1975). Responses of single olfactory cells in the pine weevil *Hylobius abietis* L. (Col.: Curculionidae). *J. Comp. Physiol.* **97**, 271–90. [21]
—— (1979). Chemoreception in bark beetles of the genus *Ips*: synergism, inhibition and discrimination of enantiomers. In *Chemical ecology: odor communication in animals* (ed. F. J. Ritter) pp. 147–58. Elsevier/N. Holland, Amsterdam. [30]
—— (1984). Olfaction. In *Chemical ecology of insects* (eds. W. J. Bell and R. T. Cardé) pp. 37–70. Chapman and Hall, London and New York. [29,30]
——, Angst, M. E. and Lanier, G. N. (1977). Responses of single receptor cells in the pine engraver beetle *Ips pini* (Say) (Coleoptera: Scolytidae) to its aggregation pheromone ipsdienol and the aggregation inhibitor, ipsenol. *J. Comp. Physiol.* **121**, 343–7. [30]
——, Angst, M. E. and Lanier, G. N. (1980). Receptor discrimination of enantiomers of the aggregation pheromone ipsdienol, in two species of *Ips*. *J. Chem. Ecol*. **8**, 689–701. [29]
Nelson, M. C. (1977). The blowfly's dance: role in the regulation of food intake. *J. Insect Physiol*. **23**, 603–12. [15]
Nettles, W. C. Jun. (1890)). Adult *Eucelatoria* sp.: response to volatiles from cotton and okra plants and from the larvae of *Heliothis virescens, Spodoptera enidania* and *Estigmene acrea. Environ. Entomol*. **9**, 759–63. [16]
Noldus, L. P. J. J. and van Lenteren, J. C. (1985a). Kairomones for the egg parasite: *Trichogramma evanescens* Westwood. I. Effect of volatile substances released by two of its hosts, *Pieris brassicae* L. and *Mamestra brassicae* L. *J. Chem. Ecol.* **11**, 781–92. [16]
—— and van Lenteren, J. C. (1985b). Kairomones for the egg parasite: *Trichogramma evanescens* Westwood. II. Effect of contact chemicals produced by two of its hosts, *Pieris brassicae* L. and *Pieris rapae* L. *J. Chem. Ecol.* **11**, 793–800. [16]
Norlund, D. A. (1981). Semiochemicals: a review of the terminology. In *Semiochemicals. Their role in pest control* (eds. D. A. Norlund, R. L. Jones and W. T. Lewis) pp. 13–28. John Wiley, New York. [30]
—— and Lewis, W. J. (1976). Terminology of chemical releasing stimuli in intraspecific and interspecific interactions. *J. Chem. Ecol*. **2**, 211–20. [16]
——, Lewis, W. J. and Gueldner, R. C. (1983). Kairomones and their use for management of entomophagous insects. XIV. Response of *Telenomus nenus* to abdominal tips of *Spodoptera frugiperda,* (*Z*)-9-tetradecene-1–01 acetate and (*Z*)-9-dodecene-1–01 acetate. *J. Chem. Ecol*. **9**, 695–701. [16]
O'Connell, R. J. (1972). Responses of olfactory receptors to the sex attractant, its synergist and inhibitor in the red-banded leafroller. In *Olfaction and taste IV* (ed. D. Schneider) pp. 180–6. Wissenschaft. Verlagsges., Stuttgart. [25,26]
—— (1975). Olfactory receptor responses to sex pheromone components in the redbanded leafroller moth. *J. Gen. Physiol.* **65**, 179–205. [25,26,29]
—— (1981). The encoding of behaviorally important odorants by insect chemosensory receptor neurons. In *Perception of behavioral chemicals* (ed. D. M. Norris) pp. 133–63. Elsevier/North Holland, Amsterdam. [25]

—— (1985). Chemical communication in invertebrates. *Experientia* in press. [25]

——, Kocsis, W. A. and Schoenfeld, R. L. (1973). Minicomputer identification and timing of nerve impulses mixed in a single recording channel. *Proc. IEEE* **61**, 1615–21. [25,31]

——, Grant, A. J., Mayer, M. S. and Mankin, R. W. (1983). Morphological correlates of differences in pheromone sensitivity in insect sensilla. *Science* **220**, 1408–10. [25,29]

Olberg, R. M. (1983a). Pheromone-triggered flip-flopping interneurones in the ventral nerve cord of the silkworm moth, *Bombyx mori*. *J. Comp. Physiol*. **152**, 297–307. [4,6,20,21,23,33]

—— (1983b). Interneurons sensitive to female pheromone in the deutocerebrum of the male silkworm moth, *Bombyx mori*. *Physiol. Entomol.*, **8**, 419–28. [21,33]

Oldfield, B. P. (1980). Accuracy of orientation in female crickets, *Teleogryllus oceanicus* (Gryllidae): dependence on song spectrum. *J. Comp. Physiol*. **141**, 93–9. [2]

Oparin, A. I. (1957). *The origin of life on Earth*. Academic Press, New York. [1]

Owens, E. D. (1982). The effects of hue, intensity, and saturation on foliar and fruit finding in the apple maggot, *Rhagoletis pomonella*. Ph.D. Thesis, Univ. Massachusetts, Amherst. [8]

Palaniswamy, P., Sviasubramania, P. and Seabrook, W. D. (1979). Modulation of sex pheromone perception in female moths of the eastern spruce budworm, *Choristoneura fumifera*, by Altosid. *J. Insect Physiol*. **25**, 571–4. [28, 29]

——, Underhill, E. W., Steck, W. F. and Chisholm, M. D. (1983). Responses of male redbacked cutworm, *Euxoa ochrogaster* (Lepidoptera: Noctuidae), to sex pheromone components in a flight tunnel. *Environ. Entomol.* **12**, 748–52. [26]

Parencia, C. R. Jun., Davis, J. W. and Cowan, C. B. Jun. (1964). Studies on the ability of overwintered boll weevils to find fruiting cotton plants. *J. Econ. Entomol*. **57**, 162. [29]

Pasquill, F. (1974). *Atmospheric diffusion* (2nd edn.) Ellis Horwood Ltd., Chichester. [3]

Payne, T. L. (1970). Electrophysiological investigations on response to pheromones in bark beetles. *Contrib. Boyce Thompson Inst*. **24**, 275–82. [11]

—— (1973). Pheromone and host odor-stimulated potentials in *Dendroctonus*. *Experientia* **30**, 509–10. [11]

—— (1975). Bark beetle olfaction. III. Antennal olfactory responsiveness of *Dendroctonus frontalis* Zimmerman and *D. brevicomis* LeConte (Coleoptera: Scolytidae) to aggregation pheromones and host tree terpene hydrocarbons. *J. Chem. Ecol*. **1**, 233–42. [11,29,30]

—— (1980). Life history and habits. In *The southern pine beetle* (eds. R. C. Thatcher, J. L. Searcy, J. E. Coster and G. D. Hertel) pp. 7–28. USFS TS Tech. Bull. 1631. [11]

——, Dickens, J. C. and Richerson, J. V. (1984). Insect predator–prey coevolution via enantiomeric specificity in a kairomone–pheromone system. *J. Chem. Ecol*. **10**, 487–92. [30]

——, Coster, J. E., Richerson, J. V., Edson, L. J. and Hart, E. R. (1978). Field response of the southern pine beetle to behavioral chemicals. *Environ. Entomol*. **7**, 578–82. [10,11]

——, Hart, E. R., Edson, L. J., McCarty, F. A., Billings, P. M. and Coster, J. E.

(1976). Olfactometer for assay of behavioral chemicals for the southern pine beetle, *Dendroctonus frontalis* (Coleoptera: Scolytidae). *J. Chem. Ecol*. **2**, 411–19. [10]

——, Richerson, J. V., Dickens, J. C., West, J. R., Mori, K., Berisford, C. W., Hedden, J. P., Vité, J. P. and Blum, M. S. (1982). Southern pine beetle: olfaction receptor and behavior discrimination of enantiomers of the attractant pheromone frontalin. *J. Chem. Ecol*. **8**, 873–81. [29]

Pearson, L. (1971). The corpora pendunculata of *Sphinx ligustri* L. and other lepidoptera: an anatomical study. *Phil. Trans. R. Soc. Lond.* (B) **259**, 477–516. [33]

Perry, J. N. and Wall, C. (1982). Practical applications. *Rep. Rothamsted Exp. Stn. for 1981*, part 1, 273–4. [9]

—— and Wall, C. (1984a). Short-term variations in catches of the pea moth, *Cydia nigricana*, in interacting pheromone traps. *Entomol. Exp. Appl.* **36**, 145–91. [9]

—— and Wall, C. (1984b). A mathematical model for the flight of pea moth to pheromone traps through a crop. *Phil. Trans. R. Soc. Lond.* B **306**, 19–48. [2,9]

—— and Wall, C. (1985). Orientation of male pea moth, *Cydia nigricana*, to pheromone traps in a wheat crop. *Entomol. Exp. Appl.* **37**, 161–7. [9]

——, Wall, C. and Greenway, A. R. (1980). Latin square designs in field experiments involving insect sex-attractants. *Ecol. Entomol.* **5**, 385–96. [9]

Person, H. L. (1931). Theory in explanation of the selection of certain trees by western pine beetle. *J. For*. **29**, 696–9. [11]

Perttunen, V. and Boman, T. (1965). Laboratory experiments on the spontaneous take-off activity of *Blastophagus piniperda* (Col., Scolytidae) in relation to temperature and light intensity at different seasons of the year. *Proc. XXII Int. Congr. Entomol*. (1964) **5**, 344–5. [10]

Pfaff, D. W. (1975). Theoretical considerations of cross-fiber pattern coding in the neural signalling of pheromones and other chemical stimuli. *Psychoneuroendocrinology* **1**, 79–93. [25]

Pfaffman, C. (1941). Gustatory afferent impulses. *J. Cell Comp. Physiol*. **17**, 243–58. [29]

Pitman, G. B. (1971). *Trans*-verbenol and *alpha*-pinene: their utility in manipulation of the mountain pine beetle. *J. Econ. Entomol*. **64**, 426–30. [10]

—— (1973). Further observations on Douglure in a *Dendroctonus pseudotsugae* management system. *Environ. Entomol*. **2**, 109–12. [10]

——, Stock, M. W. and McKnight, R. C. (1978). Pheromone application in mountain pine beetle–lodgepole pine management: theory and practice. In *Theory and practice of mountain pine beetle management in lodgepole pine forests*. (eds. A. A. Berryman, G. D. Amman and R. W. Stark) pp. 165–73. Univ. Idaho, Moscow, and USDA, Forest Service, Washington DC. [10]

——, Vité, J. P., Kinzer, G. W. and Fentiman, A. F., Jr. (1968). Bark beetle attractants: *trans*-verbenol isolated from *Dendroctonus*. *Nature* **218**, 168–9. [10]

Preiss, R. (1980). Anemotaxis im Lauf und Flung beim Schwammspinner. Versuche zur Aufklärung des Wirkungsgefüges. Diss. Univ. München. [7]

—— and Kramer, E. (1983). Stabilization of altitude and speed in tethered flying gypsy moth males: influence of (+)- and (−)-disparlure. *Physiol. Entomol.* **8**, 55–68. [5,7,12,21,23]

—— and Kramer, E. (1984). Control of flight speed by minimization of the apparent ground pattern movement. In *Localization and orientation in biology and engineering* (eds. Varjú/Schnitzler) pp. 140–2. Springer-Verlag, Berlin–Heidelberg. [7,21]

Prestwich, G. D., Golec, F. A. and Anderson, N. H. (1984). Synthesis of a highly tritiated photoaffinity labelled pheromone analog for the moth *Antheraea polyphemus J. Labelled Cmpd. Radiopharm.* **21**, 593–601. [23]

Price, P. W. (1970). Trail odors: recognition by insects parasitic on cocoons. *Science* **170**, 546–7. [16]

—— (1981). Semiochemicals in evolutionary time. In *Semiochemicals: their role in pest control*. (eds. D. A. Norlund, R. L. Jones and W. J. Lewis) pp. 251–79. John Wiley, New York. [16]

Priesner, E. (1973). Artspezifität und Funktion einiger Insekten-pheromone. *Fortschr. Zool.* **22**, 49–135. [21]

—— (1979a). Specificity studies on pheromone receptors of noctuid and tortricid Lepidoptera. In *Chemical ecology: odour communication in animals* (ed. F. J. Ritter) pp. 57–71. Elsevier/North Holland, Amsterdam. [26,30]

—— (1979b). Progress in the analysis of pheromone receptor systems. *Ann. Zool. Ecol. Anim.* **11**, 533–46. [26]

—— (1980a). Sensory encoding of pheromone signals and related stimuli in male moths. In *Insect neurobiology and insecticide action* (Neurotox 1979) pp. 359–66. Soc. Chem. Ind., London. [21,26]

—— (1980b). Sex attractant system in *Polia pisi* L. (Lepidoptera: Noctuidae) *Z. Naturforsch.* **35**, 990–4. [25,26]

—— (1983). Receptors for di-unsaturated pheromone analogues in the male summerfruit tortrix moth. *Z. Naturforsch.* **38**, 874–7. [25,26]

—— (1984a). Analysis of a sex-attractant system in the noctuid moth *Rhyacia baja* Schiff. *Z. Naturforsch.* **39c**, 845–8. [26]

—— (1984b). The pheromone receptor system of male *Eulia ministrana* L., with notes on other Cnephasiini moths. *Z. Naturforsch.* **39c**, 849–52. [26]

—— and Bogenschütz, H. (1982). Weitere Feldversuche zur Spezifität synthetischer Sexuallockstoffe bei *Choristoneura murinana. Z. Ang. Entomol.* **94**, 331–51. [26]

—— and Schroth, M. (1983). Supplementary data on the sex attractant system of *Panolis flammea. Z. Naturforsch.* **38c**, 870–3. [26]

——, Bogenschütz, H. and Arn, H. (1980). A sex attractant for the European fir budworm moth, *Choristoneura murinana. Z. Naturforsch.* **35c**, 390–26. [25,26]

Prillinger, L. (1981). Postembryonic development of the antennal lobes in *Periplaneta americana* L. *Cell Tissue Res.* **215**, 563–75. [33]

Prokopy, R. J. (1982). Getting to know a fruit fly. *J. Ga. Entomol.*, 2nd suppl. 30–8. [8]

—— and Owens, E.D. (1983). Visual detection of plants by herbivorous insects. *Ann. Rev. Entomol.* **28**, 337–64. [8]

—— and Roitberg, B. D. (1984). Foraging behavior of true fruit flies. *Amer. Sci.* **72**, 41–9. [8,20]

——, Collier, R. H. and Finch, S. (1983a). Leaf color used by cabbage root flies to distinguish among host plants. *Science* **221**, 190–2. [8]

——, Collier, R. H. and Finch, S. (1983b). Visual detection of host plants by cabbage root flies. *Entomol. Exp. Appl.* **43**, 85–9. [8]

Quinn, W. G., Sziber, P. P. and Booker, R. (1978). The *Drosophila* memory mutant *amnesiac. Nature* **277**, 212–4. [1]

Raffa, K. F. and Berryman, A. A. (1983). The role of host plant resistance in the colonization behavior and ecology of bark beetles (Coleoptera: Scolytidae). *Ecol. Monographs* **53**, 27–49. [10]

Rees, C. J. C. (1970). Age dependency of response in an insect chemoreceptor sensillum. *Nature* **227**, 740–2. [29]

Renou, M., Lalanne-Cassou, B., Frerot, B., Gallois, M. and Descoins, C. (1981). Composition de la sécrétion phéromonale émise par les femelles vierges de *Mamestra (Polia) pisi* (L). (Lepidoptére, Noctuidae, Hadeninae). *Cr. Acad. Sc. Paris* **292D**, 1117–19. [26]

Renwick, J. A. A. and Vité, J. P. (1969). Bark beetle attractants: mechanism of colonization by *Dendroctonus frontalis. Nature* **224**, 1222–3. [10,11]

—— and Vité, J. P. (1970). System of chemical communication in *Dendroctonus. Contrib. Boyce. Thompson Inst.* **24**, 283–92. [10]

Rheinlaender, J. and Blatgen, G. (1982). The precision of auditory localization in the cricket, *Gryllus bimaculatus. Physiol. Entomol.* **7**, 209–18. [2]

——, Gerhard, H. C., Yager, D. D. and Capranica, R R. (1979). Accuracy of phonotaxis by the green treefrog (*Hyla cinera*). *J. Comp. Physiol.* **133**, 247–55. [2]

Rodrigues, V. and Siddigi, O. (1978). Genetic analysis of chemosensory pathway. *Proc. Indian Acad. Sci.* **87**(B), 147–60. [1]

Roelofs, W. L. (1978). Threshold hypothesis for pheromone perception. *J. Chem. Ecol.* **4**, 658–99. [24]

—— and Cardé, R. T. (1977). Responses of Lepidoptera to synthetic sex pheromone chemicals and their analogues. *Ann. Rev. Entomol.* **22**, 377–405. [4,26]

—— and Comeau, A. (1971). Sex pheromone perception: electroantennogram responses of the redbanded leafroller moth. *J. Insect Physiol.* **17**, 1967–82. [4]

——, Hill, A. and Cardé, R. T. (1975). Sex pheromone components of the redbanded leafroller, *Argyrotaenia velutinana. J. Chem. Ecol.* **1**, 83–9. [26]

Romanes, J. G. (1882). *Animal intelligence*. Appleton, New York. [20]

Rosen, R. (1976). On rate-sensitive chemoreceptor systems and the origin of the immune response. *J. Theor. Biol.* **62**, 429–45. [1]

Rospars, J. P. (1983). Invariance and sex-specific variations of the glomerular organization in the antennal lobes of a moth, *Mamestra brassicae*, and a butterfly, *Pieris brassicae. J. Comp. Neurol.* **220**, 80–96. [32]

Ross, H. H. (1955). The taxonomy and evolution of the sawfly genus *Neodiprion. For. Sci.* **1**, 196–209. [13]

Rudinsky, J. A. (1966). Scolytid beetles associated with Douglas-fir: response to terpene. *Science* **152**, 218–19. [29]

—— (1973). Multiple function of Douglas-fir beetle pheromone 3-methyl-2-cyclohexen-1-one. *Environ. Entomol.* **2**, 580–5. [29]

—— and Ryker, L. C. (1977). Olfactory and auditory signals mediating behavioral patterns of bark beetles. In *Comportment des Insectes et Milieu Trophique* (ed. V. Lebeyrie) pp. 195–209. Coll. Intn. CNRS **265**. [10,29]

——, Kinzer, G. W., Fentiman, A. F. Jun., and Foltz, R. L. (1974). Additional components of the Douglas-fir beetle (Col.: Scolytidae) aggregative pheromone and their possible utility in pest control. *Z. Ang. Entomol.* **76**, 65–77. [29]

Rumbo, E. R. (1980). Separation of interleaved nerve spike trains in a noisy channel. *IEEE Trans. Biomed. Engin.* **27**, 470–2. [31]

—— (1981). Study of single sensillum responses to pheromone in the light-brown apple moth, *Epiphyas postvittana*, using an averaging technique. *Physiol. Entomol.* **6**, 87–98. [19]

—— (1983). Differences between single cell responses to different components of the sex pheromone in males of the light-brown apple moth (*Epiphyas postvittana*). *Physiol. Entomol.* **8**, 195–201. [19]

Ryker, L. C. and Libbey, L. M. (1982). Frontalin in the male mountain pine beetle. *J. Chem. Ecol.* **8**, 1399–409. [10]

—— and Rudinsky, J. A. (1982). Field bioassay of exo- and endo-brevicomin with *Dendroctonus ponderosae* in lodgepole pine. *J. Chem. Ecol.* **8**, 701–7. [10]

—— and Yandell, K. L. (1983). Effect of verbenone on aggregation of *Dendroctonus ponderosae* Hopkins (Coleoptera: Scolytidae) to synthetic attractant. *Z. Ang. Entomol.* **96**, 452–9. [10]

——, Libbey, L. M. and Rudinsky, J. A. (1979). Comparison of volatile compounds and stridulation emitted by the Douglas-fir beetle from Idaho and western Oregon populations. *Environ. Entomol.* **8**, 789–98. [29]

Sabelis, M. W. and Schippers, P. (1984). Variable wind directions and anemotactic strategies of searching for an odour plume. *Oecologia*, in press. [5]

Salt, G. (1935). Experimental studies in insect parasitism. III. Host selection. *Proc. R. Soc. London* Series B. **117**, 413–35. [16]

Sanders, C. J. (1981). Spruce budworm: effects of different blends of sex pheromone components on disruption of male attraction. *Experientia* **37**, 1176–8. [12]

—— (1982a). Behaviour of spruce budworm male moths in pheromone permeated air in a wind tunnel. In *Les médiateurs chimiques agissant sur le comportement des insectes.* Symp. Int. Versailles, 16–20 Nov. 1981. INRA Publ. 1982. [12]

—— (1982b)Disruption of male spruce budworm orientation to calling female in a wind tunnel by synthetic pheromone. *J. Chem. Ecol.* **8**, 493–503. [12]

—— (1984). Sex pheromone of the spruce budworm (Lepidoptera: Tortricidae): evidence for a missing component. *Can. Entomol.* **116**, 93–100. [12]

——, Lucuik, G. S. and Fletcher, R. M. (1981). Responses of male spruce budworm to different concentrations of sex pheromone as measured in a sustained-flight wind tunnel. *Can. Entomol.* **113**, 943–8. [12]

Sanders, Von W. (1983). Untersuchungen über das Verhalten des Kupferstechers *Pityogenes chalcographus* L. während der Flugphase. *Z. Ang. Entomol.* **96**, 125–31. [10]

Sandlan, K. (1980). Host location by *Coccigomimus turionella* (Hymenoptera: Ichneumonidae). *Entomol. Exp. Appl.* **27**, 233–45. [16]

Sanes, J. R. and Hildebrand, J. G. (1976a). Structure and development of antennae in a moth, *Manduca sexta* (Lepidoptera: Sphingidae). *Devel. Biol.* **51**, 282–99. [32]

—— and Hildebrand, J. G. (1976b). Origin and morphogenesis of sensory neurons in an insect antenna. *Devel. Biol.* **51**, 300–19. [32]

——, Prescott,D. J. and Hildebrand, J. G. (1977). Cholinergic neurochemical development of normal and deafferented antennal lobes in the brain of the moth, *Manduca sexta. Brain Res.* **119**, 398–402. [32]

Sass, H. (1976). Zur nervosen Codierung von Geruchsreizen bei *Periplaneta americana. J. Comp. Physiol.* **107**, 49–65. [21]

Sato, Y. (1979). Experimental studies on parasitization by *Apanteles glomeratus*. IV. Factors leading a female to the host. *Physiol. Entomol.* **4**, 63–70. [16]

Sawford, B. L. (1983). The effect of Gaussian particle-pair distribution functions in the statistical theory of concentration fluctuations in homogeneous turbulence. *Quart, J. Royal Meteorol. Soc.* **109**, 339–54. [3]

Schaefer, G. W. (1976). Radar observations of insect flight. *Symp. R. Entomol. Soc. Lond.* **7**, 157–97. [5]

Schal, C. (1982). Intraspecific vertical stratification as a mate-finding mechanism in tropical cockroaches. *Science* **215**, 1405–7. [20]

—— and Cardé, R. T. (1985). Rhythmic extrusion of pheromone gland elevates pheromone release rate. *Experientia* in press. [20]

——, Tobin, T. R., Surber, J. L., Vogel, G., Tortellot, M. K., Leban, R. A., Sizemore, R. and Bell, W. J. (1983). Search strategy of sex pheromone-stimulated male German cockroaches. *J. Insect Physiol.* **27**, 575–9. [15]

Schilderger, K. (1983). Local interneurones associated with the mushroom bodies and the central body in the brain of *Acheta domesticus. Cell Tissue Res.* **230**, 573–86. [33]

—— (1984). Multimodal interneurones in the cricket brain: properties of identified extrinsic mushroom body cells. *J. Comp. Physiol.* **154**, 71–9. [33]

Schmidt, G. T. (1974). Host acceptance behavior of *Campoletis sonorensis* toward *Heliothis zea. Ann. Entomol. Soc. Amer.* **67**, 835–44. [16]

Schmitz, R. F. and Taylor, A. R. (1969). An instance of lightning damage and infestation of ponderosa pines by the pine engraver beetles in Montana. *USDA For. Serv. Res. Note.* INT-88. 8 pp. [11]

Schneider, D. (1957). Elektrophysiologische Untersuchungen von Chem- und Mechanorezeptoren der Antenne des Seidenspinners *Bombyx mori* L. *Z. Verg. Physiol.* **40**, 8–41. [21,25]

—— (1969). Insect olfaction: deciphering system for chemical messages. *Science* **163**, 1031–7. [26,32]

—— (1975). Pheromone communication in moths and butterflies. In *Sensory physiology and behavior* (eds. R. Galun *et al.*) pp. 173–93. Plenum, New York. [21]

—— (1980). Pheromone von Insekten: Produktion—Reception—Inaktivierung. *Nova acta Leopoldina* N.F. **51**, no. 237, 249–78. [21]

—— (1983). Kommunikation durch chemische Signale bei Insekten: alte and neue Beispiele von Lepidopteren. *Verh. Dtsch. Zool. Ges.* 1983, 5–16. [21]

—— (1984). Insect olfaction—our research endeavour. In *Foundations of sensory sciences* (eds. W. W. Dawson and E. J. Enoch) pp. 381–418. Springer-Verlag, Heidelberg. [21]

—— and Boeckh, J. (1962). Rezeptorpotential und Nervenimpulse einzelner olfaktorischer Sensillen der Insekten-antenne. *Z. Verg. Physiol.* **45**, 405–12. [21]

——, Lacher, V. and Kaissling, K.-E. (1964). Die Reaktionsweise und das Reaktionsspektrum von Riechzellen bei *Antheraea pernyi* (Lepidoptera, Saturniidae). *Z. Verg. Physiol.* **48**, 632–62. [21]

——, Block, B. C., Boeckh, J. and Priesner, E. (1967). Die Reaktion der männlichen Seidenspinner auf Bombykol und seine Isomeren: Electroantennogram und Verhalten. *Z. Verg. Physiol.* **54**, 192–209. [22]

——, Boppré, M., Zweig, J., Horsley, St. B., Bell, T. B., Meinwald, J., Hansen, K. and Diehl, E. W. (1982). Scent organ development in *Creatonotos* moths: regulation by pyrrolizidine alkaloids. *Science* **215**, 1264–5. [21]

Schneider, P. (1965). Vergleichende Untersuchungen zur Steuerung der Fluggeschwindigkeit bei *Calliphora vicina* Rob.-Desvoidy (Diptera). *Z. Wiss. Zool.* **173**, 114–73. [7]

Schneiderman, A. M., Hildebrand, J. G. and Jacobs, J. J. (1983). Computer-aided morphometry of developing and mature antennal lobes in the moth *Manduca sexta*. *Soc. Neurosci. Abstr*. **9**, 834. [32]

——, Matsumoto, S. G. and Hildebrand, J. G. (1982). Trans-sexually grafted antennae influence development of sexually dimorphic neurones in moth brain. *Nature* **298**, 844–6. [32]

Schöne, H. (1984). *Spatial orientation. The spatial control of behavior in animals and man*. Princeton University Press, Princeton, N.J. [2]

Schurig, V., Weber, R., Nicholson, G. J., Oehlschlager, A. C., Pierce, H. Jun., Pierce, A. M., Borden, J. H. and Ryker, L.C. (1983). Enantiomer composition of natural exo- and endo-brevicomin by complexation gas chromatography/selected ion mass spectrometry. *Naturwissenschaften* **70**, 92–3. [30]

Schwartz, R. M. and Dayhoff, M. O. (1978). Origins of prokaryotes, eukaryotes, mitochondria, and chloroplasts. *Science* **199**, 395–403. [1]

Schweitzer, E. S., Sanes, J. R. and Hildebrand, J. G. (1976). Ontogeny of electro-antennogram responses in the moth, *Manduca sexta*. *J. Insect Physiol*. **22**, 955–60. [33]

Schwink, I. (1958). A study of olfactory stimuli in the orientation of moths. *Proc. X Int. Cong. Entomol*. **2**, 577–81. [3]

Seabrook, W. D., Hirai, K., Shorey, H. H. and Gaston, L. K. (1979). Maturation and senescence of an insect chemosensory response. *J. Chem. Ecol*. **5**, 587–94. [29]

Selzer, R. (1979). Morphological and physiological identification of food odour specific neurones in the deutocerebrum of *Periplaneta americana*. *J. Comp. Physiol*. **134**, 159–63. [33]

—— (1981). The processing of a complex food odor by antennal olfactory receptors of *Periplaneta americana*. *J. Comp. Physiol*. **144**, 509–19. [21]

Sharples, A. (1933). Injury to coconut palms by lightning. *J. Malayan Agric*. **21**, 295–6. [11]

Shepherd, R. F. (1966). Factors influencing the orientation and rates of activity of *Dendroctonus ponderosae* Hopkins (Coleoptera: Scolytidae). *Can. Entomol*. **98**, 507–18. [10]

Shrimpton, D. M. (1972). Variation in the extractives from lodgepole pine sapwood and heartwood. *Can. For. Serv., N. For. Res. Cen. Inf. Rep*. NOR-X-**18**. [10]

Siegel, S. (1956). *Nonparametric statistics for the behavioral sciences*. McGraw-Hill Book Co., New York. [25]

Silverstein, R. M. (1979). Enantiomeric composition and bioactivity of chiral semiochemicals in insects. In *Odour communication in animals* (ed. F. J. Ritter) pp. 133–46. Elsevier/North Holland, Amsterdam. [10,13,30]

Smith, D. V., Bowdan, E. and Dethier, V. G. (1984). Information transmission in tarsal sugar receptors of the blowfly. *Chemical Senses* **1**, 81–101. [1]

Smith, F. B. and Blackall, R. M. (1979). The application of field experimental data to the paramaterisation of the dispersion of plumes from ground level and elevated sources. In *Mathematical modelling of turbulent diffusion in the environment* (ed. C. J. Harris) pp. 201–37. Academic Press, London. [3]

Smith, G. L., Cleveland, T. C. and Clark, J. C. (1965). Boll weevil movement from hibernation sites to fruiting cotton. *J. Econ. Entomol*. **58**, 357–8. [29]

Smith, J. N. M. (1974). The food searching behavior of two European thrushes. I. Description and analysis of search paths. *Behaviour* **48**, 276–302. [2]

Smits, P. H. (1982). The influence of kairomones of *Mamestra brassicae* on the searching behavior of *Trichogramma evanescens*. *Les Trichogrammes, Antibes, Les Colloques de l'INRA* **9**, 139–50. [16]

Sorensen, K. and Bell, W. J. (1985). Orientation responses of the isopod, *Armadillidium nasatum*, to temporal changes in relative humidity: simulations of a 'humid zone' and of a 'humid plume'. *J. Insect Physiol.* in press. [15]

Spielman, A. (1974). Effect of synthetic juvenile hormone on ovarian diapause of *Culex pipiens* mosquitoes. *J. Med. Entomol.* **11**, 223–5. [28]

Städler, E. and Hanson, F. E. (1975). Olfactory capabilities of the 'gustatory' chemoreceptors of the tobacco hornworm larvae. *J. Comp. Physiol.* **104**, 97–102. [31]

Stanton, M. (1983). Spatial patterns of the plant community and their effects upon insect search. In *Herbivorous insects: host seeking behavior and mechanisms*. (ed. S. Ahmad) pp. 125–57. Academic Press, New York. [8]

Starrat, A. N., Dahm, K. H., Allen, N., Hildebrand, J. G., Payne, T. L. and Roller, H. (1979). Bombykal, a sex pheromone of the sphinx moth *Manduca sexta*. *Z. Naturforsch.* **34c**, 9–12. [32]

Stocker, R. F., Singh, R. N., Schorderet, M. and Siddiqui, O. (1983). Projection patterns of different types of antennal sensilla in the antennal glomeruli of *Drosophila melanogaster*. *Cell Tissue Res.* **232**, 237–48. [32]

Strausfeld, N. J. (1976). *Atlas of an insect brain*. Springer-Verlag, Berlin. [32,33]

——, Bassemir, U., Singh, R. N. and Bacon, J. P. (1984). Organizational principles of outputs from dipteran brains. *J. Insect Physiol.* **30**, 73–93. [33]

Steck, W. F., Underhill, E. W. and Chisholm, M. D. (1980). Trace components in lepidopterous sex attractants. *Environ. Entomol.* **9**, 583–5. [26]

——, Underhill, E. W., Bailey, B. K. and Chisholm, M. D. (1982). A 4-component sex attractant for male moths of the armyworm, *Pseudaletia unipuncta*. *Entomol. Exp. Appl.* **32**, 302–4. [26]

Steinbrecht, R. A. (1980). Cryofixation without cryoprotectants. Freeze substitution and freeze etching of an insect olfactory receptor. *Tiss. Cell* **12**, 73–100. [23]

—— and Müller, B. (1971). On the stimulus conducting structures in insect olfactory receptors. *Z. Zellforsch.* **117**, 570–5. [23]

Stephen, F. M. and Dahlsten, D. L. (1976). The arrival sequence of the arthropod complex following attack of *Dendroctonus brevicomis* in ponderosa pine. *Can Entomol.* **108**, 282–304. [30]

Sternlicht, M. (1973). Parasitic wasps attracted by the sex pheromone of their coccid host. *Entomophaga* **18**, 339–42. [16]

Stewart, T. E., Plummer, E. L., McCandless, L. L., West, J. R. and Silverstein, R. M. (1977). Determination of enantiomer composition of several bicylic ketal insect pheromone components. *J. Chem. Ecol.* **3**, 27–43. [29]

Storebø, P. B., Bjorvatten, T., Hønnåshagen, K., Lillegraven, A., Jones, C. D. and Van Buijtenen, C. J. P. (1983). Tracer experiments with turbulently dispersed air ions. *Boundary-layer Meteorology* **26**, 127–39. [3]

Suzuki, H. (1975a). Convergence of olfactory inputs from both antennae in the brain of the honeybee. *J. Exp. Biol.* **62**, 11–26. [33]

—— (1975b) Antennal movements induced by odour and central projection of the antennal neurones in the honeybee. *J. Insect Physiol.* **21**, 831–47. [33]

—— and Tateda, H. (1974). An electrophysiological study of olfactory interneurones in the brain of the honey-bee. *J. Insect Physiol.* **20**, 2287–99. [32,33]

——, Tateda, H. and Kuwabara, M. (1976). Activities of antennal and ocellar interneurones in the protocerebrum of the honey-bee. *J. Exp. Biol.* **64**, 405–18. [33]

Tanimura, T., Isono, K., Takamura, T. and Shimada, I. (1982). Genetic dimorphism in the taste sensitivity to trehalose in *Drosophila melanogaster. J. Comp. Physiol.* **147**, 433–7. [1]

Taylor, L. R. (1965). Flight behaviour and aphid migration. *Proc. N. Central Branch, Entomological Society of America* **20**, 9–19. [2]

—— and Taylor, R. A. J. (1983). Insect migration as a paradigm for survival by movement. In *The ecology of animal movement* (eds. I. R. Swingland and P. L. Greenwood) pp. 181–214. Clarendon Press, Oxford. [2]

Thompson, A. C. and Mitlin, N. (1979). Biosynthesis of the sex pheromone of the male boll weevil from monoterpene precursors. *Insect Biochem.* **9**, 293–4. [29]

——, Hanny, B. W., Hedin, P. A. and Gueldner, R. C. (1971). Phytochemical studies in the family Malvaceae. I. Comparison of essential oils of six species by gas–liquid chromatography. *Amer. J. Bot.* **58**, 803–7. [29]

——, Wright, B. J., Hardee, D. D., Gueldner, R. C. and Hedin, P. A. (1970). Constituents of the cotton bud. XVI. The attractancy response of the essential oils of a group of host and non-host plants. *J. Econ. Entomol.* **63**, 751–3. [29]

Thompson, S. N. and Bennett, R. B. (1971). Oxidation of fat during flight of male Douglas-fir beetles, *Dendroctonous pseudotsugae. J. Insect Physiol.* **17**, 1555–63. [10]

Thorpe, W. H. (1954). Some concepts of ethology. *Nature* **174**, 101–5. [2]

—— (1963). *Learning and instinct in animals.* Methuen, London. [2]

—— and Jones, F. G. W. (1937). Olfactory conditioning in a parasitic insect and its relation to the problem of host selection. *Proc. R. Soc. London*, Series B, **124**, 56–81. [16]

Thorson, J., Weber, T. and Huber, F. (1982). Auditory behavior of the cricket. II. Simplicity of calling-song recognition in *Gryllus*, and anomalous phonotaxis at abnormal carrier frequencies. *J. Comp. Physiol.* **146**, 361–78. [2]

Thorsteinson, A. J. (1960). Host selection in phytophagous insects. *Ann. Rev. Entomol.* **5**, 193–218. [2]

Tobin, T. R. (1981). Pheromone orientation: role of internal control mechanisms. *Science* **214**, 1147–9. [2]

—— and Bell, W. J. (1982). Guidance system for pheromone orientation in moths. *Nature* **295**, 263. [2,4]

—— and Bell, W. J. (1985). Chemo-orientation of *Trogoderma variabile* in a simulated sex pheromone plume. *J. Comp. Physiol.* in press. [15]

Tolbert, L. G. and Hildebrand, J. G. (1981). Organization and synaptic ultrastructure of glomeruli in the antennal lobes of the moth *Manduca sexta*: a study using thin sections and freeze-fracture. *Proc. R. Soc. London*, Series B **213**, 279–301. [32]

——, Matsumoto, S. G. and Hildebrand, J. G. (1983). Development of synapses in the antennal lobes of the moth *Manduca sexta* during metamorphosis. *J. Neurosci.* **3**, 1158–75. [32]

Tømmerås, B. A. (1983). Reseptorcellers spesialisering hos granbarkbillen (*Ips typographus*) og dens predator, maurbillen (*Thanasimus formicarius*) for vertsdufter og barkbilleferomoner. Thesis, Dept. Zoology, University of Trondheim. [30]

—— and Mustaparta, H. (1984). Enhanced attraction of *Ips typographus* to pheromones by adding *exo*-brevicomin. *Naturwissenschaften* **71**, 375–7. [30]

——, Mustaparta, H. and Gregoire, J.-Cl. (1984). Receptor cells in *Ips typographus* and *Dendroctonus micans* specific to pheromones of the reciprocal genus. *J. Chem. Ecol.* **10**, 759–69. [30]

Tompkins, L., Cadosa, M. J., White, F. V. and Sanders, T. G. (1979). Isolation and analysis of chemosensory behavior mutants in *Drosophila melanogaster*. *Proc. Nat. Acad. Sci. USA* **76**, 884–7. [1]

Traynier, R. M. M. (1968). Sex attraction in the Mediterranean flour moth, *Anagaster kuhniella*: location of the female by the male. *Can. Entomol.* **100**, 5–10. [5]

Trivers, R. L. (1972). Parental investment and sexual selection. In *Sexual selection and the descent of man, 1871–1971*. pp. 136–79. Aldine, Chicago. [20]

Tully, T. and Hirsch, J. (1982). Behavior-genetic analysis of *Phormia regina* II. Detection of a single major-gene effect from behavioral variation for central excitatory state (CES) using hybrid crosses. *Anim. Behav.* **30**, 1193–1202. [1]

Tumlinson, J. H., Klein, M. G., Doolittle, R. E., Ladd, T. L. and Proveaux, A. T. (1977). Identification of the female Japanese beetle sex pheromone: inhibition of male response by an enantiomer. *Science* **197**, 780–92. [29]

——, Hardee, D. D., Gueldner, R. C., Thompson, A. C., Hedin, P. A. and Minyard, J. P. (1969). Sex pheromones produced by male weevils: isolation, identification and synthesis. *Science* **166**, 1010–12. [29]

Van Der Pers, J. N. C. (1982). Comparison of single cell responses of antennal sensilla trichodea to sex attractants in nine European small ermine moths (*Yponomeuta* spp.). *Entomol. Exp. Appl.* **31**, 255–64. [27]

—— and Den Otter, C. J. (1978). Single cell responses from olfactory receptors of small ermine moths to sex-attractants. *J. Insect Physiol.* **24**, 337–43. [27]

VanderSar, T. J. D. and Borden, J. H. (1977). Visual orientation of *Pissodes strobi* (Peck) (Coleoptera: Curculionidae) in relation to host selection behaviour. *Can. J. Zool.* **55**, 2042–9. [10]

Vareschi, E. (1971). Duftunterscheidung bei der Honigbiene-Einzelzell-Ableitungen und Verhaltensreaktion. *Z. Verg. Physiol.* **75**, 143–73. [21]

Varley, G. C. (1962). A plea for a new look at Lepidoptera with special reference to the scent distributing organs of male moths. *Trans. Soc. Brit. Entom.* **15**, 29–40. [21]

Vet, L. E. M. (1983). Host habitat location through olfactory cues by *Leptopilina clavipes* (Hartig)(Hym.: Eucoilidae), a parasitoid of fungivorous *Drosophila*: the influence of conditioning. *Neth. J. Zool.* **330**, 225–48. [16]

Vinson, S. B. (1984). Parasitoid–host relationship. In *Chemical ecology of insects*. (eds. W. J. Bell and R. T. Cardé) pp. 205–33. Chapman and Hall, London and New York. [16]

——, Jones, R. L., Sonnet, P. E., Bierl, B. A. and Beroza, M. (1975). Isolation identification, and synthesis of host-seeking stimulants for *Cardiochiles nigriceps*, a parasitoid of tobacco budworm. *Entomol. Exp. Appl.* **18**, 443–50. [16]

Vité, J. P. and Bakke, A. (1979). Synergism between chemical and physical stimuli in host colonization by an ambrosia beetle. *Naturwissenschaften* **66**, 528. [10]

—— and Francke, W. (1976). The aggregation pheromones of bark beetles: progress and problems. *Naturwissenschaften* **63**, 550–5. [10]

—— and Williamson, D. L. (1970). *Thanasimus dubius*: prey perception. *J. Insect Physiol.* **16**, 233–9. [30]

——, Gara, R. I. and von Scheller, H. D. (1964). Field observations on the response to attractants of bark beetles infesting southern pines. *Contrib. Boyce Thompson Inst.* **22**, 261–70. [11]

——, Klimetzek, D., Loskant, G., Hedden, R. and Mori, K. (1976). Chirality of insect pheromones: response interruption by inactive antipodes. *Naturwissenschaften* **63**, 582. [7]

Vogt, R. G. and Riddiford, L. M. (1981a). Pheromone deactivation by antennal proteins of Lepidoptera. In *Regulation of insect development and behaviour* (eds. F. Sehnal, A. Zabza, J. J. Menn and B. Cymborowski) pp. 955–67. Polytechnical University of Wroclaw Press, Wroclaw. [23]

—— and Riddiford, L. M. (1981b). Pheromone binding and inactivation by moth antennae. *Nature* **293**, 161–3. [23]

——, Riddiford, L. M. and Prestwich, G. D. (1985). The kinetic properties of a pheromone degrading enzyme: the sensillar esterase of *Antheraea polyphemus. Proc. Nat. Acad. Sci. USA* in press. [23]

Von Uexüll, J. (1909). *Umelt und Innenwelt der Tiere.* Springer-Verlag, Berlin. [29]

Waage, J. K. (1978). Arrestment responses of the parasitoid *Nemeritis canescens* to a contact chemical produced by its host, *Plodia interpunctella. Physiol. Entomol.* **3**, 135–46. [15,16]

Wadhams, L. J., Angst, M. E. and Blight, M. M. (1982). Responses of the olfactory receptors of *Scolytus scolytus* (F.) (Coleoptera: Scolytidae) to the stereoisomers of 4-methyl-3-heptanol. *J. Chem. Ecol.* **8**, 477–92. [29]

Waldow, U. (1975). Multimodale Neurone im Deutocerebrum von *Periplaneta americana. J. Comp. Physiol.* **101**, 329–41. [32,33]

—— (1977). CNS units in cockroach (*Periplaneta americana*): specificity of response to pheromones and other odour stimuli. *J. Comp. Physiol.* **116**, 1–17. [32,33]

Wall, C. and Perry, J. N. (1978). Interactions between pheromone traps for the pea moth, *Cydia nigricana* (F.). *Entomol. Exp. Appl.* **24**, 155–62. [9]

—— and Perry, J. N. (1980). Effects of spacing and trap number on interactions between pea moth pheromone traps. *Entomol. Exp. Appl.* **28**, 313–21. [9]

—— and Perry, J. N. (1981). Effects of dose and attractant on interactions beween pheromone traps for the pea moth, *Cydia nigricana. Entomol. Exp. Appl.* **30**, 26–30. [9]

—— and Perry, J. N. (1982). The behaviour of moths responding to pheromone sources in one field: a basis for discussion. In *Les Colloques de l'INRAS* **7**, 171–88. [9]

—— and Perry, J. N. (1983). Further observations on the responses of male pea moth, *Cydia nigricana*, to vegetation previously exposed to sex-attractant. *Entomol. Exp. Appl.* **33**, 112–6. [9,20]

——, Sturgeon, D. M., Greenway, A. R. and Perry, J. N. (1981). Contamination of vegetation with synthetic sex-attractant released from traps for the pea moth, *Cydia nigricana. Entomol. Exp. Appl.* **30**, 111–115. [9,20]

Waloff, Z. (1972). Orientation of flying locusts (*Schistocerca gregaria* Forsk.) in migrating swarms. *Bull. Entomol. Res.* **62**, 1–72. [5]

Wehner, R. (1981). Spatial vision in arthropods. In *Handbook of sensory physiology*. (ed. H. Autrum) pp. 287–616. Springer-Verlag, New York. [8]

Weiss, M. J. (1974). Neuronal connections and the function of the corpora pedunculus in the brain of the American cockroach, *Periplaneta americana*. *J. Morph*. **142**, 21–69. [33]

Wellington, W. G. (1977). Returning the insect to insect ecology: some consequences for pest management. *Environ. Entomol.* **6**, 1–8. [10]

Whittaker, R. H. and Feeny, P. P. (1971). Allelochemics: chemical interactions between species. *Science* **171**, 757–70. [30]

Wigglesworth, V. B. (1972). *The principles of insect physiology* (7th ed.). Chapman and Hall, London and New York. [21]

Willis, M. A. and Baker, T. C. (1984). Effects of intermittent and continuous pheromone stimulation on the flight behavior of the oriental fruit moth, *Grapholitha molesta*. *Physiol. Entomol.* **9**, 341–58. [2,4,20]

Wilson, E. O. (1962). Chemical communication among workers of the fire ant *Solenopsis saevissima* (Fr. Smith). I. The organization of mass-foraging. *Anim. Behav*. **10**, 134–47. [14]

Witzgall, P. and Priesner, E. (1984). Behavioural responses of *Coleophora laricella* male moths to synthetic sex-attractant, (*Z*)-5-decenol, in the field. *Z. Ang. Entomol*. **98**, 15–32. [2]

Wood, D. L. (1972). The selection and colonization of ponderosa pine by bark beetles. In *Insect/plant relationships* (ed. H. F. van Emden) pp. 101–17. Blackwell Scientific Publications, Oxford. [11]

—— (1976). Host selection by bark beetles in the mixed-conifer forests of California (summary only). *Symp. Biol. Hung*. **16**, 301. [11]

—— (1982). The role of pheromones, kairomones, and allomones in the host selection and colonization behavior of bark beetles. *Ann. Rev. Entomol*. **27**, 411–46. [11]

—— and Bushing, R. W. (1963). The olfactory responses of *Ips confusus* (LeConte) (Coleoptera: Scolytidae) to the secondary attraction in the laboratory. *Can. Entomol*. **95**, 1066–78. [10]

——, Browne, L. E., Bedard, W. D., Tilden, P. E., Silverstein, R. M. and Rodin, J. O. (1968). Response of *Ips confusus* to synthetic sex pheromones in nature. *Science* **159**, 1373–4. [30]

——, Cobb, F. W. Jun., Cohen, D. J., Browne, L. E., Moeck, H. A. and Stark, R. W. (1974). Host selection by bark beetles (Coleoptera: Scolytidae) in the mixed conifer forests of California. *Symp. on the host plant in relation to insect behavior and production.* June 11–14, 1974. Tihany, Hungary. [11]

——, Browne, L. E., Ewing, B., Lindahl, K., Bedard, W. D., Tilden, P. E., Mori, K., Pitman, G. B. and Hughes, P. R. (1976). Western pine beetle: specificity among enantiomers of male and female components of an attractant pheromone. *Science* **192**, 896–8. [29]

Wright, R. H. (1958). The olfactory guidance of flying insects. *Can. Entomol*. **90**, 81–9. [3,4,20]

—— (1962). How insects follow a scent. *New Scientist* **14**, 339–41. [4]

—— (1964). *The science of smell*. Allen and Unwin, London. [4]

Yamada, M. (1971). A search for odour encoding in the olfactory lobe. *J. Physiol*. **214**, 127–143. [32,33]

——, Ishii, S. and Kuwahara, Y. (1970). Odour discrimination: 'sex pheromone specialists' in the olfactory lobe of the cockroach. *Nature* **227**, 855. [33]

Zack, C. (1979). Sensory adaptation in the sex pheromone receptor cells of saturnid moths. Diss. Ludwig-Maximilians-Universität. München. [6]

Index